This textbook is a general introduction to the dynamics of astrophysical fluids for students with knowledge of basic physics at undergraduate level. No previous knowledge of fluid dynamics or astrophysics is required because the author develops all new concepts in context. The first four chapters cover classical fluids, relativistic fluids, photon fluids, and plasma fluids, with many cosmic examples being included. The remaining six chapters deal with astrophysical applications, stars, stellar systems, astrophysical plasmas, cosmological applications, and the large-scale structure of the Universe.

Astrophysical fluid dynamics is a promising branch of astronomy, with wide applicability. This textbook considers the role of plasma and magnetism in planets, stars, galaxies, the interplanetary, interstellar and intergalactic media, as well as the Universe at large.

Astrophysical Fluid Dynamics

Astrophysical Fluid Dynamics

E. BATTANER
Departamento de Física Teórica y del Cosmos
University of Granada, Spain

CAMBRIDGE
UNIVERSITY PRESS

CAMBRIDGE UNIVERSITY PRESS
Cambridge, New York, Melbourne, Madrid, Cape Town, Singapore, São Paulo

Cambridge University Press
The Edinburgh Building, Cambridge CB2 2RU, UK

Published in the United States of America by Cambridge University Press, New York

www.cambridge.org
Information on this title: www.cambridge.org/9780521431668

First published 1996

A catalogue record for this publication is available from the British Library

Library of Congress Cataloguing in Publication data
Battaner, E.
Astrophysical fluid dynamics / E. Battaner.
   p.   cm.
ISBN 0 521 43166 2.–ISBN 0 521 43747 4 (pbk.)
1. Fluid dynamics. 2. Astrophysics. I. Title.
QB466.F58B38 1996
523.01–dc20   95-22253 CIP

ISBN-13 978-0-521-43166-8 hardback
ISBN-10 0-521-43166-2 hardback

ISBN-13 978-0-521-43747-9 paperback
ISBN-10 0-521-43747-4 paperback

Transferred to digital printing 2006

TO EDUARDEJO

Contents

Preface

This small book is intended as a general introduction to astrophysical fluid dynamics. The reader is presumed to possess a knowledge of basic physics, namely, classical physics, elements of relativity, and introductory ideas about quantum mechanics. No previous knowledge of fluid dynamics or of astrophysics is required, these topics being introduced in the book. Although fluid dynamics may constitute a complementary, original, natural, fecund, unexplored, simple, and enjoyable way to introduce astrophysics, the topic of astrophysical fluid dynamics is a promising, distinct, and particularly wide branch of astrophysics at the present time.

The first part of the book (Chapters 1–4) deals with basic fluid dynamics. Although it could also be used for non-astrophysical purposes, it was written with the former in mind. It often includes cosmic examples that are mainly related to a stationary, static, and stratified atmosphere. These conditions provide the greatest simplification while maintaining a high degree of astrophysical interest.

Following the first chapter on classical fluids, Chapter 2 is devoted to relativistic fluids. The early introduction of relativistic fluids is necessary, as many cosmic fluids, and the cosmic fluid itself, are relativistic. One important advantage is that radiative transfer can be developed as transport in a relativistic fluid, thereby avoiding the usual classical mis-interpretation of the radiative Boltzmann equation. Plasmas and magnetohydrodynamics are also included because of their growing interest in the field of astrophysics. The important role played by magnetic fields in a large sample of cosmic systems is only now beginning to be appreciated.

The second part of the book deals with astrophysical applications. The fluid in a star (stellar interiors), the fluid of stars (dynamics of stellar systems), astrophysical plasma fluids (solar magnetic phenomena, interstellar gas dynamics, accretion discs and jets . . .) and three chapters devoted to cosmology and large-scale structure formation (the Newtonian cosmic fluid, the relativistic cosmic fluid, and the fluid of galaxies) are the main topics discussed. The list is reasonably complete. To a greater, or lesser, extent we have considered planets, stars, and galaxies, interplanetary, interstellar, and intergalactic media, as well as the Universe as a whole.

As elementary particle theories and unification models are neither assumed nor introduced here, the history of the fluid that composes our Universe can only be traced back to a time shortly before annihilation. This poses a strict space-time limit on our objectives: $z \approx 10^{10}$.

The development of each chapter, and even of each section, is obsessively similar, reproducing the same logical scheme: the microscopic Boltzmann equation followed by its 'daughter' equations of macroscopic interest, continuity, motion, and energy balance. Only in those chapters with a relativistic treatment are the last two equations merged. This unified presentation not only has logical and aesthetic grounds, but is also didactic in that it attempts to demonstrate the fruitfulness of a systematic hydrodynamical approach.

E. Battaner
January 1995

Acknowledgements

Glenn Harding tried, unsuccessfully, to keep me from repeating the same mis-spelled words and incorrect sentences, but was then able to transform them into understandable English.

Throughout the 1992–1994 course my students experimented with an unusual didactic method consisting of finding errors in a manuscript containing plenty. In particular, I am very grateful to J.A. Vacas, A. Sarsa, M. Pérez-Victoria, A. Moñino, C. López-Sánchez, L.F. Jiménez-Fuentes, M.J. Carrillo, P. Casado, and A. Delgado.

My colleagues at the Astrophysics and Particle Physics Group permitted me to quit the real world from time to time and R. Rebolo and C. Abia revised part of the manuscript.

Professor John Beckman played a very special role in the production of the final version. He was asked to undertake a general revision of the scientific English, which he did with generosity and punctiliousness. More importantly, he contributed valuable scientific suggestions and maintained fruitful discussions with me, which have improved the book considerably. Only a philanthropic and bilingual astronomer could have carried out this task at his level of perfection.

Estrella Florido gave me her time and her science. She has been my muse, and my censor.

1 Classical fluids

1.1 The distribution function

Macroscopic equations of fluids can be obtained either from macroscopic principles or from the Boltzmann microscopic approach. The first method is more direct and intuitive, but when the mathematical difficulties of the second were eventually overcome, this acquired a higher theoretical interest, and even predicted unknown effects. For astrophysical applications the microscopic approach is preferable, because we do not possess an intuitive perception of many astrophysical fluid systems, with such extreme values of thermodynamic parameters that are clearly beyond human concepts of orders of magnitude. The microscopic approach has provided a powerful tool with which to study many different cosmic problems using a unique and systematic technique.

As is usual, let us introduce the one-particle probability distribution function $f(\vec{r}, \vec{p}, t)$ (where $\vec{r}$ is the position vector, $\vec{p}$ the particle's momentum and t the time), which when multiplied by the six-dimensional phase space volume element $d\tau = d\tau_r \, d\tau_p = dx \, dy \, dz \, dp_x \, dp_y \, dp_z$ gives the total number of particles contained in this volume element.

Our objective is the determination of f. Once f is determined, any macroscopic quantity or property of a fluid system will be easy to obtain.

1.1.1 Number density and mean quantities

Firstly, a one-component fluid, with all molecules having the same mass and properties, will be considered. The number density of particles per spatial volume element can be obtained by

$$n = \int_p f \, d\tau_p \tag{1.1}$$

where

$$\int_p \equiv \int_{p_x=-\infty}^{\infty} \int_{p_y=-\infty}^{\infty} \int_{p_z=-\infty}^{\infty} \tag{1.2}$$

The number density has an obvious macroscopic importance. If m is the mass of a particle, the density is

$$\rho = mn \tag{1.3}$$

Let G be a quantity defined for each particle, function of the mass, momentum, and energy. The mean value of G is defined by

$$\langle G \rangle = \frac{1}{n} \int_p Gf \, d\tau_p \tag{1.4}$$

where $\langle G \rangle$, n, and any other macroscopic quantities are functions of $\vec{r}$ and t, but not of $\vec{p}$. A particular function of obvious macroscopic interest is the mean velocity; taking $G \equiv \vec{v}$:

$$\vec{v}_0 = \langle \vec{v} \rangle = \frac{1}{n} \int_p \vec{v}f \, d\tau_p \tag{1.5}$$

where $\vec{v}$ is the velocity of a particle, equal to $\frac{\vec{p}}{m}$.

The peculiar velocity of a particle is defined as

$$\vec{V} = \vec{v} - \vec{v}_0 \tag{1.6}$$

so that the velocity of a particle $\vec{v}$ can be decomposed into a mean velocity $\vec{v}_0$, identical for all particles in the volume element, plus a peculiar velocity $\vec{V}$ characterizing the random thermal motion of each particle. It is obvious that $\langle \vec{V} \rangle = 0$, and this can easily be confirmed:

$$\langle \vec{V} \rangle = \frac{1}{n} \int_p (\vec{v} - \vec{v}_0)f \, d\tau_p = \frac{1}{n} \int_p \vec{v}f \, d\tau_p - \frac{1}{n} \int_p \vec{v}_0 f \, d\tau_p = \vec{v}_0 - \frac{1}{n} \vec{v}_0 n = 0 \tag{1.7}$$

The temperature is also of obvious macroscopic interest. Though the thermodynamic definition of temperature is very restrictive and in particular requires thermodynamic equilibrium, a kinetic temperature, representative of the magnitude of thermal motions, is introduced here:

$$\frac{3}{2} kT = \frac{1}{2} m \langle V^2 \rangle \tag{1.8}$$

where k is Boltzmann's constant. This definition is only valid for monatomic gases. If the gas is diatomic, $\frac{3}{2}$ must be replaced by $\frac{5}{2}$, and so on. This definition of temperature permits us to use this concept in a system which is not in thermodynamic equilibrium.

1.1.2   Transport fluxes

Let us assume a surface element $d\vec{S}$ inside the fluid, through which molecules freely flow. The number of molecules having a given $\vec{v}$ passing through $d\vec{S}$ is calculated by $f d\tau_r \, d\tau_p$, where $d\tau_r = (\vec{v} \, dt) \cdot d\vec{S}$. Therefore $f\vec{v} \cdot d\vec{S} \, dt \, d\tau_p$ molecules with momentum between $\vec{p}$ and $\vec{p} + d\vec{p}$ will cross the surface element in dt. Let G again be a mechanical quantity defined for each molecule. Then $Gf\vec{v} \cdot d\vec{S} \, dt \, d\tau_p$ is the magnitude of G transported through $d\vec{S}$, with a given momentum, in dt. The magnitude of G transported through $d\vec{S}$ in dt by any particle, with any $\vec{p}$, is

$$\int_p Gf\vec{v} \cdot d\vec{S} \, dt \, d\tau_p = n \langle G\vec{v} \rangle \cdot d\vec{S} \, dt \tag{1.9}$$

and per surface and time element $n\langle G\vec{v}\rangle \cdot \vec{u}$, where $\vec{u}$ is the unitary vector with direction $d\vec{S}$. This is the projection of the vector $n\langle G\vec{v}\rangle$ along $\vec{u}$. This vector is called the transport flux $\vec{\phi}$ of G:

$$\vec{\phi}(G) = n\langle G\vec{v}\rangle \tag{1.10}$$

and has the property that when projected along any direction, it yields the magnitude of G transported per unit area per unit time in this direction.

Let us give some examples of transport fluxes of quantities that can be transported. If G is the mass of a particle, its flux is the vector

$$\vec{\varphi} = n\langle m\vec{v}\rangle = mn\langle \vec{v}\rangle = \rho\vec{v}_0 \tag{1.11}$$

Even if there is no net mass flux, G can be transported:

$$\vec{\phi}(G) = n\langle G\vec{v}\rangle = n\langle G(\vec{v}_0 + \vec{V})\rangle = n\vec{v}_0\langle G\rangle + n\langle G\vec{V}\rangle \tag{1.12}$$

If $\vec{\varphi} = 0$, then $\vec{v}_0 = 0$ and the first term is zero, but the second one, $n\langle G\vec{V}\rangle$, may be non-vanishing.

$\vec{\phi}$ has been decomposed into two fluxes: a flux of the fluid carrying the mean value of G and another flux of G which is present even if the fluid is at rest.

As a second important example let G be the momentum. When G is a scalar, $\vec{\phi}$ is a vector. When $\vec{G}$ is a vector, the flux becomes a second-order tensor. The flux of $\vec{p}$ is

$$\mathcal{R} = n\langle \vec{p}\vec{v}\rangle = \rho\langle \vec{v}\vec{v}\rangle \tag{1.13}$$

Products such as $\vec{p}\vec{v}$ or $\vec{v}\vec{v}$ (without a point) are diadic or external products. Second-order tensors will in general be represented by script capital letters. We now use the relation $\vec{v} = \vec{v}_0 + \vec{V}$ again:

$$\begin{aligned}
\mathcal{R} &= \rho\langle (\vec{v}_0 + \vec{V})(\vec{v}_0 + \vec{V})\rangle \\
&= \rho\langle \vec{v}_0\vec{v}_0\rangle + \rho\langle \vec{v}_0\vec{V}\rangle + \rho\langle \vec{V}\vec{v}_0\rangle + \rho\langle \vec{V}\vec{V}\rangle \\
&= \rho\vec{v}_0\vec{v}_0 + \rho\langle \vec{V}\vec{V}\rangle
\end{aligned} \tag{1.14}$$

since $\vec{v}_0$ is a macroscopic quantity, constant in the volume element and can be taken outside the brackets, and $\langle \vec{V}\rangle = 0$. The first term contains $\vec{\varphi} = \rho\vec{v}_0$, so that it represents the momentum transport associated with mass flow. The second one represents an internal momentum transport that is present even if the fluid is at rest. This important tensor is called the pressure tensor:

$$\mathcal{P} = \rho\langle \vec{V}\vec{V}\rangle \tag{1.15}$$

which is, by definition, a symmetric second-order tensor.

To enhance familiarity with the pressure tensor, suppose that the fluid is in thermodynamic equilibrium. Then no privileged direction exists; so we can write (1.15) in the form

$$\mathcal{P} = p\delta \tag{1.16}$$

where $\mathcal{P}$ is triply degenerate, δ is the Kroenecker tensor, and p the eigenvalue; p is called the hydrostatic pressure. All off-diagonal components of $\mathcal{P}$ are zero because for any molecule with $V_1 V_2$ a molecule with $-V_1 V_2$ can always be found, so the mean of $V_1 V_2$ is zero.

The use of 'hydrostatic pressure', a well-known concept, must be justified. We will do this using two particularly simple examples.

First, suppose that a fluid is contained in a vessel. The walls are pushed perpendicularly with a force per surface element of the wall, given by the momentum which would have been transported through it in dt if the wall had been suppressed, that is, $\mathcal{P} \cdot \vec{u}$, with $\vec{u}$ perpendicular to the wall. The component of this force along $\vec{u}$, perpendicular to the surface is $(\mathcal{P} \cdot \vec{u}) \cdot \vec{u}$. But because of the threefold degeneracy of $\mathcal{P}$ in equilibrium we have

$$(\mathcal{P} \cdot \vec{u}) \cdot \vec{u} = (p\delta \cdot \vec{u}) \cdot \vec{u} = p\vec{u} \cdot \vec{u} = p \tag{1.17}$$

Hence, under the equilibrium condition, p is the force per surface element perpendicular to the wall.

Second, assume an ideal monatomic gas in equilibrium. The trace of $\mathcal{P}$ will be $\mathcal{P}_{ii} = 3p = \rho\langle V^2 \rangle$ and, recalling the definition (1.8) of the temperature of a monatomic gas,

$$p = nkT \tag{1.18}$$

which is the equation of state of an ideal gas.

Therefore, at least in equilibrium, the trace of the pressure tensor is three times the hydrostatic pressure. In general, this is adopted as the definition of the hydrostatic pressure.

Suppose, now, that $G \equiv \frac{1}{2}mv^2$, the kinetic energy of a molecule. If we restrict ourselves to a monatomic gas we can obtain the energy flux

$$\begin{aligned}
\vec{\Phi} &= n\left\langle \frac{1}{2}mv^2\vec{v} \right\rangle = \frac{1}{2}\rho\langle v^2\vec{v} \rangle \\
&= \frac{1}{2}\rho\langle (\vec{v}_0 + \vec{V}) \cdot (\vec{v}_0 + \vec{V})(\vec{v}_0 + \vec{V}) \rangle \\
&= \frac{1}{2}\rho v_0^2\vec{v}_0 + \frac{1}{2}\rho\vec{v}_0 \cdot \langle \vec{V}\vec{V} \rangle \\
&\quad + \frac{1}{2}\rho\langle \vec{V} \cdot \vec{v}_0\vec{V} \rangle + \frac{1}{2}\rho\langle V^2 \rangle\vec{v}_0 + \frac{1}{2}\rho\langle V^2\vec{V} \rangle \\
&= \frac{1}{2}\rho v_0^2\vec{v}_0 + \mathcal{P} \cdot \vec{v}_0 + \left(\frac{3}{2}kT \right)n\vec{v}_0 + \vec{q}
\end{aligned} \tag{1.19}$$

where we have used some tensor properties $((\vec{A}\vec{B}) \cdot \vec{C} = \vec{A}(\vec{B} \cdot \vec{C}) = \vec{A}\vec{B} \cdot \vec{C}$ for any three vectors; the symmetry of $\mathcal{P}$ has also been used). The vector $\vec{q}$, the conduction flux, is defined as

$$\vec{q} = \frac{1}{2}\rho\langle V^2\vec{V} \rangle \tag{1.20}$$

which is again present even if the fluid is at rest. The first term is the macroscopic kinetic energy transported by the fluid. The third term is microscopic kinetic energy transported by the fluid. $\mathcal{P} \cdot \vec{v}_0$ is an interesting term which will be discussed later.

1.1.3 Multicomponent fluids

In a fluid with more than one component, the ith component will have its own distribution function f_i, so that $f_i \, d\tau_r \, d\tau_p$ will give the number of molecules of type i contained in the volume element $d\tau_r \, d\tau_p$ at $\vec{r}$ and $\vec{p}$ in six-dimensional phase space. Then

$$f = \sum_i f_i \tag{1.21}$$

There is a number density n_i for each type of molecule:

$$n_i = \int_p f_i \, d\tau_p \tag{1.22}$$

and

$$n = \sum_i n_i \tag{1.23}$$

and a partial density

$$\rho_i = m_i n_i \tag{1.24}$$

The density of the mixture is

$$\rho = \sum_i \rho_i \tag{1.25}$$

The equivalent mass of the mixture is defined as

$$m = \frac{\rho}{n} \tag{1.26}$$

which is a weighted average of m_i taking n_i as the weighting factor. For different reasons, not all mean quantities are defined with the same weight for the calculation of the average. The mean velocity $\vec{v}_0$ of the mixture is defined with the partial mass density as the weighting factor. If $\vec{v}_i = \frac{\vec{p}}{m_i}$,

$$\vec{v}_0 = \frac{\sum_i m_i n_i \langle \vec{v}_i \rangle}{\sum_i m_i n_i} \tag{1.27}$$

where, obviously,

$$\langle \vec{v}_i \rangle = \frac{1}{n_i} \int_p f_i \vec{v}_i \, d\tau_p \tag{1.28}$$

is the mean velocity of the ith component. The peculiar velocity of a particle is again defined as

$$\vec{V}_i = \vec{v}_i - \vec{v}_0 \tag{1.29}$$

but now the mean value of $\vec{V}_i$ for the ith component may be non-vanishing, even where the mean value for the fluid as a whole is zero:

$$\langle \vec{V}_i \rangle = \frac{1}{n_i} \int_p f_i \vec{V}_i \, d\tau_p = \langle \vec{v}_i \rangle - \vec{v}_0 \tag{1.30}$$

This quantity is macroscopically interesting and is called the diffusion velocity of the ith component. The different diffusion velocities must compensate one another; to be precise:

$$\sum_i m_i n_i \langle \vec{V}_i \rangle = \sum_i m_i n_i \langle \vec{v}_i - \vec{v}_0 \rangle$$

$$= \sum_i m_i n_i \langle \vec{v}_i \rangle - \left(\sum_i m_i n_i \right) \vec{v}_0 \qquad (1.31)$$

$$= \left(\sum_i m_i n_i \right) \vec{v}_0 - \left(\sum_i m_i n_i \right) \vec{v}_0 = 0$$

There is a physical reason for defining $\vec{v}_0$ as in (1.27) instead of adopting n_i as a weighting factor. If we used n_i we would deduce $\vec{v}_0 = 0$, when the fluid is in fact flowing. Suppose, for instance, a massive component such as uranium is moving to the right with the same speed as a second lighter component such as hydrogen is moving to the left, the numbers of particles of both components being equal. If the weighting factor were n_i we would infer that the fluid was at rest, which would be true for numbers of particles, but not for the mass of the fluid as a whole.

The microscopic velocity $\vec{v}_i$ of a particle can now be written as

$$\vec{v}_i = \vec{v}_0 + \langle \vec{V}_i \rangle + \vec{V}_i' \qquad (1.32)$$

$\vec{v}_0$ and $\langle \vec{V}_i \rangle$ are macroscopic. $\vec{V}_i'$ now represents the individual chaotic thermal velocities. It can easily be found that $\langle \vec{V}_i' \rangle = 0$ for each component. In general, $\langle \vec{V}_i \rangle$ is much lower than typical values of $\vec{V}'$ and will be considered negligible when compared with the chaotic velocities.

The pressure tensor of an individual component is

$$\mathcal{P}_i = \rho_i \langle \vec{V}_i' \vec{V}_i' \rangle \qquad (1.33)$$

To define the temperature of the mixture, an average with the number density as a weighting factor is needed. For monatomic gases:

$$\frac{3}{2} k T_i = \frac{1}{2} m_i \langle V_i'^2 \rangle \qquad (1.34)$$

and

$$T = \frac{1}{n} \sum_i n_i T_i \qquad (1.35)$$

The physical reason for defining T as in (1.35) is that, under equilibrium conditions, all molecules, irrespective of their mass, have the same energy.

The total flux of the mixture is defined as the sum of the ith fluxes. In particular,

$$\vec{\varphi} = \sum_i \vec{\varphi}_i = \sum_i n_i (\vec{v}_0 + \langle \vec{V}_i \rangle) \qquad (1.36)$$

$$\mathcal{P} = \sum_i \mathcal{P}_i = \sum_i \rho_i \langle \vec{V}_i' \vec{V}_i' \rangle \qquad (1.37)$$

$$p = \sum_i p_i = \frac{1}{3} \sum_i \sum_j \mathcal{P}_{ijj} \qquad (1.38)$$

$$\vec{q} = \sum_i \vec{q}_i = \frac{1}{2} \sum_i \rho_i \langle V_i'^2 \vec{V}_i' \rangle \qquad (1.39)$$

where the definitions of $\vec{\varphi}_i$, $\mathcal{P}_i$, p_i, $\vec{q}_i$ are obvious generalizations.

1.1.4 Boltzmann's equation

The objective now is to deduce the differential equations for f_i and integrate them. Once each f_i is known, all functions of macroscopic interest may be calculated from their defining integrals. In particular, we will be able to calculate n_i, n, $\vec{v}_0$, $\langle \vec{V}_i \rangle$, T_i, and T.

Let us first assume a one-component fluid. Let $\vec{F}$ be the force acting on a particle, excluding the short-range particle–particle interaction force in a collision. To clarify this concept, we can take a self-gravitating fluid. $\vec{F}$ will be the gravitational force arising from the whole ensemble of particles, but not the gravitational force induced on a particle by a single neighbour particle. $\vec{F}$ can be an external force or it can be produced by the fluid itself.

Particles in the six-dimensional phase space cell $[\vec{r}, \vec{p}]$ (i.e. having their position vector $\vec{r}$ and their momentum $\vec{p}$ in a given volume element $d\tau_r\, d\tau_p$) would travel after dt to the cell $[\vec{r} + \vec{v}\, dt, \vec{p} + \vec{F}\, dt]$ if there were no collisions. The number of particles in each cell, in the absence of collisions, would then be invariant:

$$f\,(\vec{r} + \vec{v}\, dt, \vec{p} + \vec{F}\, dt, t + dt)d\tau_r\, d\tau_p = f\,(\vec{r}, \vec{v}, t)\, d\tau_r\, d\tau_p \tag{1.40}$$

Using a series expansion on the left-hand side:

$$\frac{\partial f}{\partial t} + \vec{v} \cdot \nabla f + \vec{F} \cdot \nabla_p f = 0 \tag{1.41}$$

where the momentum-gradient ∇_p of f is the vector $\frac{\partial f}{\partial p_i}$. Some collisions will have the effect of propelling particles into the cell $[\vec{r} + \vec{v}\, dt, \vec{p} + \vec{F}dt]$ which were not previously contained in $[\vec{r}, \vec{p}]$. Others will have the effect of removing particles that were in $[\vec{r}, \vec{p}]$, preventing them from reaching $[\vec{r} + \vec{v}\, dt, \vec{p} + \vec{F}\, dt]$. Equation (1.41) must be modified by a term that takes into account the effect of collisions. This term is conventionally called Γ and the result is Boltzmann's equation:

$$\frac{\partial f}{\partial t} + \vec{v} \cdot \nabla f + \vec{F} \cdot \nabla_p f = \Gamma \tag{1.42}$$

For a multicomponent fluid we have

$$\frac{\partial f_i}{\partial t} + \vec{v}_i \cdot \nabla f_i + \vec{F}_i \cdot \nabla_p f_i = \Gamma_i \tag{1.43}$$

This equation was first derived by Boltzmann in 1872, but the integration was not performed until 1916 by Chapman and Enskog (independently), and even then not for the most general case (only for a dilute monatomic gas not very far from thermodynamic equilibrium). In the absence of collisions and under specified conditions, it was later integrated by Tonks and Langmuir in 1929, Chandrasekhar in 1942 for the fluid comprising galactic stars, Landau in 1946, and others.

Of course, in order to integrate Boltzmann's equation, we should write Γ explicitly as a function of our independent variables $\vec{r}$, $\vec{p}$, and t, and of the distribution function f. However, no attempt to describe the integration of this equation in detail will be made here, as it would take up too much space. The reader is referred to classical texts such as those by Chapman and Cowling (1970) or Hirschfelder, Curtiss and Bird (1954). Instead, macroscopic equations consequences of Boltzmann's equation will be derived, which do not require an explicit knowledge of the collision term. These macroscopic equations are the fluid dynamic equations. They provide less information than (1.42), as they do not provide $\langle \vec{V}_i \rangle$, $\mathcal{P}$, and $\vec{q}$, and to overcome this problem and complete the set of differential equations, some indirect arguments will be invoked in each case, bearing in mind that a formal integration underpins our conclusions.

1.2   Macroscopic implications

In order to obtain macroscopic equations in which the collision term is not present, the following procedure may be adopted:

$$\sum_i \int_p (Boltzmann's\ equation)_i G_i\ d\tau_p \tag{1.44}$$

Now G_i is a function defined for each particle which has the property of collisional invariance, such as mass, momentum, energy, and any combination of these quantities. Then the collision term gives

$$\sum_i \int_p \Gamma_i G_i\ d\tau_p = 0 \tag{1.45}$$

because it represents the change in G in a given time element due to collisions between molecules in a space volume element, with any momentum and belonging to any constituent. It is necessary to perform the operation $\sum_i$ because the momentum (for instance) gained by one type of particle will be lost by another type.

In this way, three macroscopic equations can be obtained, corresponding to the conservation of mass, momentum, and energy.

1.2.1   Continuity equations

Let us consider mass as the collisional invariant. In this exceptional case, the final sum over all constituents in (1.44) is not necessary because not only is the mass of the whole mixture conserved, but the mass of each individual constituent must also be conserved.

From the first term of Boltzmann's equation, applying (1.44),

$$\int_p \frac{\partial f_i}{\partial t} m_i\ d\tau_p = m_i \frac{\partial}{\partial t} \int_p f_i\ d\tau_p = m_i \frac{\partial n_i}{\partial t} \tag{1.46}$$

Note that in this six-dimensional phase space t, $\vec{r}$, and $\vec{p}$ are independent variables and can have any value assigned by the observer. (For instance, $\vec{p} = m_i \frac{d\vec{r}}{dt}$ cannot be written; $\vec{r}$ and t are independent; and $\frac{d\vec{r}}{dt}$ would be zero.) Due to this fact $\frac{\partial}{\partial t}$ is taken outside the integral. The second term in Boltzmann's equation gives

$$\int_p m_i \vec{v}_i \cdot \nabla f_i\ d\tau_p = m_i \nabla \cdot \int_p f_i \vec{v}_i\ d\tau_p = m_i \nabla \cdot (n_i \langle \vec{v}_i \rangle) \tag{1.47}$$

and the third term gives

$$\int_p m_i \vec{F}_i \cdot \nabla_p f_i \, d\tau_p = m_i F_{i1} \int_{p_1} \int_{p_2} \int_{p_3} \frac{\partial f_i}{\partial p_1} \, dp_1 \, dp_2 \, dp_3 + \dots$$

$$= m_i F_{i1} \int_{p_2} \int_{p_3} [f_i]_{-\infty}^{\infty} \, dp_2 \, dp_3 + \dots = 0 \tag{1.48}$$

because $f_i(\infty)$, as well as $f_i(-\infty)$, must vanish, as in any distribution function; otherwise n would be infinite. Another assumption used in obtaining (1.48) is that $\vec{F}_i$ is assumed to be independent of $\vec{p}$ (i.e. independent of $\vec{v}_i$), as is true for most types of force. This is not always the case and the most important exception is the Lorentz force. This force acts on charged particles; electromagnetic forces and plasmas are considered later in Chapter 4. $\vec{F}_i$ will now be considered independently of $\vec{p}$ and the results obtained will be revised when this is required.

Finally, we obtain

$$\frac{\partial n_i}{\partial t} + \nabla \cdot (n_i \langle \vec{v}_i \rangle) = 0 \tag{1.49}$$

which form the continuity equations for each constituent i. They are usually written as

$$\frac{\partial n_i}{\partial t} + \nabla \cdot (n_i (\vec{v}_0 + \langle \vec{V}_i \rangle)) = 0 \tag{1.50}$$

A continuity equation for the mixture is obtained when these equations are multiplied by m_i and added together, taking (1.31) into account:

$$\frac{\partial \rho}{\partial t} + \nabla \cdot (\rho \vec{v}_0) = 0 \tag{1.51}$$

Note that $\vec{r}$ and t are independent variables. We are normally interested in knowing the density at a position chosen by us, and at a time chosen by us; there is no sense in calculating $\frac{d\vec{r}}{dt}$. The meaning of $\frac{\partial \rho}{\partial t}$ is obviously the time variation observed in the density, at a fixed point, without any spatial displacement. If the total derivative $\frac{d\rho}{dt}$ were calculated, as $\frac{dx_i}{dt} = 0$ we would obtain $\frac{d\rho}{dt} = \frac{\partial \rho}{\partial t}$. However, there is another time derivative, called the convective derivative, which can be denoted by $\frac{d\rho}{dt}$ without any risk of confusion. In general, $d\rho/dt$ will be given by

$$\frac{d\rho}{dt} = \frac{\partial \rho}{\partial t} + \frac{\partial \rho}{\partial x} \frac{dx}{dt} + \frac{\partial \rho}{\partial y} \frac{dy}{dt} + \frac{\partial \rho}{\partial z} \frac{dz}{dt} \tag{1.52}$$

Now, we accept that there will be spatial displacements in dt, and that these will actually be taking place macroscopically in the fluid, that is, we accept $\frac{dx}{dt} = v_{ox}$, and so on. Then we write, as a definition of the new symbol $\frac{d}{dt}$,

$$\frac{d\rho}{dt} = \frac{\partial \rho}{\partial t} + \vec{v}_0 \cdot \nabla \rho \tag{1.53}$$

To calculate $\frac{d\rho}{dt}$, we follow the fluid motion $\vec{r} = \vec{v}_0 t$: therefore $\frac{d}{dt}$ is the time derivative that would be measured by an observer travelling with the fluid. This observer, who has a velocity $\vec{v}_0(\vec{r}, t)$ at each point, will be called the 'wet' observer, in contrast with the inertial observer – the 'dry' observer – who measures time variations at a fixed point.

Equation (1.53) is valid not only for the density, but also for any other quantity, so that the relation between operators can be written as

$$\frac{d}{dt} \equiv \frac{\partial}{\partial t} + \vec{v}_0 \cdot \nabla \tag{1.54}$$

It can also be applied to vectors, and even to $\vec{v}_0$ itself:

$$\frac{d\vec{v}_0}{dt} = \frac{\partial \vec{v}_0}{\partial t} + \vec{v}_0 \cdot \nabla \vec{v}_0 \tag{1.55}$$

The use of the dry observer's derivative is preferable for integrating the equations, but the calculation of $\frac{d}{dt}$, the wet observer's time derivative, or convective derivative, sometimes provides interesting interpretative insights. This is true, for example, for the Milky Way star fluid or for the Universe as a whole, in which we are wet observers.

Another possible expression of the continuity equation might be

$$\frac{d\rho}{dt} + \rho \nabla \cdot \vec{v}_0 = 0 \tag{1.56}$$

which is in fact of less practical interest than (1.51).

The continuity equation is the macroscopic, fluid dynamic form of mass conservation.

1.2.2   The equation of motion

Now $G_i \equiv \vec{p}$, and the operations implied in (1.44) will be carried out. However, $\sum_i$ will be delayed until the end of this section (1.67). Let us define

$$\vec{A}_i = \int_p \Gamma_i \vec{p} \, d\tau_p \tag{1.57}$$

We know, of course, that

$$\sum_i \vec{A}_i = 0 \tag{1.58}$$

From the first term of Boltzmann's equation we obtain

$$\int_p \vec{p} \frac{\partial f_i}{\partial t} \, d\tau_p = m_i \frac{\partial}{\partial t} \left(n_i (\vec{v}_0 + \langle \vec{V}_i \rangle) \right) \tag{1.59}$$

From the second term we obtain

$$\int_p \vec{p}(\vec{v}_i \cdot \nabla f_i) \, d\tau_p = m_i \int_p \vec{v}_i(\vec{v}_i \cdot \nabla f_i) \, d\tau_p = \nabla \cdot (\rho_i \langle \vec{v}_i \vec{v}_i \rangle)$$
$$= \nabla \cdot (\rho_i \langle (\vec{v}_0 + \langle \vec{V}_i \rangle + \vec{V}_i')(\vec{v}_0 + \langle \vec{V}_i \rangle + \vec{V}_i') \rangle) \tag{1.60}$$
$$\approx \nabla \cdot (\mathcal{P}_i + \rho_i \vec{v}_0 \vec{v}_0)$$

neglecting $\langle \vec{V}_i \rangle$ compared to the mean values of $\vec{V}_i'$, and noting that $\langle \vec{V}_i' \rangle = 0$. Remember that the divergence of a second-order tensor τ is the vector $(\nabla \cdot \tau)_i = \frac{\partial}{\partial x_j}(\tau_{ji})$, that is, the contraction of its gradient.

It will be assumed not only that f_i goes to zero at infinity, but that the distribution moments $[f_i p^\alpha]_{-\infty}^\infty = 0$ for any value of α (1, 2, 3, . . .). We will now therefore assume that $[f_i \vec{p}]_{-\infty}^{\infty} = 0$. This is a common property of any physically useful distribution function. Then

$$\int_p \vec{p}\left(\vec{F}_i \cdot \nabla_p f_i\right) d\tau_p = \int_p \vec{p}\left((\nabla_p f_i) \cdot \vec{F}_i\right) d\tau_p = \left(\int_p (\vec{p}\nabla_p f_i) d\tau_p\right) \cdot \vec{F}_i \qquad (1.61)$$

The jth component of this vector would be

$$\left(\int_p p_j \frac{\partial f_i}{\partial p_k} d\tau_p\right) F_{ik} = \left(\int_p \frac{\partial}{\partial p_k} (f_i p_j) d\tau_p\right) F_{ik} - \left(\int_p f_i \frac{\partial p_j}{\partial p_k} d\tau_p F_{ik}\right) \qquad (1.62)$$

The subindex i is reserved for the ith constituent. Subindices j and k denote vector or tensor components.

The first integral on the right-hand side for $k = 1, j = 2$, takes the form

$$\int_p \frac{\partial}{\partial p_1} (f_i p_2) \, dp_1 \, dp_2 \, dp_3 = \int_{p2} \int_{p3} [f_i p_2]_{-\infty}^{\infty} \, dp_2 \, dp_3 = 0 \qquad (1.63)$$

The second integral is

$$-\int_p f_i \frac{\partial p_j}{\partial p_k} d\tau_p = -\int_p f_i \delta_{jk} \, d\tau_p = -\delta_{jk} \int_p f_i \, d\tau_p = -\delta_{jk} n_i \qquad (1.64)$$

Therefore, in (1.61),

$$\int_p \vec{p}(\vec{F}_i \cdot \nabla_p f_i) \, d\tau_p = -n_i \delta \cdot \vec{F}_i = -n_i \vec{F}_i \qquad (1.65)$$

and, finally,

$$\vec{A}_l = m_i \frac{\partial}{\partial t} \left(n_i(\vec{v}_0 + \langle \vec{V}_i \rangle) \right) + \nabla \cdot (\mathcal{P}_i + m_i n_i \vec{v}_0 \vec{v}_0) - n_i \vec{F}_i \qquad (1.66)$$

This formula will be needed later. It is now time to carry out the last sum $\sum_i$, and use (1.31):

$$\frac{\partial}{\partial t} (\rho \vec{v}_0) + \nabla \cdot (\mathcal{P} + \rho \vec{v}_0 \vec{v}_0) - \sum_i n_i \vec{F}_i = 0 \qquad (1.67)$$

Though this is a possible way of presenting the equation of motion, it is not the most usual. Combining (1.67) and the continuity equation it can easily be found that

$$\rho \frac{\partial \vec{v}_0}{\partial t} + \rho \vec{v}_0 \cdot \nabla \vec{v}_0 + \nabla \cdot \mathcal{P} - \sum_i n_i \vec{F}_i = 0 \qquad (1.68)$$

This is called the equation of motion because it is the macroscopic, fluid dynamic expression of Newton's second law. The first term is of the type *mass* × *acceleration*, and therefore the other terms are to be interpreted as forces per volume element. The term $\vec{v}_0 \cdot \nabla \vec{v}_0$ is the difference between the accelerations seen by the wet and the dry observers, and is therefore called the inertial term. It introduces complications in the mathematical treatment, especially for the peculiar turbulent behaviour often present in fluids. These complications arise from the non-linear character of this term. In equilibrium, $\nabla \cdot \mathcal{P} = \nabla p$ is a force arising from the hydrostatic pressure gradient, called the gradient force. Away from equilibrium, $\nabla \cdot \mathcal{P}$ also includes viscosity forces, as will be described later (1.4.2). The latter term has an obvious interpretation, as it represents the action of real forces.

Gravity (as well as other forces) can be written as $\vec{F}_i = m_i\vec{g}$, where $\vec{g}$ has no subindex. This type of force, which includes all types of inertial force, is called non-diffusive. The force term for them takes the form

$$\sum_i n_i\vec{F}_i = \sum_i m_i n_i \vec{g} = \left(\sum_i m_i n_i\right)\vec{g} = \rho\vec{g} \tag{1.69}$$

The equation of motion for a multicomponent fluid is in this case exactly the same as the equation for a single component fluid.

1.2.3   The energy balance equation

Let $G_i \equiv \frac{1}{2}m_i v_i^2$, the kinetic energy of an atom of a monatomic gas. We know that

$$\sum_i \int_p \frac{1}{2}m_i v_i^2 \Gamma_i \, d\tau_p = 0 \tag{1.70}$$

The derivation of the heat balance equation is similar to the previous derivation, so we will not go into detail. The results are

$$\textit{First term} \quad \frac{\partial}{\partial t}\left(\frac{3}{2}nkT + \frac{1}{2}\rho v_0^2\right)$$

$$\textit{Second term} \quad +\nabla\cdot(\vec{q} + \mathcal{P}\cdot\vec{v}_0 + \frac{3}{2}nkT\vec{v}_0 + \frac{1}{2}\rho v_0^2\vec{v}_0) \tag{1.71}$$

$$\textit{Third term} \quad -\sum_i\left(n_i\vec{F}_i\cdot(\vec{v}_0 + \langle\vec{V}_i\rangle)\right) = 0$$

However, another form of this equation is more often used, obtained by combining (1.71), the equation of motion, and the equations of continuity:

$$\frac{3}{2}nk\left(\frac{\partial T}{\partial t} + \vec{v}_0\cdot\nabla T\right) + \nabla\cdot\vec{q} + P_{kj}\frac{\partial v_{ok}}{\partial x_j} - \sum_i(n_i\vec{F}_i\cdot\langle\vec{V}_i\rangle) = 0 \tag{1.72}$$

The first set of parentheses can be shortened using the wet observer time derivative, $\frac{\partial T}{\partial t} + \vec{v}_0\cdot\nabla T = \frac{dT}{dt}$, although that does not help in integrating the equation. Though we do not derive it here, we can use the fact that, for monatomic gases,

$$c_v = \frac{3}{2}\frac{k}{m} \tag{1.73}$$

where c_v is the specific heat of constant volume. So $\frac{3}{2}nk$ in (1.72) can be replaced by $c_v\rho$. The expression obtained is in fact not restricted to monatomic gases, but is more general. We can also use the expression for the specific internal energy per particle for monatomic gases:

$$u = \frac{3}{2}kT \tag{1.74}$$

to obtain a general formula, although in fact we will deal here with monatomic gases, unless otherwise specified. The first term in (1.72) thus has some useful alternative forms:

$$\frac{3}{2}nk\left(\frac{\partial T}{\partial t}+\vec{v}_0\cdot\nabla T\right)=c_v\rho\left(\frac{\partial T}{\partial t}+\vec{v}_0\cdot\nabla T\right)=n\left(\frac{\partial u}{\partial t}+\vec{v}_0\cdot\nabla u\right) \tag{1.75}$$

$\frac{dT}{dt}$ is a heating (cooling when negative) term, so the other terms must represent different means of heating the fluid. One of them, $\nabla\cdot\vec{q}$, is a divergent heat conduction flux. The term $\mathcal{P}_{kj}\frac{\partial\vec{v}_{ok}}{\partial x_j}$ is the trace of the internal product $\mathcal{P}\cdot\nabla\vec{v}_0$ and its physical meaning needs further study. We will treat it in detail later (1.3.2 and 1.4.5) but in anticipation, we can say that it includes adiabatic and viscous heating. The last term is not present in single-component fluids and can generally be ignored.

The final balance equation is the macroscopic, fluid dynamic expression of the first law of thermodynamics.

1.2.4   Fluid dynamic equations

The macroscopic fluid dynamic equations are
(a) continuity equation for the *i*th component:

$$\frac{\partial n_i}{\partial t}+\nabla\cdot\left(n_i(\vec{v}_0+\langle\vec{V}_i\rangle)\right)=0 \tag{1.76}$$

(b) continuity equation for the mixture:

$$\frac{\partial\rho}{\partial t}+\nabla\cdot(\rho\vec{v}_0)=0 \tag{1.77}$$

(c) equation of motion:

$$\rho\frac{\partial\vec{v}_0}{\partial t}+\rho\vec{v}_0\cdot\nabla\vec{v}_0+\nabla\cdot\mathcal{P}-\sum_i n_i\vec{F}_i=0 \tag{1.78}$$

(d) heat balance equation:

$$\frac{3}{2}nk\left(\frac{\partial T}{\partial t}+\vec{v}_0\cdot\nabla T\right)+\nabla\cdot\vec{q}+\mathcal{P}_{kj}\frac{\partial v_{ok}}{\partial x_j}-\sum_i\left(n_i\vec{F}_i\cdot\langle\vec{V}_i\rangle\right)=0 \tag{1.79}$$

This is a system of $N+3+1$ equations, N being the number of constituents. Equation (1.77) is not counted as it is obtained from (1.76) and is not independent. As unknowns, we have $(n_i, \vec{v}_0, T)$ plus $(3N+9+3)$, corresponding to $\langle\vec{V}_i\rangle$, $\mathcal{P}$ and $\vec{q}$. As $\mathcal{P}$ is symmetrical we have in practice six unknowns $\mathcal{P}_{kj}$ instead of nine. But in any case the large number of unknowns is clear. If it were possible to find the transport fluxes, $\langle\vec{V}_i\rangle$, $\mathcal{P}$, and $\vec{q}$ as functions of $\vec{r}$, t, n_i, $\vec{v}_0$, and T, the system of differential equations would become closed. In subsequent sections these functions will be investigated under different conditions.

1.3   The zeroth-order approximation. Perfect fluids

Euler's equations are the fluid dynamic equations in which the distribution function is the Maxwell–Boltzmann distribution corresponding to thermodynamic equilibrium

$$f=n(2\pi mkT)^{-\frac{3}{2}}e^{-\frac{(\vec{p}-m\vec{v}_0)^2}{2mkT}} \tag{1.80}$$

At first glance, thermodynamic equilibrium is incompatible with motion, and the existence of gradients and time variations of p, T, ρ, etc. However, our assumption will not imply full thermodynamic equilibrium, but local thermodynamic equilibrium, LTE. If a given volume element is observed, a Maxwellian velocity distribution is obtained for a certain temperature, because the system is very close to equilibrium. For larger scales, gradients of T may exist, but these scales are so large that, locally, the equilibrium is not perturbed and LTE is a good approximation. Additionally, time derivatives of the temperature may exist, but for such a long characteristic time that the instantaneous local equilibrium is not perturbed. The same may be assumed for ∇p and $\frac{\partial p}{\partial t}$, and so on. Local thermodynamic equilibrium is an idealization that enables us to use the fluid equations (1.76)–(1.79) and the distribution (1.80) to calculate transport fluxes.

Local thermodynamic equilibrium is the equivalent for fluids to the zeroth-order approximation, and is also equivalent to the perfect fluid idealization. Perfect fluids obey Euler's equations.

1.3.1 Euler's equations

Using (1.80) $\langle \vec{V}_i \rangle$, $\mathcal{P}$ and $\vec{q}$ are easily calculated. It is even possible to avoid calculating the defining integrals by invoking an equilibrium property: the directional isotropy of $\vec{V}_i$. This property has already been used to determine $\mathcal{P} = p\delta$ in equilibrium. Now we can prove that under this condition $\langle \vec{V}_i \rangle$ and $\vec{q}$ vanish. For a molecule with $\vec{V}_i$ another is always found with $-\vec{V}_i$, so that they are compensated and the average is zero. Then, it is easy to show that

$$\nabla \cdot \mathcal{P} = \nabla p \tag{1.81}$$

and

$$\mathcal{P}_{kj} \frac{\partial v_{0k}}{\partial x_j} = p\nabla \cdot \vec{v}_0 \tag{1.82}$$

so that Euler's equations can be formulated:

$$\textit{Continuity} \quad \frac{\partial \rho}{\partial t} + \nabla \cdot (\rho \vec{v}_0) = 0 \tag{1.83}$$

$$\textit{Motion} \quad \rho \frac{\partial \vec{v}_0}{\partial t} + \rho \vec{v}_0 \cdot \nabla \vec{v}_0 + \nabla p - \sum_i n_i \vec{F}_i = 0 \tag{1.84}$$

$$\textit{Heat balance} \quad \frac{3}{2} nk \left(\frac{\partial T}{\partial t} + \vec{v}_0 \cdot \nabla T \right) + p\nabla \cdot \vec{v}_0 = 0 \tag{1.85}$$

Note that there are five differential equations and six unknowns ($\rho, \vec{v}_0, p, T$). We have not counted n_i because in the absence of diffusion $\frac{n_i}{n}$ must remain constant for the wet observer. In the absence of diffusion the multicomponent fluid behaves like a single-component fluid.

For non-diffusive forces we also bring in (1.69). As we have a number of unknowns greater than the number of equations, to solve the system we need the equation of state $p = p(\rho, T)$.

1.3.2 Basic astrophysical applications

To become familiar with Euler's equations we now consider some examples of astrophysical and geophysical interest, such as the concepts of scale height and plasmapause, adiabatic cooling in ascension and geostrophic winds.

Hydrostatics. Scale height. Plasmaspheres

A system is said to be stationary if $\frac{\partial}{\partial t} = 0$, and static if $\vec{v}_0 = 0$. The simplest case to begin with is a stationary static fluid. We further assume that $\vec{F}$ is non-diffusive. Then, if $\vec{F} = \rho\vec{g}$ the hydrostatic equation is obtained, in the form

$$\nabla p = \rho\vec{g} \tag{1.86}$$

or

$$\nabla p + \rho\nabla\mathcal{F} = 0 \tag{1.87}$$

where $\mathcal{F}$ is the gravitational potential.

If the fluid is incompressible, i.e. if ρ is a constant, then $p + \rho\mathcal{F} = $ constant. The surfaces $\mathcal{F} = $ constant are usually defined as horizontal. We then deduce that equipotential surfaces are isobars and are horizontal. A liquid surface must be an isobar, and is thus horizontal. Similarly consider a not so trivial classical problem in fluid dynamics: the rotating glass. A glass rotates with an axis coinciding with its symmetry axis; what is the equation of the liquid surface? For a non-inertial observer rotating with the glass $\mathcal{F} = gz - \frac{1}{2}\omega^2 r^2$, as the centrifugal potential must be included. ω is the rotation velocity. We then deduce that the required equation for $p = $ constant is a paraboloid, $gz - \frac{1}{2}\omega^2 r^2 = $ constant.

Consider now a stratified atmosphere of a planet or star. Stratification implies $\frac{\partial}{\partial x} = \frac{\partial}{\partial y} = 0$ which greatly simplifies our equation. Then, (1.87) gives

$$dp = -\rho g \, dz \tag{1.88}$$

for a planetary atmosphere. We can assume that the atmosphere is an ideal gas, for which $p = nkT$. Then,

$$\frac{dp}{p} = -\frac{mg}{kT}dz = -\frac{dz}{H} \tag{1.89}$$

where H is termed the scale height

$$H = \frac{kT}{mg} \tag{1.90}$$

It may depend on z, because T, g and m may depend on z. (As m is the equivalent mass, it depends on the chemical composition which may be a function of z.) The integration of (1.89) requires a knowledge of the relation between T and p. Consider the simplest case $H = $ constant, which in practice requires the constancy of T, g and m. For an isothermal atmosphere, with g and m also constants

$$p = p_0 e^{-\frac{z-z_0}{H}} \tag{1.91}$$

$$\rho = \rho_0 e^{-\frac{z-z_0}{H}} \tag{1.92}$$

$$n = n_0 e^{-\frac{z-z_0}{H}} = n_0 e^{-\frac{\mathcal{F}}{kT}} \tag{1.93}$$

The subindex 0 denotes the reference height z_0. In particular (1.93) is the Maxwell distribution function, as would be expected for an isothermal atmosphere. The exponential functions (1.91)–(1.93) permit the interpretation of the scale height as a characteristic vertical length in which significant variations of atmospheric parameters take place. Another way to interpret the concept of scale height is to use

$$\int_{z_0}^{\infty} \rho \, dz = \rho_0 H \tag{1.94}$$

This formula, which is easily deduced, tells us that the scale height is the height above the base z_0, of an atmosphere of uniform density (that at z_0) with the same column density as the real atmosphere. For instance, if our atmosphere were compressed to the surface density, it would be 8 km high.

The interpretation of (1.93) as a Maxwell distribution enables us to derive the vertical distribution of atoms in an atmosphere at those high altitudes where the gravitational acceleration can no longer be assumed constant, nor is the centrifugal force negligible. For points in the equatorial plane, for an observer rotating with the planet, we must consider both the gravitational potential and a centrifugal one, so that

$$\mathcal{F} = -\frac{GM}{r} - \frac{1}{2}\Omega^2 r^2 + \text{constant} \tag{1.95}$$

where r is the distance from the centre of the planet, GM is equal to $R_0^2 g_0$, R_0 is the planetary radius, g_0 the surface gravitational acceleration and Ω the rotational velocity of the planet. The constant can be chosen as $\mathcal{F}(R_0) = 0$, thus

$$\mathcal{F} = g_0 R_0 \left(1 - \frac{R_0}{r}\right) - \frac{1}{2}\Omega^2 (r^2 - R_0^2) \tag{1.96}$$

and the vertical number density distribution is given by the right-hand equality of (1.93), where $\mathcal{F}$ is now given by (1.96).

For moderate altitudes the centrifugal potential is negligible, so n decreases with altitude. However, for a certain altitude the preceding formula predicts an implausible increase of n with r. Above a certain radius, called the plasmapause radius, the force due to (1.95) becomes positive (i.e. directed outwards) and no equilibrium can be expected. In the absence of other forces, such as magnetic forces, atoms will simply escape into space. The equatorial position of the plasmapause is calculated by setting the force to zero, i.e. $\frac{\partial \mathcal{F}}{\partial r} = 0$.

$$r_p = \left(\frac{g_0 R_0^2}{\Omega^2}\right)^{\frac{1}{3}} = \left(\frac{GM}{\Omega^2}\right)^{\frac{1}{3}} \tag{1.97}$$

At this plasmapause radius there is a sudden decrease in the number density. On the Earth r_p is about six Earth radii, whilst this ratio is 50 on Venus and two on Jupiter. We note that above the plasmapause on some planets there is another planetary region, the magnetosphere, which we will discuss below.

Adiabatic cooling during ascent. Adiabatic atmosphere

As a second application of Euler's equations, consider a stationary but non-static situation in a nearly stratified atmosphere, in which the term $p\nabla \cdot \vec{v}_0$ plays an important role. Consider a rising column of air surrounded by static air. The air rises slowly so that it is possible to assume that the pressure height distribution is not perturbed and that it is still given by (1.89). In this sense the atmosphere is stratified, but it is clear that rising air implies a different vertical velocity for different x, y, and we speak of quasi-stratification. Inside the column, the continuity equation is

$$\frac{\partial}{\partial z}(nv_0) = 0 \tag{1.98}$$

noting that $v_{0x} = v_{0y} = 0$. The heat balance equation is

$$\frac{3}{2}nkv_0\frac{\partial T}{\partial z} + p\frac{\partial v_0}{\partial z} = 0 \tag{1.99}$$

It is simple to obtain

$$\frac{5}{2}\frac{1}{T}\frac{\partial T}{\partial z} + \frac{1}{H} = 0 \tag{1.100}$$

The coefficient $\frac{5}{2}$ reminds us of the constant pressure specific heat $c_p = \frac{5}{2}\frac{k}{m}$. Therefore

$$\frac{\partial T}{\partial z} = -\frac{g}{c_p} = -\Gamma \tag{1.101}$$

The wet observer notices a heating of

$$\frac{dT}{dt} = \frac{\partial T}{\partial t} + v_0\frac{\partial T}{\partial z} = -v_0\frac{g}{c_p} = -v_0\Gamma \tag{1.102}$$

If $v_0 > 0$, $\frac{dT}{dt} < 0$, the air is cooled during the ascent. Clearly, descending air would become hotter.

For instance in the Earth's atmosphere we have the approximate values, $c_p \approx 0.24$ cal/K gr, $\Gamma \approx 10^{-4}$ K cm^{-1}. A 100-m-high column of rising air would have a temperature difference between the bottom and the top of about 1 K. Therefore, this effect is important in the heat balance of any atmosphere.

When an air bubble is rising and cooling, it can become cooler than the surrounding gas; then it will be denser than its surroundings and will fall again. The atmosphere is then said to be stable. If, on the other hand, the bubble is hotter than the surroundings after rising, it will rise further and the atmosphere is said to be unstable. Suppose that the atmospheric region surrounding the bubble has a structural temperature gradient $\frac{\partial T}{\partial z} = -\alpha$. Then the condition for an atmosphere to be stable is

$$\alpha < \Gamma \tag{1.103}$$

If not, convection cells will develop. But convection cells transport energy upwards, as we will describe in detail below. This positive vertical energy flux modifies the structural temperature gradient, reducing the value of α. It may even happen that the critical value $\alpha = \Gamma$ is reached, and then convection ceases. Therefore, an unstable atmosphere tends to become stable by means of internal convective processes and has a gradient $\frac{\partial T}{\partial z} = -\frac{g}{c_p}$. The atmosphere is then said to be adiabatic. In an adiabatic atmosphere there is therefore a known relation $T(z)$. As it has been shown that there is a known pressure height distribution $p(z)$, there must be a known relation $p = p(T)$ in an adiabatic atmosphere. And therefore the relations $n(T)$, $n(p)$ also characterize an adiabatic atmosphere. What is this relation? Or, in particular, what is the (T, ρ) relation? From (1.89), the equation of state of ideal gases and the differential equation describing adiabatic atmospheres $\frac{\partial T}{\partial z} = -\frac{g}{c_p}$, it is easily calculated that

$$T\rho^{1-\gamma} = \text{constant} \tag{1.104}$$

where $\gamma = \frac{c_p}{c_v}$ (equal to 5/3 if our gas is monatomic). Equation (1.104) is clearly the equation of adiabatic processes in an ideal gas. Equation (1.104) is therefore a pleasant and perhaps unexpected result.

When α is negative, $\frac{\partial T}{\partial z} > 0$, and the atmosphere is always stable. If α is positive, $\frac{\partial T}{\partial z} < 0$, and the atmosphere is either stable or unstable, depending on its absolute value. In an unstable atmosphere, convective or turbulent vertical transport may be very important and must be included in the equations. If the atmosphere is stable, only a static mechanism of energy transport, such as conduction or radiation, need be considered.

The sign of $\frac{\partial T}{\partial z}$ is therefore closely related to the stability of a planetary atmosphere or a region of an atmosphere. It has been taken as one of the criteria for the classification of atmospheric layers and the assignation of their names. Following this classification, in the terrestial atmosphere we have (in order of increasing height) the troposphere ($\frac{\partial T}{\partial z} < 0$), the stratosphere ($\frac{\partial T}{\partial z} > 0$), the mesosphere ($\frac{\partial T}{\partial z} < 0$), the thermosphere ($\frac{\partial T}{\partial z} > 0$), and the exosphere ($\frac{\partial T}{\partial z} = 0$). The troposphere and the mesosphere are the most unstable layers, both producing cloud formation. Other planets also have these layers but they do not exhibit the temperature maximum at the stratopause (region where the stratosphere ceases). This maximum at the terrestrial stratopause is due to near ultraviolet radiation absorption by ozone. Chemical reactions produce this ozone from molecular oxygen, and thus our particular terrestrial maximum is associated with the existence of life on Earth. In other planets there is also an exosphere, where a high conduction coefficient ensures temperature homogeneity, a thermosphere, where the solar ultraviolet radiation is mainly absorbed in the upper layers producing ($\frac{\partial T}{\partial z} > 0$), and a troposphere with ($\frac{\partial T}{\partial z} < 0$) because the whole atmosphere is transparent to visible radiation, which then heats the surface and this in turn heats the lower layers, producing ($\frac{\partial T}{\partial z} < 0$).

The geostrophic wind

Consider now the atmospheric wind of a rotating planet or star. In addition to gravity, other inertial forces such as the centrifugal and the Coriolis forces must be included in the equation of motion. On the Earth, usually neither the centrifugal force nor the horizontal component of gravity are included as forces. The Earth is an ellipsoidal-like planet: the equatorial radius is larger than the polar radius and therefore there is a horizontal (constant distance to the centre) component of gravity directed polewards. On the other hand, the centrifugal force has an equatorwards horizontal component.

Both forces should be included in precise calculations, but they almost compensate one another and either both are included or both are neglected in most practical calculations. This equality has not arisen by chance but because the primitive fluid Earth adopted its present ellipsoidal shape as a compromise between centrifugal and gravitational forces. The same can be said for other rotating planets such as Jupiter and Saturn and even for some stars which are fluids at present.

However, the horizontal component of the Coriolis force cannot be neglected, because it is of high importance in wind dynamics. The three-dimensional equation of motion will then be

$$\rho\frac{\partial \vec{v}_0}{\partial t} + \rho\vec{v}_0 \cdot \nabla\vec{v}_0 + 2(\vec{\Omega} \times \vec{v}_0)\rho + \nabla p - \rho\vec{g} = 0 \tag{1.105}$$

where $\vec{\Omega}$ is the rotation angular velocity of the planet. Attention must be paid to the sign of the Coriolis force because of its pseudovectorial nature. The directions of the axes must be specified, and here we adopt, x to the east, y to the north, z upwards. We also assume a stationary situation and that the inertial force is negligible. As only horizontal winds are now being considered, $v_{0z} = 0$. Geostrophic winds are idealized winds, explaining reasonably well actual wind patterns in rotating planets which are the result of a balance between two horizontal forces: the Coriolis force and the pressure gradient force. By projecting (1.105) into the horizontal plane, we obtain

$$-2\Omega v_{0y} \sin\lambda + \frac{1}{\rho}\frac{\partial p}{\partial x} = 0 \tag{1.106}$$

$$2\Omega v_{0x} \sin\lambda + \frac{1}{\rho}\frac{\partial p}{\partial y} = 0 \tag{1.107}$$

where λ is the latitude. The wind speed is

$$v_0 = (v_{0x}^2 + v_{0y}^2)^{\frac{1}{2}} = \frac{|\nabla p|}{2\Omega\rho\sin\lambda} \tag{1.108}$$

A high horizontal pressure gradient produces a high wind speed, as expected. It is rather unexpected, even though we may be familiar with weather maps, to find $\vec{v}_0$ perpendicular to ∇p.

The wind does not follow the direction from high to low pressures, but turns clockwise around high pressure zones and anticlockwise around low pressure zones. These results are valid only for the northern hemispheres, the sense of rotation being inverted in the southern hemispheres. (Of all the solar system bodies only Venus rotates in the inverse sense, and its rotation velocity is too small to produce geostrophic winds.)

When integration over time intervals that are long compared with the rotation period is considered, the equator, which receives more heat, becomes a higher pressure zone. Then the pressure gradient is from pole to equator and consequently winds along the parallels prevail. This result, known as prevailing zonal winds, is suggested by the familiar images of the outer planets, which rotate very rapidly.

Geostrophic winds are also present in the terrestrial troposphere and stratosphere. At higher altitudes, another force, the ion-drag force, predominates and winds are no longer geostrophic.

1.4   **First-order approximation. Imperfect fluids**

Perfect fluids are in local thermodynamic equilibrium and are subject to Euler's equations. Perfect fluids do not permit any diffusion, do not transport momentum other than $p\delta$, and heat conduction is not present. Imperfect fluids can be influenced by diffusion, viscosity, and heat conduction. Viscosity is any kind of momentum transport other than hydrostatic pressure transport. To calculate general expressions for $\langle \vec{V}_i \rangle$, $\mathcal{P}$, and $\vec{q}$, large and complicated procedures are available, which become relatively simple for first-order approximation. In this approximation, the fluid equations are called the Navier–Stokes equations, which are now our main objective.

Transport fluxes are expressed as the product of a transport coefficient and a potential which may be either a vector or a tensor quantity. Potentials are functions of gradients or other causes producing transport fluxes. They are variable depending on the physical situation of the fluid at any time. Transport coefficients indicate the capacity of the fluid to produce fluxes from potentials. They are not constants but may depend on the state of the fluid, that is, on the temperature, pressure and composition. They are either tabulated or are obtainable from the theory of Chapman and Enskog. Let us first obtain the diffusion, viscosity, and heat conduction potentials, and then simplified expressions for diffusion, viscosity, and heat conduction coefficients.

1.4.1   Diffusion

The diffusion velocity $\langle \vec{V}_i \rangle$ can be calculated by solving the individual equations of motion. These were in fact obtained in (1.66), where the left-hand side is unknown. $\vec{A}_i$ was defined in (1.57) with the restriction (1.58), and represents the net increase in momentum by particles of type i due to collisions with particles of other types. Let us, then, adopt the assumption that

$$\vec{A}_i = \sum_{j \neq i} \vec{A}_{ij} \tag{1.109}$$

$$\vec{A}_{ij} = s_{ij} n_i n_j (\langle \vec{V}_j \rangle - \langle \vec{V}_i \rangle) \tag{1.110}$$

$\vec{A}_{ij}$ will be the momentum gain per time and volume elements by particles of type i, due to collisions with particles of type j. It is assumed that $\vec{A}_{ij}$ is a linear function of the relative velocity of the two constituents i and j (note that the relative velocity $\langle \vec{v}_j \rangle - \langle \vec{v}_i \rangle = \langle \vec{V}_j \rangle - \langle \vec{V}_i \rangle$) and is proportional to both number densities. s_{ij} is a coefficient with a value depending on the adopted intermolecular potential. For instance, for the elastic collision model between two rigid spheres

$$s_{ij} = \frac{16}{3} \sqrt{\mu_{ij}} \sigma_{ij}^2 \sqrt{\frac{\pi k T}{2}} \tag{1.111}$$

where μ_{ij} is the reduced mass and σ_{ij} is the sum of the radii of both intervening spheres. Let us define ν_i, the momentum transfer collision frequency by

$$m_i \nu_i = \sum_{j \neq i} n_j s_{ij} \tag{1.112}$$

which in fact has dimensions of frequency, but this name will be fully justified later.

Then the system of equations to be solved is

$$m_i \frac{\partial}{\partial t}\left(n_i(\vec{v}_0 + \langle \vec{V}_i \rangle)\right) + \nabla \cdot (\mathcal{P}_i + m_i n_i \vec{v}_0 \vec{v}_0) - n_i \vec{F}_i$$
$$= n_i \sum_j s_{ij} n_j (\langle \vec{V}_j \rangle - \langle \vec{V}_i \rangle) \tag{1.113}$$

In order to obtain a better insight into the diffusion phenomenon, let us assume some simplifying hypotheses. Diffusion probably behaves similarly when the fluid is in motion to when it is at rest. Therefore we adopt $\vec{v}_0 = 0$. A diffusion acceleration $\frac{\partial \langle \vec{V}_i \rangle}{\partial t}$ and the force $\frac{\partial}{\partial t}\left(n_i \langle \vec{V}_i \rangle\right)$ are neglected, and no individual viscosity is considered, that is, $\mathcal{P}_i = p_i \delta$. Then

$$\nabla p_i - n_i \vec{F}_i = n_i \sum_{j \neq i} s_{ij} n_j (\langle \vec{V}_j \rangle - \langle \vec{V}_i \rangle) \tag{1.114}$$

There are only $N - 1$ equations where is N the number of constituents because their sum is the equation of motion of the mixture, which is considered separately. But there are also $N - 1$ unknowns, because one of them is not independent, but is related by (1.31). $\langle \vec{V}_i \rangle$ must be obtained by solving (1.114) using standard methods, but quite often an iterative method is preferred. From (1.114) we obtain

$$\langle \vec{V}_i \rangle = \frac{\sum_{j \neq i} s_{ij} n_j \langle \vec{V}_j \rangle}{\sum_{j \neq i} s_{ij} n_j} - \frac{\frac{\nabla p_i}{n_i} - \vec{F}_i}{\sum_{j \neq i} n_j s_{ij}} = \vec{V}_{Fi} - \frac{1}{\nu_i}\left(\frac{\nabla p_i}{\rho_i} - \frac{\vec{F}_i}{m_i}\right) \tag{1.115}$$

where the friction velocity is defined by

$$\vec{V}_{Fi} = \frac{\sum_{j \neq i} s_{ij} n_j \langle \vec{V}_j \rangle}{m_i \nu_i} \tag{1.116}$$

The iterative procedure consists of calculating $\langle \vec{V}_i \rangle$ by using $\vec{V}_{Fi}$ using $\langle \vec{V}_i \rangle$ obtained in the previous step. For moderate rates of diffusion the procedure converges rapidly. Often $\vec{V}_{Fi}$ can be ignored, but this must be checked in each particular case. Equation (1.115) is usually written in the form

$$\langle \vec{V}_i \rangle = -\mathcal{D}_i \left(\nabla \ln(n_i kT) - \frac{\vec{F}_i}{kT}\right) \tag{1.117}$$

where $\mathcal{D}_i$ is one of the transport coefficients, called the diffusion coefficient:

$$\mathcal{D}_i = \frac{kT}{m_i \nu_i} \tag{1.118}$$

In (1.117) $\vec{V}_{Fi}$ was neglected. The diffusion potential is identified with the expression inside the parenthesis.

It is clear that $\mathcal{D}_i$ is not a constant as it depends on T and on ν_i. From its definition it is also clear that ν_i increases when ρ increases, so $\mathcal{D}_i$ is bigger in tenuous media. When $\mathcal{D}_i$ is large, the system quickly reaches so-called diffusive equilibrium, $\langle \vec{V}_i \rangle = 0$

$$\nabla(\ln n_i kT) = \frac{\vec{F}_i}{kT} \tag{1.119}$$

As a simple case, consider a binary mixture, one of the components being a minor constituent, $n_1 \ll n_2$. Then $n \approx n_2$ is approximately constant. If in addition the mixture remains isothermal, $\mathcal{D}_i$ is a constant. Suppose also that $\vec{F}_i = 0$. Then the diffusion flux of the minor constituent is

$$n_1 \langle \vec{V}_1 \rangle = -\mathcal{D}_1 \nabla n_1 \tag{1.120}$$

known as the Fick's first law. The diffusion equation is coupled with the continuity equation, and the net action of diffusion is observed when both equations are considered:

$$\frac{\partial n_1}{\partial t} = -\nabla \cdot (n_1 \langle \vec{V}_1 \rangle) = \mathcal{D}_1 \nabla^2 n_1 \tag{1.121}$$

which is Fick's second law. For steady state conditions, n_1 is distributed according to Laplace's equation $\nabla^2 n_1 = 0$.

Atmospheric molecular diffusion
This kind of diffusion is often called molecular diffusion, in contrast with eddy diffusion, which will be considered later. Consider a stratified atmosphere belonging to a planet or a star. Suppose again that $\vec{V}_{Fi} = 0$. Only the vertical component of $\langle \vec{V}_i \rangle$, called simply $\langle V_i \rangle$, is non-vanishing. From (1.117) we find after a short derivation

$$\langle V_i \rangle = -\mathcal{D}_i \left(\frac{1}{n_i} \frac{\partial n_i}{\partial z} + \frac{1}{T} \frac{\partial T}{\partial z} + \frac{1}{H_i} \right) \tag{1.122}$$

an equation commonly used in atmospheric studies. It is mainly used for minor constituents because then $\mathcal{D}_i$ is constant for a given height. Remember that it is only valid as a first approximation. Otherwise, other effects, such as thermal diffusion, introduce corrective terms. In (1.122), the individual scale height H_i is defined by

$$H_i = \frac{kT}{m_i g} \tag{1.123}$$

Molecular diffusion processes are more important in upper atmospheric layers, where n is low and $\mathcal{D}_i$ large. In the terrestrial thermosphere, there is an upward flux of molecular oxygen reaching the upper layers where it is photodissociated by UV solar radiation. A downward flux of atomic oxygen produced via O_2 photodissociation reaches the lower layers where the high density promotes recombination, producing O_2 again. In still higher layers diffusion is such a fast process that diffusive equilibrium is easily reached. The n_i distribution with altitude is given by (1.122) with $\langle V_i \rangle = 0$, or directly by (1.119)

$$n_i = \frac{n_{i0} T_0}{T} \exp \left(-\int_{z_0}^z \frac{dz}{H_i} \right) \tag{1.124}$$

where n_{i0} and T_0 are n_i and T for $z = z_0$, a reference altitude where boundary conditions are adopted. Light constituents, such as atomic hydrogen, have low m_i, high H_i, and therefore decrease slowly with altitude and are dominant in the upper layers (in practice, atomic hydrogen even escapes into space; it does not reach equilibrium). More massive molecules such as N_2 have low H_i, decrease in density faster with height, and are absent in the upper layers. Therefore, above a certain layer, called the homopause or the turbopause, when molecular diffusion becomes important, there is a diffusive separation of constituents. Above the turbopause, the atmosphere is called the heterosphere, because each layer has a different chemical composition. The terrestrial turbopause is situated at around 90 km above sea level.

1.4.2 Viscosity and heat conduction

It has been seen that thermodynamic equilibrium implies $\mathcal{P} = p\delta$, where p is one-third of the trace of $\mathcal{P}$. In the absence of equilibrium a new tensor must be added to $p\delta$. This tensor must take into account viscosity, that is, internal friction. Our objective now is to find an expression for this tensor, taking different physical constraints into account.

Friction forces are usually assumed to be of the form $-kv$, v being a body's speed and k a constant of proportionality. By analogy we can express the tensor we need by $-\eta S$, where η is the viscosity coefficient and S is the shear tensor. The viscosity coefficient is one of the transport coefficients and the shear tensor is the potential for viscous momentum transport. Then

$$\mathcal{P} = p\delta - \eta S \tag{1.125}$$

Viscosity is friction between a given volume of the fluid and adjacent volumes, so that it is not $\vec{v}_0$ that determines S, but $\nabla \vec{v}_0$. The shear tensor must be a function of $\nabla \vec{v}_0$, and it must itself be a second-order tensor. There are not many second-order tensors obtainable from $\nabla \vec{v}_0$. The first candidate could be $\nabla \vec{v}_0$ itself. Is $S = \nabla \vec{v}_0$?

A physical requirement which strongly restricts the function $S(\nabla \vec{v}_0)$ is that $S = 0$ for solid-body rotation. In solid-body rotation, the fluid rotates like a solid, the velocity field being given by $\vec{v}_0 = \vec{\Omega} \times \vec{r}$, where $\vec{\Omega}$ is a constant. Then no friction between adjacent volumes is present. Assume first that $S = \nabla \vec{v}_0$:

$$\nabla \vec{v}_0 = \nabla(\vec{\Omega} \times \vec{r}) = \nabla(\hat{\Omega} \cdot \vec{r}) = (\nabla \vec{r}) \cdot (\hat{\Omega})^+ = \delta \cdot (\hat{\Omega})^+ = (\hat{\Omega})^+ = -\hat{\Omega} \tag{1.126}$$

which is not zero. We will introduce some tensor nomenclature that is employed generally in this book. The tensor $\hat{\Omega}$ is the antisymmetric tensor associated with the vector $\vec{\Omega}$. It is defined so that $\vec{\Omega} \times \vec{A} = \hat{\Omega} \cdot \vec{A}$, $\vec{A}$ being any vector. This tensor permits the conversion of vector products into scalar products, which are easier to handle in tensor algebra. If $\mathcal{R}$ is a tensor, $(\mathcal{R})^+$ is the transpose, so that $(\mathcal{R}_{ij})^+ = \mathcal{R}_{ji}$. We see then that S cannot be $\nabla \vec{v}_0$, but (1.126) gives us an idea. If we transpose (1.126):

$$(\nabla \vec{v}_0)^+ = \hat{\Omega} \tag{1.127}$$

and therefore

$$\nabla \vec{v}_0 + (\nabla \vec{v}_0)^+ = 0 \tag{1.128}$$

corresponds to solid-body rotation. But there is another possibility. If we calculate the trace in (1.126):

$$\nabla \cdot \vec{v}_0 = 0 \tag{1.129}$$

hence

$$(\nabla \cdot \vec{v}_0)\delta = 0 \tag{1.130}$$

There are no other tensors with the specified characteristics, so that the shear tensor must be a combination of (1.128) and (1.129):

$$S = (\nabla \vec{v}_0 + (\nabla \vec{v}_0)^+) + x(\nabla \cdot \vec{v}_0)\delta \tag{1.131}$$

where x is to be determined. To do this, we recall that p is one-third of the trace of $\mathcal{P}$. The interpretations given for the hydrostatic pressure in (1.17) and (1.18) suggest that this property may persist even under non-equilibrium conditions. Then the trace of S should vanish so that x must have a value of $\frac{-2}{3}$, and we finally derive

$$S = \nabla \vec{v}_0 + (\nabla \vec{v}_0)^+ - \frac{2}{3}(\nabla \cdot \vec{v}_0)\delta \tag{1.132}$$

It is probable that our result of $x = -\frac{2}{3}$ is not completely satisfactory, and that a new tensor of the form $\kappa(\nabla \cdot \vec{v}_0)\delta$ should be added to produce a more general expression. κ is another transport coefficient, called the bulk viscosity coefficient, and it is indeed often negligible. Finally, we can write

$$\mathcal{P} = p\delta - \eta\left(\nabla \vec{v}_0 + (\nabla \vec{v}_0)^+ - \frac{2}{3}(\nabla \cdot \vec{v}_0)\delta\right) - \kappa(\nabla \cdot \vec{v}_0)\delta \tag{1.133}$$

which is a rather general expression, confirmed by the Chapman–Enskog theory.

When this expression for $\mathcal{P}$ is inserted into the equation of motion, it is often necessary to calculate

$$\nabla \cdot \left(\nabla \vec{v}_0 + (\nabla \vec{v}_0)^+ - \frac{2}{3}(\nabla \cdot \vec{v}_0)\delta\right) = \nabla^2 \vec{v}_0 + \frac{1}{3}\nabla\nabla \cdot \vec{v}_0$$

$$= \frac{4}{3}\nabla^2 \vec{v}_0 + \frac{1}{3}\nabla \times \nabla \times \vec{v}_0 \tag{1.134}$$

$$= \frac{4}{3}\nabla\nabla \cdot \vec{v}_0 - \nabla \times \nabla \times \vec{v}_0$$

For the heat conduction flux the usual hypothesis

$$\vec{q} = -\lambda\nabla T \tag{1.135}$$

will be assumed here. The Chapman–Enskog theory provides higher-order terms, which are negligible in most astrophysical applications. λ is another transport coefficient, called the coefficient of thermal conductivity.

1.4.3   Approximate values of transport coefficients

For neutral gases, the mean free length l, was found by Clausius to be of the form

$$l = (n\Sigma)^{-1} \tag{1.136}$$

where Σ is the collision cross-section. For hydrogen, the major constituent in most astrophysical systems, $\Sigma = \pi a_0^2$, where a_0 can be identified with the radius of the first Bohr orbit, $a_0 \approx 0.5 \times 10^{-8}$ cm. As a typical value of a particle's speed, we usually take the root-mean-square velocity

$$V = \left(\frac{3kT}{m}\right)^{\frac{1}{2}} \tag{1.137}$$

for a Maxwellian velocity distribution. The mean-free-time between collisions is $\tau \approx lV^{-1}$ or, more exactly, for a Maxwellian distribution

$$\tau = l \left(\frac{2m}{\pi kT} \right)^{\frac{1}{2}} \tag{1.138}$$

Observe that the frequency between collisions $\nu = \tau^{-1}$, that is,

$$\nu = n\Sigma \left(\frac{3kT}{m} \right)^{\frac{1}{2}} \tag{1.139}$$

does not differ very much in order of magnitude from the value given in (1.112) with (1.111), which justifies the name collision frequency given to ν_i in (1.112). We already know how to calculate the diffusion coefficient using (1.118), (1.112), and (1.111). Inserting (1.139) into (1.118), the same order of magnitude is obtained:

$$\mathcal{D}_i \approx \sqrt{kT} \frac{1}{\Sigma_i} \frac{1}{n} \frac{\sqrt{m}}{m_i} \tag{1.140}$$

where m is the mean molecular mass and Σ_i the collision-cross section for molecules of type i with other molecules. More exact formulae require a knowledge of the intermolecular potentials and their derivation would be very lengthy.

Maxwell first found the viscosity coefficient to be

$$\eta = mnlV \approx \frac{\sqrt{3mkT}}{\Sigma} \tag{1.141}$$

We see that it is independent of the density. This is due to a compensation: when n is high, the mean-free-path is short, and the probability for a molecule to interchange momentum at a great distance decreases.

The coefficient of thermal conduction is given by

$$\lambda = \frac{5}{2} kT \frac{l}{V} n \left(\frac{3}{2} \frac{k}{m} \right) \approx \frac{5}{4} \sqrt{\frac{3kT}{m}} \frac{k}{\Sigma} \tag{1.142}$$

again independent of the density.

1.4.4 Navier–Stokes equations

These are the first-order approximations of the fluid dynamic equations. They are equations (1.76)–(1.79), where $\langle \vec{V}_i \rangle$, $\mathcal{P}$, and $\vec{q}$ are taken from (1.117), (1.133), and (1.135).

1.4.5 Entropy variations of an imperfect fluid

Let σ be the entropy per particle. Assume that particles are conserved; sometimes σ is called the entropy per baryon, especially in cosmological calculations. Let u be the internal energy per particle; for instance, for a monatomic gas $u = \frac{3}{2} kT$. Taking into account (1.56), (1.85) and (1.133), when a single-component fluid is considered for simplicity, for a wet observer

$$nT\frac{d\sigma}{dt} = n\frac{du}{dt} - \frac{p}{n}\frac{dn}{dt} = n\frac{du}{dt} + p\nabla \cdot \vec{v}_0 = -\nabla \cdot \vec{q} - P_{ij}\frac{\partial v_{0i}}{\partial x_j} + p\nabla \cdot \vec{v}_0$$

$$= -\nabla \cdot \vec{q} - p\nabla \cdot \vec{v}_0 + \eta\left(\frac{\partial v_{0j}}{\partial x_i} + \frac{\partial v_{0i}}{\partial x_j}\right)\frac{\partial v_{0i}}{\partial x_j}$$

$$+ \left(-\frac{2}{3}\eta + \kappa\right)\frac{\partial v_{0k}}{\partial x_k}\delta_{ij}\frac{\partial v_{0i}}{\partial x_j} + p\nabla \cdot \vec{v}_0 \tag{1.143}$$

$$= -\nabla \cdot \vec{q} + \eta\left(\frac{\partial v_{0j}}{\partial x_i} + \frac{\partial v_{0i}}{\partial x_j}\right)\frac{\partial v_{0i}}{\partial x_j} - \frac{2}{3}\eta(\nabla \cdot \vec{v}_0)^2 + \kappa(\nabla \cdot \vec{v}_0)^2$$

One can show that

$$\left(\frac{\partial v_{0j}}{\partial x_i} + \frac{\partial v_{0i}}{\partial x_j}\right)\frac{\partial v_{0i}}{\partial x_j} - \frac{2}{3}(\nabla \cdot \vec{v}_0)^2$$

$$= \frac{2}{3}\left(\frac{\partial v_{01}}{\partial x_1} - \frac{\partial v_{02}}{\partial x_2}\right)^2 + \frac{2}{3}\left(\frac{\partial v_{01}}{\partial x_1} - \frac{\partial v_{03}}{\partial x_3}\right)^2 + \frac{2}{3}\left(\frac{\partial v_{02}}{\partial x_2} - \frac{\partial v_{03}}{\partial x_3}\right)^2 \tag{1.144}$$

$$+ \left(\frac{\partial v_{01}}{\partial x_2} + \frac{\partial v_{02}}{\partial x_1}\right)^2 + \left(\frac{\partial v_{01}}{\partial x_3} + \frac{\partial v_{03}}{\partial x_1}\right)^2 + \left(\frac{\partial v_{02}}{\partial x_3} + \frac{\partial v_{03}}{\partial x_2}\right)^2 = Q$$

where we have defined Q, which is always positive. This fact is important because, in the absence of bulk viscosity, and of heating effects other than viscosity, we would obtain

$$\frac{du}{dt} = \eta Q \tag{1.145}$$

which means that the wet observer always measures viscous heating, never cooling. Then

$$nT\frac{d\sigma}{dt} = -\nabla \cdot \vec{q} + \eta Q + \kappa(\nabla \cdot \vec{v}_0)^2 \tag{1.146}$$

The time variation of the total entropy of the fluid contained in a finite volume τ (either real or imaginary) is

$$\frac{\partial S}{\partial t} = \frac{\partial}{\partial t}\int_\tau n\sigma \, d\tau = \int_\tau \frac{\partial}{\partial t}(n\sigma) \, d\tau \tag{1.147}$$

But

$$\frac{\partial}{\partial t}(n\sigma) = n\frac{\partial\sigma}{\partial t} + \sigma\frac{\partial n}{\partial t}$$

$$= n\frac{d\sigma}{dt} - n\vec{v}_0 \cdot \nabla\sigma - \sigma\nabla \cdot (n\vec{v}_0) = n\frac{d\sigma}{dt} - \nabla \cdot (n\sigma\vec{v}_0) \tag{1.148}$$

$$= -\frac{1}{T}\nabla \cdot \vec{q} + \frac{\eta Q}{T} + \frac{\kappa}{T}(\nabla \cdot \vec{v}_0)^2 - \nabla \cdot (n\sigma\vec{v}_0)$$

Inserting (1.148) in (1.147) and calculating some of the integrals:

$$-\int_\tau \frac{1}{T}\nabla \cdot \vec{q} \, d\tau = -\int_\tau \nabla \cdot \left(\frac{\vec{q}}{T}\right) d\tau + \int_\tau \vec{q} \cdot \nabla\left(\frac{1}{T}\right) d\tau$$

$$= -\int\int_s \frac{\vec{q}}{T} \cdot d\vec{S} + \int_\tau \lambda\frac{(\nabla T)^2}{T^2} \, d\tau \tag{1.149}$$

where Gauss' theorem was used to obtain a surface integral along the surface surrounding τ. We further assume that λ is a constant. Similarly,

$$-\int_\tau \nabla \cdot (n\sigma\vec{v}_0)\, d\tau = -\iint_s n\sigma\vec{v}_0 \cdot d\vec{S} \tag{1.150}$$

Therefore,

$$\frac{\partial S}{\partial t} = -\iint_s n\sigma\vec{v}_0 \cdot d\vec{S} - \iint_s \frac{\vec{q}}{T} \cdot d\vec{S}$$

$$+ \int_\tau \lambda \frac{(\nabla T)^2}{T^2}\, d\tau + \int_\tau \frac{1}{T}\eta Q\, d\tau + \int_\tau \frac{\kappa}{T}(\nabla \cdot \vec{v}_0)^2 d\tau \tag{1.151}$$

where there are two surface integrals and three volume integrals. The surface integrals represent the increase (decrease) in the total entropy of the fluid in τ, via entropy fluxes passing through the surrounding surface. The first surface integral is the entropy flux associated with the matter flux. The second surface integral is the entropy flux associated with heat conduction flux, and can be present even if no matter flux passes through the boundary surface. It is clear that the total entropy of the fluid will increase either if matter is added or if it is heated from outside. These surface integrals can be either positive or negative. They vanish if the fluid is isolated, for instance when the boundary surface has a sufficiently large radius.

The three volume integrals involve sources of entropy which are produced in processes taking place within the fluid itself. The first one represents the entropy production due to heat conduction from one volume element to another. The two last volume integrals represent the entropy production due to viscous heating. Experimentally we know that λ, η, and κ are always positive. All other quantities in the volume integrals are positive. Therefore, internal heat conduction and viscosity always produce a net increase of entropy, in agreement with the second law of thermodynamics. In fact, the argument may be inverted: as entropy in isolated systems always increases, λ, η, and κ must be positive coefficients.

1.5 Turbulence

Most cosmic fluids manifest turbulent flow. Turbulence is a complex phenomenon responsible for the majority of complex structures in the Universe. For instance, any terrestrial map is complex, hence it suggests turbulence, which in this case is hidden from direct observation, in the athenosphere. Gravity alone would act with perfect spherical symmetry, rendering terrestrial maps as inexpressive white sheets. Galaxies are complex as a result of the turbulence of the interstellar gas, preventing a simple classification. The Sun's surface convective cells are turbulent, tropospheres are highly turbulent, and there are many other examples of cosmic turbulence.

1.5.1   Hierarchy of turbulent motions

The vorticity equation gives a good insight into the process under study. Vorticity is defined as

$$\vec{\omega} = \nabla \times \vec{v}_0 \tag{1.152}$$

and is a quantity representing the local rotation magnitude, that is, the eddy magnitude. Let us assume an incompressible fluid $\rho = constant$, hence $\nabla \cdot \vec{v}_0 = 0$. We will consider for the moment only incompressible fluids, as the equations are shorter and easier to assess, so it is easier to concentrate on the most important properties of turbulence. Let us apply $\nabla \times$ to both sides of the equation of motion, taking the tensor relation into account:

$$\frac{1}{2}\nabla \vec{v}_0^2 = \vec{v}_0 \times (\nabla \times \vec{v}_0) + \vec{v}_0 \cdot \nabla \vec{v}_0 \tag{1.153}$$

(only valid for incompressible fluids), noting that operators $\frac{\partial}{\partial t}$ and $\nabla \times$ commute, and assuming a perfect fluid without viscosity:

$$\frac{\partial \vec{\omega}}{\partial t} = \nabla \times (\vec{v}_0 \times \vec{\omega}) \tag{1.154}$$

which is the vorticity equation for incompressible perfect fluids. It clearly shows the intrinsic instability of a perfect fluid. If $\vec{\omega} = 0$ initially, then $\frac{\partial \vec{\omega}}{\partial t} = 0$ and $\vec{\omega}$ will remain zero. But if there exists a very small perturbation at the beginning, consisting of a very small vorticity, it will create vorticity, which in turn will create vorticity, and so on. The fluid becomes unpredictable and chaotic as a result of the fact that vorticity generates vorticity. The right-hand side in (1.154) arises directly from the inertial term $\vec{v}_0 \cdot \nabla \vec{v}_0$ in the equation of motion. This term is non-linear, so that the mathematical complexity in the equation corresponds to physical complexity in the motion. Perfect fluids are unstable.

As not all fluids in nature are unstable, the absence of turbulence must be associated with an imperfect fluid condition. Heat conduction is not present in the equation of motion, so we infer that viscosity is the process preventing chaotic motion. Two effects oppose one another: the inertial term $\rho \vec{v}_0 \cdot \nabla \vec{v}_0$ which produces turbulence, and viscosity $-\eta \nabla^2 \vec{v}_0$ which dissipates it. The first has a typical order of magnitude of $\rho v_0^2 L^{-1}$, where L is the maximum characteristic length in which significant changes in the velocity v_0, or any other quantity representative of the fluid's large-scale properties, takes place. The L in laboratory experiments are usually interpreted as the dimensions of the obstacles that produce the velocity field on the largest scale. In cosmic problems there are in general no solid obstacles and L is of the order of the atmospheric scale length, or the thickness of the disc of a spiral galaxy, etc., depending on the physical system under study. On the other hand, viscous forces are of the order of $\eta v_0 L^{-2}$. Turbulence will be present if inertial forces are much greater than viscous ones, that is, if

$$\mathcal{R} = \frac{\rho v_0^2 L^{-1}}{\eta v_0 L^{-2}} = \frac{\rho v_0 L}{\eta} \tag{1.155}$$

is much larger than unity. $\mathcal{R}$ is the well-known dimensionless quantity called the Reynolds number. If $\mathcal{R} \ll 1$, there will be no turbulence. If $10 > \mathcal{R} > 0.1$, we cannot be sure, as the calculation is an order of magnitude estimate.

A mean flux with a certain viscosity generates eddies on a smaller scale. On this smaller scale the small eddies become a mean flux, which contains smaller eddies on a still smaller scale, and so on. Therefore, the flux interpretable as fluctuating on a given scale is interpreted as a mean flux at the next scale down. Mean and fluctuating flows are therefore relative concepts depending on the scale length considered. There is a hierarchy of eddies, as first envisaged by Taylor. The largest scale is denoted by L, and the others are represented by the continuous variable l. We know when there is turbulence on the maximum scale L. The question is, what is the range of l over which there is turbulence? Down to what value of l, called l_{min}, does the mean flux at scale l generate eddies of a lower l?

The velocity v_l of the mean flow on scale l will almost certainly be a function of l. Based on everyday experience the assumption that v_l is an increasing function of l is entirely credible.

On a scale l there will be turbulence if the Reynolds number on this scale, defined as

$$\mathcal{R}_l = \frac{\rho v_l l}{\eta} \qquad (1.156)$$

is much larger than unity. Starting from L and going to lower values of l, we see that $\mathcal{R}_l$ will also fall to lower and lower values. ρ and η are assumed to be constants, and v_l increases with l, therefore $\mathcal{R}_l$ increases with l. The minimum value of l with turbulence will be $l = l_{min}$ so that $\mathcal{R}_{l_{min}}$ reaches a value of about unity. At this scale l_{min}, viscosity is no longer negligible; it dissipates eddies and eddies on smaller scales are not produced. Turbulence can only exist for scales in the range $L \geq l \geq l_{min}$. Note that no particular function $v_l = v_l(l)$ has been adopted; in fact we would like to obtain such a function.

1.5.2 The mean flow equations

When we are interested in the flow on scale L, we may ignore the highly variable eddies on lower scales. But the fluid on scale L moves differently and with different energy when underlying turbulence is present on smaller scales. To find the mean flow hydrodynamic equations, any quantity is split into a mean quantity and a fluctuating one, for example

$$\vec{v}_0 = \vec{v}_1 + \vec{v}' \qquad (1.157)$$

$\vec{v}'$ varies in such a way that when averaged over a scale larger than the eddies, or over times longer than characteristic times of turbulent oscillation, the result is zero. Let us denote this kind of average by $[\ ]$, different from the symbol $\langle \rangle$ which denotes molecular mean values. Taking averages in (1.157)

$$[\vec{v}_0] = \vec{v}_1 \qquad (1.158)$$

$$[\vec{v}'] = 0 \qquad (1.159)$$

With (1.157) and (1.158), both $\vec{v}_1$ and $\vec{v}'$ are defined. In a similar way to (1.157) we can decompose any quantity

$$p = p_1 + p' \tag{1.160}$$

Using this form and calculating [ ] in the resulting equation, the mean flow equations will be obtained. We will do this only for the scale L, but as mean flow and fluctuating flow are relative concepts, the equations derived will be valid for the mean flow at any intermediate scale l.

The continuity equations. Eddy diffusion

Let us first assume a one-component fluid. The continuity equation, $\nabla \cdot \vec{v}_0$, gives

$$\nabla \cdot (\vec{v}_1 + \vec{v}') = 0 \tag{1.161}$$

and taking averages, if [ ] commutes with either ∇ or $\frac{\partial}{\partial t}$

$$\nabla \cdot \vec{v}_1 = 0 \tag{1.162}$$

This is the mean flow equation of continuity, which as we see does not differ in form from the original expression. The continuity equation of the fluctuating flow is obtained by subtracting (1.162) from (1.161), obtaining

$$\nabla \cdot \vec{v}' = 0 \tag{1.163}$$

With respect to the continuity equation, the behaviour of the fluid does not depend on the presence or absence of internal turbulence.

Let us now assume that the fluid is multicomponent. n is a constant, but n_i may not be, so turbulent diffusion is possible. Let us start with equation (1.50), but neglecting $\langle \vec{V}_i \rangle$ compared to $\vec{v}_0$:

$$\frac{\partial n_{i1}}{\partial t} + \frac{\partial n_i'}{\partial t} + \nabla \cdot (n_{i1}\vec{v}_1) + \nabla \cdot (n_i'\vec{v}_1) + \nabla \cdot (n_{i1}\vec{v}') + \nabla \cdot (n_i'\vec{v}') = 0 \tag{1.164}$$

Taking [ ]:

$$\frac{\partial n_{i1}}{\partial t} + \nabla \cdot (n_{i1}\vec{v}_1) + \nabla \cdot [n_i'\vec{v}'] = 0 \tag{1.165}$$

We see that $[\vec{v}'] = 0$, $[p'] = 0$, but $[n_i'\vec{v}']$, the average of a product of two fluctuating quantities, does not disappear and in practice is often very large. The mean individual continuity equation is different from the original continuity equation. There is a new term, which after defining the eddy diffusion flux $\vec{\varphi}_{ti}$ as

$$\vec{\varphi}_{ti} = [n_i'\vec{v}'] \tag{1.166}$$

enables us to rewrite

$$\frac{\partial n_{i1}}{\partial t} + \nabla \cdot (n_{i1}\vec{v}_1) + \nabla \cdot \vec{\varphi}_{ti} = 0 \tag{1.167}$$

$\nabla \cdot (n_{i1}\vec{v}_1)$ is $\vec{v}_1 \cdot \nabla n_{i1}$, but it is not a new term; it was present in the original expression. Even if molecular diffusion $\langle \vec{V}_i \rangle$ is negligible, there is an eddy diffusion flux $\vec{\varphi}_{ti}$.

The equation of motion

The equation of motion for our incompressible fluid, after decomposition into mean flow and fluctuating quantities is

$$\frac{\partial \vec{v}_1}{\partial t} + \frac{\partial \vec{v}'}{\partial t} + \vec{v}_1 \cdot \nabla \vec{v}_1 + \vec{v}_1 \cdot \nabla \vec{v}' + \vec{v}' \cdot \nabla \vec{v}_1 + \vec{v}' \cdot \nabla \vec{v}'$$
$$+ \frac{\nabla p_1}{\rho} + \frac{\nabla p'}{\rho} - \frac{\eta}{\rho} \nabla^2 \vec{v}_1 - \frac{\eta}{\rho} \nabla^2 \vec{v}' - \vec{g} = 0 \tag{1.168}$$

Taking averages:

$$\frac{\partial \vec{v}_1}{\partial t} + \vec{v}_1 \cdot \nabla \vec{v}_1 + [\vec{v}' \cdot \nabla \vec{v}'] + \frac{\nabla p_1}{\rho} - \frac{\eta}{\rho} \nabla^2 \vec{v}_1 - \vec{g} = 0 \tag{1.169}$$

This is the same equation as we originally had, with the exception of $[\vec{v}' \cdot \nabla \vec{v}']$, which because of (1.163) can be written as $\nabla \cdot [\vec{v}'\vec{v}']$. The Reynolds tensor is defined as

$$\tau_t = \rho[\vec{v}'\vec{v}'] \tag{1.170}$$

The mean flow equation of motion is

$$\rho \frac{\partial \vec{v}_1}{\partial t} + \rho \vec{v}_1 \cdot \nabla \vec{v}_1 + \nabla \cdot \tau_t + \nabla p_1 - \eta \nabla^2 \vec{v}_1 - \vec{g} = 0 \tag{1.171}$$

τ_t is therefore interpreted as a turbulent flux of momentum, its divergence producing a force in the fluid. The Reynolds tensor is not negligible, and is even a dominant flux of momentum.

The energy balance equation

The heat balance equation can now be written as

$$n\left(\frac{\partial u}{\partial t} + \vec{v}_0 \cdot \nabla u\right) - \eta\left(\frac{\partial v_{0i}}{\partial x_j} + \frac{\partial v_{0j}}{\partial x_i}\right)\frac{\partial v_{0j}}{\partial x_i} = 0 \tag{1.172}$$

and by the standard procedure used before, the mean flow heat-balance equation, is derived to be

$$n\left(\frac{\partial u_1}{\partial t} + \vec{v}_1 \cdot \nabla u_1\right) + n[\vec{v}' \cdot \nabla u']$$
$$- \eta\left(\frac{\partial v_{1i}}{\partial x_j} + \frac{\partial v_{1j}}{\partial x_i}\right)\frac{\partial v_{1j}}{\partial x_i} - \eta\left[\left(\frac{\partial v_i'}{\partial x_j} + \frac{\partial v_j'}{\partial x_i}\right)\frac{\partial v_j'}{\partial x_i}\right] = 0 \tag{1.173}$$

(1.173) differs from (1.172) by two terms. If the turbulent heat flux is defined as

$$\vec{q}_t = \rho[u'\vec{v}'] \tag{1.174}$$

one of them can be rewritten as

$$n[\vec{v}' \cdot \nabla u'] = n[\nabla \cdot (u'\vec{v}')] = n\nabla \cdot [u'\vec{v}'] = \frac{1}{m}\nabla \cdot \vec{q}_t \tag{1.175}$$

The evaluation of $\vec{q}_t$ will be considered later. The other different term is called the energy dissipation rate, ϵ:

$$\epsilon = \eta \left[\left(\frac{\partial v_i'}{\partial x_j} + \frac{\partial v_j'}{\partial x_i} \right) \frac{\partial v_j'}{\partial x_i} \right] \tag{1.176}$$

and plays an important role in current theories of turbulence.

The energy dissipation rate ϵ is always positive. The derivation (1.144) could now be confirmed or, more specifically for incompressible fluids, it can easily be shown that

$$\sum_{ij} \left(\frac{\partial v_i'}{\partial x_j} + \frac{\partial v_j'}{\partial x_i} \right) \frac{\partial v_j'}{\partial x_i} = \frac{1}{2} \sum_{ij} \left(\frac{\partial v_i'}{\partial x_j} + \frac{\partial v_j'}{\partial x_i} \right)^2 \tag{1.177}$$

It has dimensions of $[\epsilon] = ML^2 T^{-3}$ as it is the energy gained per particle and per unit time. The final form of the mean flow heat balance equation is

$$n \left(\frac{\partial u_1}{\partial t} + \vec{v}_1 \cdot \nabla u_1 \right) + \frac{1}{m} \nabla \cdot \vec{q}_t - \eta \left(\frac{\partial v_{1i}}{\partial x_j} + \frac{\partial v_{1j}}{\partial x_i} \right) \frac{\partial v_{1j}}{\partial x_i} - \epsilon = 0 \tag{1.178}$$

Thus we see that the internal energy is enhanced by an amount ϵ. What kind of internal energy is gained? It cannot be thermal energy, because turbulence has indeed been produced on the largest scale. This implies that on this scale viscosity as a dissipation mechanism can be neglected. The answer can only be that ϵ is an internal energy consisting of eddies on smaller scales. ϵ is used either in creating small eddies or in maintaining them when the turbulence reaches statistical equilibrium. If the fluctuating heat balance equation were written down, by subtracting (1.178) from (1.172), ϵ would appear as an energy loss.

The important fact is, in view of the relativity of the concepts of mean and fluctuating flow, that (1.178) is valid not only for the mean flow on the largest scale L, but on any scale l greater than l_{min}. When stationary turbulence is reached, in the absence of any dissipation, ϵ produces eddies on lower scales, through the whole hierarchy of scales, from L to l_{min}. The chain stops at l_{min}, where viscosity is no longer negligible and ϵ contributes to heating the fluid, thereby increasing its temperature. Finally, at l_{min}, the term $\eta (\frac{\partial v_{1i}}{\partial x_j} + \frac{\partial v_{1j}}{\partial x_i}) \frac{\partial v_{1j}}{\partial x_i}$ in (1.178) becomes important.

The energy dissipation rate ϵ cannot depend on η, as it is transferred on scales where the fluid is perfect, and η has no influence on the motion. To put it rather crudely, at $l \gg l_{min}$ the fluid does not know that viscosity exists. ϵ cannot depend on l because the same value is transferred on any scale, irrespective of the value of l, nor can ϵ depend on v_l because $v_l(l)$ is a function of l only. From its definition in (1.176), however, in order of magnitude terms:

$$\epsilon \approx \text{constant} \times \eta \frac{v_l^2}{l^2} \tag{1.179}$$

where this constant is of the order of unity and dimensionless. If $v_l = v_l(l)$ and neither l nor v_l is present in (1.179), the only possibility is

$$v_l = \text{constant} \times l \tag{1.180}$$

which is a reasonable conclusion on intermediate scales $L \gg l \gg l_{min}$, where the turbulent process can be considered as statistically homogeneous and isotropic. On this range of scales the physical behaviour becomes simplified, as shown in the theory of

Kolmogorov and Oboukov. The constant in (1.179) must also depend on η in such a way that the final expression is itself independent of η.

To find the equation which gives ϵ we must perhaps abandon (1.179). ϵ may depend on n and, possibly, on the quantities on the largest scale, v_0 and L, as they are characteristic of the process which generates turbulence. The particle mass cannot be excluded. A relation between quantities ϵ, m, n, v_0, L is then sought. Dimensional analysis states that the general equation relating these quantities is

$$\epsilon = m \frac{v_0^3}{L} \psi(nL^3) \tag{1.181}$$

where ψ is an unknown function. But nL^3 is proportional to the total number of particles in the total volume, $\sim L^3$, which cannot be considered as a variable parameter in our problem. Then $\psi(nL^3)$ is a constant, and m is a constant. Therefore,

$$\epsilon \propto \frac{v_0^3}{L} \tag{1.182}$$

The dissipation rate does not depend on the density, is proportional to the cube of the mean velocity, and is lower for large systems. ϵ may be taken as representative of the violence of the turbulent motion.

1.5.3   Turbulent transport fluxes

The preceding equations do not form a closed system, as $\vec{\varphi}_{ti}$, τ_t, and $\vec{q}_t$ are unknown. Turbulence models must provide these functions, taking their definitions (1.166), (1.170) and (1.174) as a starting point. Turbulence is, however, such a complicated field that even mathematicaly sophisticated models cannot provide these fluxes. Therefore, indirect and intuitive hypotheses will be assumed here which have been of practical interest in several cosmic problems, and which are in part empirically supported.

For $\vec{\varphi}_{ti}$ we can take

$$\vec{\varphi}_{ti} = -Kn\nabla \frac{n_i}{n} \tag{1.183}$$

since the relative concentration $\frac{n_i}{n}$ is the potential producing eddy diffusion. When complete mixing is reached eddy diffusion ceases. Clearly, $\vec{\varphi}_{ti}$ must have the opposite sign to $\nabla\left(\frac{n_i}{n}\right)$. $\vec{\varphi}_{ti}$ will be larger if n is larger. K is called the eddy diffusion coefficient, and depends on the magnitude of the turbulence.

Stellar and planetary atmospheres provide simple applications of our formula (1.183). Under conditions of stratification it is easy to obtain

$$\varphi_{ti} = -Kn_i\left(\frac{1}{n_i}\frac{\partial n_i}{\partial z} + \frac{1}{T}\frac{\partial T}{\partial z} + \frac{1}{H}\right) \tag{1.184}$$

This is Lettau's formula. φ_{ti} is now the vertical component of $\vec{\varphi}_{ti}$. We can see that (1.184) is very similar to the equation giving the molecular diffusion flux, equation (1.122). The important difference, however, is that in (1.184) the atmospheric scale height H appears in the expression of the flux instead of the individual scale height

H_i. Molecular diffusion tends to separate the atmospheric constituents according to their molecular masses. Eddy diffusion, by contrast, tends to mix them and provide a uniform chemical composition. Mixing equilibrium, or eddy diffusion equilibrium, with $\varphi_{ti} = 0$ is given by

$$n_i = n_{i0} \frac{T_0}{T} \exp\left(-\int_{z_0}^{z} \frac{dz}{H}\right) \tag{1.185}$$

The right-hand side has no subindex i (except for the boundary value n_{i0}). Therefore, the vertical variation is the same for all constituents. Troposphere were shown to be unstable and must have a large value of K. Thermospheres on the other hand have low densities, therefore $\mathcal{D}_i$ must be very large. It is therefore to be expected that molecular diffusion will be dominant in the upper layers and eddy diffusion in the lower ones. In the homosphere, eddy diffusion provides complete mixing. The homosphere ends at the homopause, or turbopause, and then begins the heterosphere where molecular diffusion provides separation of constituents.

A similar hypothesis could be proposed for the Reynolds tensor:

$$\tau_t = -K\nabla(\rho\vec{v}_0) \tag{1.186}$$

as well as for turbulent heat conduction:

$$\vec{q}_t = -K\nabla(nu) = -K\nabla(c_v\rho T) \tag{1.187}$$

In a stratified atmosphere:

$$q_t = -Kc_v\frac{\partial}{\partial z}(\rho T) = -Kc_v\frac{\partial}{\partial z}\left(\frac{pm}{k}\right) = \frac{Kc_v m}{k}\frac{p}{H} = K\rho g\frac{c_v m}{k} \tag{1.188}$$

For a diatomic molecule, for instance $\frac{c_v m}{k} = \frac{5}{2}$, then

$$q_t = \frac{5}{2}k\rho g \tag{1.189}$$

which is always positive. In an atmosphere, turbulent energy transport is directed upwards. This effect is also important in stellar atmospheres and interiors, in those layers where convective processes are stronger than in radiative ones.

2 Relativistic fluids

2.1 The distribution function

As in classical fluid dynamics, a distribution function f is defined such that $f d\tau_r d\tau_p$ gives the total number of particles in a given volume element $d\tau_r$, having a momentum contained in a momentum volume element $d\tau_p$, that is, the number of particles per volume element in six-dimensional phase space.[1]

Two different observers, one moving with $\vec{v}_0$ with respect to the other, with $\Gamma = 1/\sqrt{1-v_0^2}$, will not agree on the values of either dx_i or dp_i which limit the volume element $d\tau_r, d\tau_p$. If particles at the cell boundary are used, however, to determine the cell size for a given observer, all of them will be consistent in the number of particles in the cell. Therefore, $f d\tau_r d\tau_p$ is a relativistic invariant. It is a number of particles, and all observers would count the same number.

Therefore $f d\tau_r d\tau_p$ is a scalar. What is the relativistic transformation of f? It is well known that $d\tau_r$ transforms so that

$$d\tau_r = \frac{d\tau_r'}{\Gamma} \tag{2.1}$$

The relativistic transformation of $d\tau_p$ requires more thought. Consider the hypersphere in the momentum space:

$$-E^2 + \sum_i p_i^2 = -m^2 \tag{2.2}$$

where m is the rest mass of a particle of fluid, E its energy, and p_i the components of its three-dimensional momentum. Remember that E, p_1, p_2, p_3 are the components of the four-dimensional momentum, which is the contravariant vector p^α. Remember also that p^α is defined by $p^\alpha = mv^\alpha$. If the modulus of p^α is calculated, taking into account that $v_\alpha v^\alpha = -1$, the result is $-m^2$, as stated in (2.2).

As usual we have adopted the subindex 0 for the time coordinate; a Latin subindex such as i or j denotes any spatial coordinate. Greek subindexes such as α or β are 0, 1, 2, or 3.

---

1 Following most astrophysical literature, c.g.s. units have been adopted through most of the book. However, relativity is usually presented in a different system of units. Here we will adopt the so-called geoetrized unit system, in which the basic universal constants G, c, k, ϵ_0 are set equal to unity, with the exception of Planck's constant h. To avoid problems in frequent changes from one system to another, an appendix containing conversion factors and dimensions will be provided at the end. Chapters 2,3, 8, 9, and 10 use this unit system.

The zeroth component of the hypersurface element $d\sigma^\alpha$ of this hypersphere is $d\tau_p = dp_1 dp_2 dp_3$. Also, $d\sigma^\alpha$ will be parallel to the position vector, p^α, in this momentum space, at each point on the hypersurface; therefore $d\sigma^\alpha \propto p^\alpha$. The zeroth component of these vectors will also be proportional, and the constant of proportionality will be the same for all points on the hypersphere. Different Lorentz frames correspond to different points on the hypersphere. By changing from one point to another, $p^0 \equiv E$ varies, and therefore $d\sigma^0$ varies in the same proportion. Therefore, the relativistic transformation for p^0 is the same as that for $d\sigma^0$, and $d\tau_p$ has the same transformation as E under Lorentz frame changes. As we know that $E = \Gamma E'$, with $E' \equiv m$, we conclude that

$$d\tau_p = \Gamma d\tau_p' \tag{2.3}$$

Then the volume element in the six-dimensional space phase is invariant. From (2.1) and (2.3):

$$d\tau_p \, d\tau_r = \Gamma \, d\tau_p' \frac{1}{\Gamma} \, d\tau_r' = d\tau_p' \, d\tau_r' \tag{2.4}$$

If $f d\tau_r \, d\tau_p$ is a scalar and $d\tau_p \, d\tau_r$ is also a scalar, then f is also a relativistic invariant. This is the great advantage of adopting this space instead of the space of positions and velocities.

2.1.1   Transport fluxes

We know that $\frac{d\tau_p}{E}$ is a relativistic invariant, so it will be used instead of $d\tau_p$ in the defining integrals of transport fluxes. In this way the flux of a true tensor quantity G, defined as

$$\phi^\alpha(G) = \int_p Gp^\alpha f \frac{d\tau_p}{E} \tag{2.5}$$

will itself be a true tensor. If G is a scalar, its flux will be a vector ϕ^α.

Assume, as a first example, that $G \equiv 1$. The flux vector ϕ^α is then termed J^α, the current four-vector

$$J^\alpha = \int_p p^\alpha f \frac{d\tau_p}{E} \tag{2.6}$$

In particular, we observe that

$$J^0 = n = \int_p f d\tau_p \tag{2.7}$$

that is, the number density n is not a scalar, but rather the zeroth component of the current vector. To interpret J^α physically, note that for a classical fluid $J^i = \int_p f v_i \, d\tau_p = n\langle v_i \rangle$, using the nomenclature of the preceding chapter.

Now let $G \equiv p^\alpha$. Using (2.5), the flux of four-momentum, called the energy-momentum tensor, $\tau^{\alpha\beta}$, is

$$\tau^{\alpha\beta} = \int_p \frac{p^\alpha p^\beta}{E} f d\tau_p \tag{2.8}$$

It is by definition a symmetric tensor. τ^{00} is called the energy density, ϵ:

$$\epsilon = \tau^{00} = \int_p Ef d\tau_p \tag{2.9}$$

The hydrostatic pressure is defined as

$$p = \frac{1}{3}\tau^{ii} \tag{2.10}$$

Neither ϵ nor p is a scalar.

2.1.2 Perfect fluids

A perfect fluid is defined as that fluid for which it is locally possible to find an observer who sees his local microstate as perfectly isotropic. This observer will again be called the wet observer, that is, the observer of a comoving Lorentz frame. For this special observer the integrals (2.6)–(2.10) are easier to obtain. Because of the condition of isotropy of the directions of the particles, we would simply obtain

$$J^\alpha = (n, 0, 0, 0) \tag{2.11}$$

$$\tau^{\alpha\beta} = \begin{bmatrix} \epsilon & 0 & 0 & 0 \\ 0 & p & 0 & 0 \\ 0 & 0 & p & 0 \\ 0 & 0 & 0 & p \end{bmatrix} \tag{2.12}$$

where n, ϵ, and p will now be called the proper number density, proper energy density, and proper hydrostatic pressure respectively. In the case of a perfect fluid:

$$\epsilon = \tau^{00} = \int_p fE\, d\tau_p = n\langle E \rangle \tag{2.13}$$

$$p = \frac{1}{3}\int_p \frac{p^2}{E}f\, d\tau_p = \frac{1}{3}n\left\langle \frac{p^2}{E} \right\rangle \tag{2.14}$$

It is certainly confusing that the hydrostatic pressure and the momentum of a particle are denoted by the same symbol p, but both quantities have aquired this right by force of tradition. In (2.14) it is clear that p^2 is the squared modulus of the three-momentum. For material particles we obtain

$$\epsilon = mn\langle \gamma \rangle \tag{2.15}$$

$$p = \frac{1}{3}n\left\langle \frac{m^2\gamma^2 v^2}{\gamma m} \right\rangle = \frac{1}{3}mn\langle \gamma v^2 \rangle \tag{2.16}$$

where

$$\gamma = \frac{1}{\sqrt{1 - v^2}} \tag{2.17}$$

should not be confused with Γ, which is a function of the mean speed of the fluid or of a Lorentz frame. It is a macroscopic quantity. On the other hand, γ is a function of the speed of a fluid particle and is a microscopic quantity.

For a cold gas with non-relativistic particles, a series development gives

$$\gamma \approx 1 + \frac{1}{2}v^2 \qquad (2.18)$$

$$\epsilon \approx mn\langle 1 + \frac{1}{2}v^2\rangle = mn + \frac{1}{2}mn\langle v^2\rangle \qquad (2.19)$$

$$\gamma v^2 \approx v^2 \qquad (2.20)$$

$$p \approx \frac{1}{3}mn\langle v^2\rangle \qquad (2.21)$$

and therefore

$$\epsilon \approx mn + \frac{3}{2}p \qquad (2.22)$$

which is the energetic equation of state of a cold gas. ϵ is the sum of the rest mass energy density plus a contribution due to thermal motion.

For a hot gas, with v close to unity,

$$p = \frac{1}{3}mn\langle \gamma v^2\rangle \approx \frac{1}{3}mn\langle \gamma\rangle = \frac{\epsilon}{3} \qquad (2.23)$$

Hence,

$$\epsilon = 3p \qquad (2.24)$$

is the energetic equation of state of a hot gas. It is well known and will be seen in the next chapter that this equation holds for photons.

From the expressions of J^α and $\tau^{\alpha\beta}$ for the comoving Lorentz frame, it is easy to obtain the general expressions for any observer (e.g. for the dry observer) by performing a Lorentz transformation. One particular case of Lorentz transformation is given by the matrix

$$\Lambda = \begin{bmatrix} \Gamma & \Gamma v_0 & 0 & 0 \\ \Gamma v_0 & \Gamma & 0 & 0 \\ 0 & 0 & 1 & 0 \\ 0 & 0 & 0 & 1 \end{bmatrix} \qquad (2.25)$$

equivalent to a displacement along the common OX-axis of one system with respect to the other. There is no loss of generality when (2.25) is used, as no special OX-axis was chosen previously. Then

$$J^\alpha = \Lambda^\alpha{}_\beta J'^\beta = n\begin{bmatrix} \Gamma & \Gamma v_0 & 0 & 0 \\ \Gamma v_0 & \Gamma & 0 & 0 \\ 0 & 0 & 1 & 0 \\ 0 & 0 & 0 & 1 \end{bmatrix}\begin{bmatrix} 1 \\ 0 \\ 0 \\ 0 \end{bmatrix} = n\begin{bmatrix} \Gamma \\ \Gamma v_0 \\ 0 \\ 0 \end{bmatrix} \qquad (2.26)$$

For a general Lorentz matrix we would have obtained a current vector which could be represented by $J^\alpha = n(\Gamma, \Gamma\vec{v}_0) = n\Gamma(1, \vec{v}_0)$, and which can therefore be written as

$$J^\alpha = nU^\alpha \qquad (2.27)$$

where U^α is the four-velocity of the fluid at a given point. In practice, (2.27) can also be understood as the definition of U^α, and is always true even for imperfect fluids. For the four-velocity it is clear, and can be checked immediately, that

$$U_\alpha U^\alpha = -1 \tag{2.28}$$

The same Lorentz transformation can be applied to (2.12) to obtain the general expression of the energy-momentum tensor of a perfect fluid for any dry observer:

$$\tau^{\alpha\beta} = \Lambda^\alpha{}_\gamma \Lambda^\beta{}_\delta \tau'^{\gamma\delta} = \begin{bmatrix} \Gamma^2(\epsilon + pv_0^2) & \Gamma^2 v_0(\epsilon + p) & 0 & 0 \\ \Gamma^2 v_0(\epsilon + p) & p + \Gamma^2 v_0^2(\epsilon + p) & 0 & 0 \\ 0 & 0 & p & 0 \\ 0 & 0 & 0 & p \end{bmatrix} \tag{2.29}$$

For any Lorentz matrix:

$$\tau^{ij} = p\delta_{ij} + (p + \epsilon)\Gamma^2 v_{0i} v_{0j} \tag{2.30}$$

$$\tau^{i0} = \Gamma^2(\epsilon + p)v_{0i} \tag{2.31}$$

$$\tau^{00} = \Gamma^2(\epsilon + pv_0^2) \tag{2.32}$$

which can be expressed by the covariant equation

$$\tau^{\alpha\beta} = p\eta^{\alpha\beta} + (p + \epsilon)U^\alpha U^\beta \tag{2.33}$$

where $\eta^{\alpha\beta}$ is the Minkowski tensor ($\eta^{00} = -1; \eta^{ii} = 1$ and all off-diagonal components are zeros). Finally, (2.33) is the energy-momentum tensor for a perfect fluid. Instead of using a particular Lorentz matrix (2.25) and then guessing the general result for any matrix, we can use a general Lorentz matrix directly, which is given by the symbolic expression

$$\Lambda = \begin{bmatrix} \Gamma & (\Gamma\vec{v}_0) \\ (\Gamma\vec{v}_0) & \left(\delta + \frac{\Gamma-1}{v_0^2}\vec{v}_0\vec{v}_0\right) \end{bmatrix} \tag{2.34}$$

2.1.3 Boltzmann's equation

First a covariant Boltzmann equation must be found. From the three-dimensional equation, the generalization is proposed in the form

$$v^\alpha \frac{\partial f}{\partial x^\alpha} + F^\alpha \frac{\partial f}{\partial p^\alpha} = \Pi \tag{2.35}$$

with v^α being the four-velocity of a particle. This is Boltzmann's equation in special relativity; we will see that it is covariant, and that it reduces to the classical expression in the classical approximation. The first term $v^\alpha \frac{\partial f}{\partial x^\alpha}$ takes the form $\gamma\frac{\partial f}{\partial t} + \gamma v_i \frac{\partial f}{\partial x_i}$ equivalent to $\gamma(\frac{\partial f}{\partial t} + \vec{v} \cdot \nabla f)$, which is the first part of the classical Boltzmann equation (1.42) multiplied by γ. The second term $F^\alpha \frac{\partial f}{\partial p^\alpha}$ contains $F^0 \frac{\partial f}{\partial E}$.

However, f does not depend explicitly on E, and therefore this contribution vanishes. Then $F^\alpha \frac{\partial f}{\partial p^\alpha} = F^i \frac{\partial f}{\partial p^i}$. Remember that $F^i = \gamma\frac{dp_i}{dt}$ and $p^i = p_i$ (only, here, p_i is the three-dimensional momentum, not a covariant vector). Therefore, $F^\alpha \frac{\partial f}{\partial p^\alpha}$ is also present in the classical Boltzmann equation (1.42), again multiplied by γ. With respect to the collision term Π we could say that it is also γ times the classical collision term in (1.42), because $\Pi\, d\tau_r\, d\tau_p\, d\tau$ (with $d\tau$ being the proper time) is a number of particles equivalent to the classical term $\Gamma\, d\tau_r\, d\tau_p\, dt$. As $d\tau = \frac{dt}{\gamma}$, then $\Pi = \gamma\Gamma$. Therefore, (2.35) reduces to (1.42) in the classical limit.

The other condition that (2.35) must meet to be the relativistic generalization of (1.42) is that it be covariant. All the quantities in (2.35) are vectors or scalars. This is obvious for v^α, $\frac{\partial f}{\partial x^\alpha}$, and F^α. Now we show that $\frac{\partial f}{\partial p^\alpha}$ is a true covariant vector. This is easily demonstrated as

$$p^\alpha = \Lambda^\alpha{}_\beta p'^\beta \tag{2.36}$$

because p^α is a contravariant vector. From this we obtain p'^β:

$$p'^\beta = \Lambda_\alpha{}^\beta p^\alpha \tag{2.37}$$

where $\Lambda_\alpha{}^\beta$ is the inverse Lorentz matrix. Then

$$\frac{\partial p'^\beta}{\partial p^\alpha} = \Lambda_\alpha{}^\beta \tag{2.38}$$

and

$$\frac{\partial f}{\partial p^\alpha} = \frac{\partial f}{\partial p'^\beta}\frac{\partial p'^\beta}{\partial p^\alpha} = \Lambda_\alpha{}^\beta \frac{\partial f}{\partial p'^\beta} \tag{2.39}$$

which is the transformation rule for covariant vectors. However, $\Pi\, d\tau_r\, d\tau_p\, d\tau$ is a scalar as it represents a number of particles; $d\tau_r\, d\tau_p$ and $d\tau$ are scalars; therefore Π is a scalar. We thus conclude that (2.35) is covariant, and that it is Boltzmann's special relativity equation.

However, two variables v^α and p^α appearing in (2.35) are related: $p^\alpha = mv^\alpha$. We will use only one of them: p^α. Then we multiply by the rest mass and obtain the final form for the relativistic Boltzmann equation:

$$p^\alpha \frac{\partial f}{\partial x^\alpha} + mF^\alpha \frac{\partial f}{\partial p^\alpha} = m\Pi \tag{2.40}$$

2.2   Macroscopic implications

As seen in Chapter 1, when this equation is multiplied by $G\frac{d\tau_p}{E}$, where G is a collisional invariant, and then integrated over all possible values of p^i, that is, performing $\int_p$, macroscopically interesting equations will be obtained which do not require an explicit knowledge of the collision term $m\Pi$. In this way we will next obtain the continuity equation and the equation of conservation of energy-momentum.

2.2.1   The continuity equation

Let $G \equiv 1$. From (2.40),

$$\int_p p^\alpha \frac{\partial f}{\partial x^\alpha}\frac{d\tau_p}{E} + \int_p mF^\alpha \frac{\partial f}{\partial p^\alpha}\frac{d\tau_p}{E} = 0 \tag{2.41}$$

The first integral is

$$\frac{\partial}{\partial x^\alpha} \int_p fp^\alpha \frac{d\tau_p}{E} = \frac{\partial}{\partial x^\alpha} J^\alpha \tag{2.42}$$

The second one, when F^α does not depend on p^α, is

$$mF^\alpha \int_p \frac{\partial f}{\partial p^\alpha} \frac{d\tau_p}{E} = 0 \tag{2.43}$$

because, as discussed in the preceding chapter, $[fp^{\alpha^x}]_{-\infty}^\infty = 0$ for any value of $x = 1, 2, 3, \ldots$. Then the continuity equation reads

$$\frac{\partial J^\alpha}{\partial x^\alpha} = 0. \tag{2.44}$$

However, the basic hypothesis made, that F^α is independent of p^α, excludes electromagnetic forces, which is probably too restrictive. Of the four natural forces, gravitation can be included as a geometrical effect, the strong and weak forces require special theories, and therefore the electromagnetic force should not be excluded.

Electromagnetic forces are of the type

$$F^\alpha = \frac{q}{m} \mathcal{F}^\alpha{}_\beta p^\beta \tag{2.45}$$

F^α is the four-force acting on a particle of electrical charge q and four-momentum p^β. $\mathcal{F}^\alpha{}_\beta$ is the contravariant–covariant form of the Faraday tensor

$$\mathcal{F}^{\alpha\beta} = \begin{bmatrix} 0 & E_1 & E_2 & E_3 \\ -E_1 & 0 & B_3 & -B_2 \\ -E_2 & -B_3 & 0 & B_1 \\ -E_3 & B_2 & -B_1 & 0 \end{bmatrix} \tag{2.46}$$

where E_1, E_2, and E_3 are the components of the three-dimensional electric field and B_1, B_2, and B_3 are the components of the three-dimensional magnetic field strength. Now the second integral in (2.41) must be revised:

$$\int_p m \frac{q}{m} \mathcal{F}^\alpha{}_\beta p^\beta \frac{\partial f}{\partial p^\alpha} \frac{d\tau_p}{E} = q\mathcal{F}^\alpha{}_\beta \int_p \frac{\partial}{\partial p^\alpha} (fp^\beta) \frac{d\tau_p}{E} - \int_p q\mathcal{F}^\alpha{}_\beta f \frac{\partial p^\beta}{\partial p^\alpha} \frac{d\tau_p}{E}$$

$$= 0 - \int_p q\mathcal{F}^\alpha{}_\beta f \delta_\alpha^\beta \frac{d\tau_p}{E} = -\int_p q\mathcal{F}^\alpha{}_\alpha f \frac{d\tau_p}{E} = 0 \tag{2.47}$$

as $\mathcal{F}^\alpha{}_\alpha = 0$, because $\mathcal{F}^\alpha{}_\beta$ is antisymmetric. Therefore the continuity equation in the form (2.44) is always valid, even when electromagnetic forces are acting on the fluid particles. The divergence of the current four vector is zero. It is also convenient to express this continuity equation in three dimensional form. If n is the proper number density

$$\frac{\partial}{\partial t} \left(\frac{n}{\sqrt{1 - v_0^2}} \right) + \nabla \cdot \left(\frac{n\vec{v}_0}{\sqrt{1 - v_0^2}} \right) = 0 \tag{2.48}$$

where the relativistic corrections are significant, when either the fluid velocity or the observer's velocity is large.

2.2.2 Equation of energy-momentum conservation

Now let $G \equiv p^\alpha$ and perform the integral $\int_p p^\alpha [Boltzmann] \frac{d\tau_p}{E}$. The first term gives

$$\int_p p^\alpha p^\beta \frac{\partial f}{\partial x^\beta} \frac{d\tau_p}{E} = \frac{\partial}{\partial x^\beta} \int_p fp^\alpha p^\beta \frac{d\tau_p}{E} = \frac{\partial}{\partial x^\beta} \tau^{\beta\alpha}. \tag{2.49}$$

To calculate the second integral, we take into account directly that the forces are electromagnetic

$$\int_p m \frac{q}{m} \mathcal{F}^\gamma{}_\beta p^\beta \frac{\partial f}{\partial p^\gamma} p^\alpha \frac{d\tau_p}{E} = q \int_p \mathcal{F}^\gamma{}_\beta \frac{\partial}{\partial p^\gamma} (fp^\beta p^\alpha) \frac{d\tau_p}{E}$$

$$- q \int_p \mathcal{F}^\gamma{}_\beta f \frac{\partial p^\beta}{\partial p^\gamma} p^\alpha \frac{d\tau_p}{E} - q \int_p \mathcal{F}^\gamma{}_\beta fp^\beta \frac{\partial p^\alpha}{\partial p^\gamma} \frac{d\tau_p}{E}$$

$$= 0 - q\mathcal{F}^\gamma{}_\beta \int_p f \delta^\beta_\gamma p^\alpha \frac{d\tau_p}{E} - q\mathcal{F}^\gamma{}_\beta \int_p fp^\beta \delta^\alpha_\gamma \frac{d\tau_p}{E} \tag{2.50}$$

$$= 0 - 0 - q\mathcal{F}^\alpha{}_\beta \int_p fp^\beta \frac{d\tau_p}{E} = -q\mathcal{F}^\alpha{}_\beta J^\beta$$

The similarity between (2.50) and (2.45) permits the obvious interpretation of (2.50). Usually the electrical current J^β_{EM} is introduced and defined as

$$J^\beta_{EM} = qJ^\beta. \tag{2.51}$$

J^β is the particle current. If it is multiplied by the charge of each particle the electrical current is obtained. In the absence of forces either electromagnetic or any other type, the equation of energy-momentum conservation simply reads

$$\frac{\partial}{\partial x^\alpha} \tau^{\alpha\beta} = 0. \tag{2.52}$$

In the case of electromagnetic forces, J^β_{EM} can be obtained from the covariant form of Maxwell's equations:

$$\frac{\partial}{\partial x^\alpha} \mathcal{F}^{\alpha\beta} = -4\pi J^\beta_{EM} \tag{2.53}$$

$$\frac{\partial}{\partial x^\alpha} \mathcal{F}_{\beta\gamma} + \frac{\partial}{\partial x^\beta} \mathcal{F}_{\gamma\alpha} + \frac{\partial}{\partial x^\gamma} \mathcal{F}_{\alpha\beta} = 0 \tag{2.54}$$

Introducing J^β_{EM} in the force expression given by (2.50):

$$\frac{\partial}{\partial x^\alpha} \tau^{\alpha\beta} + \frac{1}{4\pi} \mathcal{F}^\beta{}_\alpha \frac{\partial}{\partial x^\gamma} \mathcal{F}^{\gamma\alpha} = 0 \tag{2.55}$$

which can be rewritten in another form that is very interesting from the point of view of interpretation. If we define the so-called electromagnetic energy-momentum tensor, $\tau^{\alpha\beta}_{EM}$, by

$$4\pi \tau^{\alpha\beta}_{EM} = \mathcal{F}^\alpha{}_\gamma \mathcal{F}^{\beta\gamma} - \frac{1}{4\pi} \eta^{\alpha\beta} \mathcal{F}_{\gamma\delta} \mathcal{F}^{\gamma\delta} \tag{2.56}$$

then (2.55) becomes

$$\frac{\partial}{\partial x^\alpha} \tau^{\alpha\beta}_{total} = 0 \tag{2.57}$$

where $\tau_{total}^{\alpha\beta}$ is the total energy-momentum tensor

$$\tau_{total}^{\alpha\beta} = \tau^{\alpha\beta} + \tau_{EM}^{\alpha\beta} \qquad (2.58)$$

Equation (2.57) is a very general equation, although it only considers electromagnetic forces. As stated above, of the four forces of nature, gravitation requires general relativity and will be dealt with below as a geometrical effect, while the strong and weak forces require a quantum description and are beyond our scope, which is restricted to classical and relativistic cosmic fluids. Therefore, electromagnetic forces are the only remaining forces to be considered. A mnemonic rule replaces (2.56):

$$\tau_{EM}^{\alpha\beta} = \frac{1}{4\pi}\left[\begin{array}{cc} 0 & (\vec{E} \times \vec{B}) \\ \left(\begin{array}{c}\vec{E} \\ \times \\ \vec{B}\end{array}\right) & (-\vec{E}\vec{E} - \vec{B}\vec{B}) \end{array}\right] + \frac{1}{8\pi}(E^2 + B^2)\delta_\beta^\alpha \qquad (2.59)$$

As a consequence it is possible to refer to an electromagnetic energy density

$$\epsilon_{EM} = \tau_{EM}^{00} = \frac{1}{8\pi}(E^2 + B^2) \qquad (2.60)$$

which is a very familiar expression. τ^{0i} reminds us of the Poynting three-vector. There is also an electromagnetic hydrostatic pressure:

$$3p_{EM} = \tau^{ii} = \frac{1}{8\pi}(E^2 + B^2) = \epsilon_{EM} \qquad (2.61)$$

Therefore, $\epsilon_{EM} = 3p_{EM}$, just as in (2.24) for a hot gas.

The equation of energy-momentum conservation is written in (2.52) in covariant form. For many purposes it may be preferable to integrate the three-dimensional equations with relativistic corrections. Directly from (2.52) we obtain

$$\frac{\partial}{\partial t}\left(\Gamma^2 \vec{v}_0(\epsilon + p)\right) + \nabla p + \nabla \cdot \left(\Gamma^2(\epsilon + p)\vec{v}_0\vec{v}_0\right) = 0 \qquad (2.62)$$

$$\frac{\partial}{\partial t}\left(\Gamma^2(\epsilon + pv_0^2)\right) + \nabla \cdot \left(\Gamma^2(\epsilon + p)\vec{v}_0\right) = 0 \qquad (2.63)$$

These four equations, together with the continuity equation (2.48) and the equation of state, are the six equations to be solved for the six unknowns $\vec{v}_0$, ϵ, p, and n. They are not, however, the most commonly used expressions. Performing some algebra [(2.62) - $\vec{v}_0$(2.63)] the equation of motion is

$$\frac{\partial \vec{v}_0}{\partial t} + \vec{v}_0 \cdot \nabla\vec{v}_0 + \frac{\nabla p + \frac{\partial p}{\partial t}\vec{v}_0}{\Gamma^2(\epsilon + p)} = 0 \qquad (2.64)$$

and the equation of energy balance is

$$\frac{\partial}{\partial t}\left(\Gamma^2(\epsilon + p)\right) + \vec{v}_0 \cdot \nabla\left(\Gamma^2(\epsilon + p)\right) = \frac{\partial p}{\partial t} - \Gamma^2(\epsilon + p)\nabla \cdot \vec{v}_0 \qquad (2.65)$$

Equation (2.64) should be compared with its classical counterpart. The acceleration and the inertial terms remain the same, $\frac{\nabla p}{\rho}$ is replaced by $\frac{\nabla p}{(\Gamma^2(\epsilon+p))}$, which is the same for a very cold gas. The term with $\frac{\partial p}{\partial t}$ is completely new. Equation (2.65) should also be compared with its classical counterpart. The internal energy nu is now replaced by $\Gamma^2(\epsilon + p)$. The adiabatic heating term is clearly identified, and again $\frac{\partial p}{\partial t}$ is a new term, which is in part matched by the first term.

2.2.3   Entropy variations of a perfect fluid

It is clear that the flux of a classical perfect fluid is isentropic. This result has been obtained by calculating $\frac{d\sigma}{dt}$, the entropy time variation with respect to the wet observer. We have already shown that the entropy variations in the interior of the fluid are due only to viscosity and heat conduction, both effects being disregarded in perfect fluids. A similar result is to be expected for a relativistic perfect fluid. Let us calculate $\frac{d\sigma}{d\tau}$, where, as before, σ is the specific entropy per particle (if the number of particles is conserved). In most cases of practical interest, σ is the entropy per baryon. Now,

$$\frac{d\sigma}{d\tau} = \frac{\partial\sigma}{\partial x^\alpha}\frac{dx^\alpha}{d\tau} = U^\alpha\frac{\partial\sigma}{\partial x^\alpha} \tag{2.66}$$

Here $\frac{d\sigma}{d\tau}$ is the time derivative for the wet observer, $\frac{\partial\sigma}{\partial x^\alpha}$ are derivatives for the dry observer, and (2.66) replaces the classical expression connecting the two time derivatives $\frac{d\sigma}{dt}$ and $\frac{\partial\sigma}{\partial t}$.

The second law of thermodynamics can now be written:

$$Td\sigma = pd\left(\frac{1}{n}\right) + d\left(\frac{\epsilon}{n}\right) \tag{2.67}$$

Therefore,

$$nT\frac{d\sigma}{d\tau} = np\frac{d}{d\tau}\left(\frac{1}{n}\right) + n\frac{d}{d\tau}\left(\frac{\epsilon}{n}\right) = nU^\alpha\left(p\frac{\partial}{\partial x^\alpha}\left(\frac{1}{n}\right) + \frac{\partial}{\partial x^\alpha}\left(\frac{\epsilon}{n}\right)\right) \tag{2.68}$$

It can now be shown that the right-hand side is equal to

$$U_\beta\frac{\partial}{\partial x^\alpha}\tau^{\alpha\beta} = 0 \tag{2.69}$$

This is easily demonstrated, taking the expression of $\tau^{\alpha\beta}$ for a perfect fluid, (2.33), into account:

$$U_\beta\frac{\partial}{\partial x^\alpha}\tau^{\alpha\beta} = U_\beta\frac{\partial p}{\partial x^\alpha}\eta^{\alpha\beta} + U_\beta\frac{\partial}{\partial x^\alpha}\left((\epsilon+p)U^\alpha U^\beta\right)$$
$$= U^\alpha\frac{\partial p}{\partial x^\alpha} + U_\beta U^\beta\frac{\partial}{\partial x^\alpha}((\epsilon+p)U^\alpha) + (\epsilon+p)U^\alpha U_\beta\frac{\partial U^\beta}{\partial x^\alpha} \tag{2.70}$$

The last term vanishes because

$$0 = \frac{\partial}{\partial x^\alpha}(-1) = \frac{\partial}{\partial x^\alpha}(U_\beta U^\beta)$$
$$= \frac{\partial U_\beta}{\partial x^\alpha}U^\beta + U_\beta\frac{\partial U^\beta}{\partial x^\alpha} = \frac{\partial U^\beta}{\partial x^\alpha}U_\beta + U_\beta\frac{\partial U^\beta}{\partial x^\alpha} = 2U_\beta\frac{\partial U^\beta}{\partial x^\alpha} \tag{2.71}$$

Therefore,

$$U_\beta\frac{\partial}{\partial x^\alpha}\tau^{\alpha\beta} = U^\alpha\frac{\partial p}{\partial x^\alpha} - \frac{\partial}{\partial x^\alpha}((\epsilon+p)U^\alpha) = U^\alpha\frac{\partial p}{\partial x^\alpha} - \frac{\partial}{\partial x^\alpha}\left(\frac{\epsilon+p}{n}nU^\alpha\right)$$
$$= U^\alpha\frac{\partial p}{\partial x^\alpha} - \frac{\partial}{\partial x^\alpha}\left(\frac{\epsilon+p}{n}\right)nU^\alpha = -nU^\alpha\left(p\frac{\partial}{\partial x^\alpha}\left(\frac{1}{n}\right) + \frac{\partial}{\partial x^\alpha}\left(\frac{\epsilon}{n}\right)\right) \tag{2.72}$$

Finally, using (2.68) and (2.69),

$$\frac{d\sigma}{d\tau} = 0 \tag{2.73}$$

A perfect relativistic fluid has an isentropic flow.

Imperfect relativistic fluids are not considered in this introductory text. Most astrophysical systems can be accounted for by using the perfect fluid approximation. Indeed, when dealing with the cosmological problem it will be shown that the Universe as a whole evolves like a perfect fluid. Some aspects of galaxy formation are influenced by viscosity and heat conduction, but they will be considered using order of magnitude arguments which do not require a formal development of imperfect fluid theory.

2.2.4 Sound

A fluid can be relativistic because either $v_0 \approx 1$, or because it is very hot. This second possibility will now be considered. We will consider sound in a perfect fluid which is sufficiently hot that $\epsilon = 3p$. For simplicity we will assume that $\vec{v}_0 = 0$. As is usual in this type of derivation, we will assume that any quantity is decomposed into a mean value and a perturbed quantity. For instance,

$$p \rightarrow p + p' \tag{2.74}$$

We do not use p_1 in order not to confuse this subindex with a covariant component subindex. The velocity will, however, not be decomposed this way; rather we will use

$$\vec{v}_0 \rightarrow 0 + \vec{v}' \tag{2.75}$$

We will further assume that the fluid is homogeneous and stationary (except for the very small variations induced by the passage of the sound wave), so that, for example, $\nabla p = 0$, $\frac{\partial p}{\partial t} = 0$. A considerable difference with respect to a similar calculation carried out for turbulence in the preceding chapter is that, here, perturbed quantities are considered to be much smaller than the principal quantities. Therefore, terms containing products of two or more perturbed quantities may be neglected.

For didactic reasons we will carry out our derivations in two different ways: first using the corrected three-dimensional equations; second using the covariant four-dimensional equation.

For a pure three-dimensional equation, from (2.64) and taking $v_0 = 0$, $\Gamma = 1$

$$\frac{\partial \vec{v}'}{\partial t} + \frac{\nabla p'}{4p} = 0 \tag{2.76}$$

and from (2.65):

$$\frac{\partial(3p')}{\partial t} + 4p\nabla \cdot \vec{v}' = 0 \tag{2.77}$$

Taking ∇ in (2.76) and $\frac{\partial}{\partial t}$ in (2.77), multiplying appropriately, and summing:

$$\nabla^2 p' = \frac{\partial^2}{\partial t^2}(3p') \tag{2.78}$$

Another similar expression is obtained for $\vec{v}'$. Equation (2.78) is the equation for wave propagation, and the speed of sound is obtained directly:

$$v_s = \frac{1}{\sqrt{3}} \approx 0.58 \tag{2.79}$$

Let us now repeat the above calculation by means of the four-dimensional equation (2.52). Now $U^\alpha \equiv (1,0,0,0)$ and $\tau^{\alpha\beta}$ is given by (2.33). It is then necessary to evaluate $pU^\alpha U^\beta$, which will become

$$pU^\alpha U^\beta \rightarrow pU^\alpha U^\beta + pU^\alpha U'^\beta + pU'^\alpha U^\beta + pU'^\alpha U'^\beta \\ + p'U^\alpha U^\beta + p'U^\alpha U'^\beta + p'U'^\alpha U^\beta + p'U'^\alpha U'^\beta \tag{2.80}$$

$$\frac{\partial}{\partial x^\alpha}\left(pU^\alpha U^\beta\right) \rightarrow 0 + p\frac{\partial U'^\beta}{\partial t} + p\frac{\partial U'^\alpha}{\partial x^\alpha}U^\beta + 0 + \frac{\partial p'}{\partial t}U^\beta + 0 + 0 + 0 \tag{2.81}$$

On the other hand, it is clear that

$$\frac{\partial}{\partial x^\alpha}\left(p'\eta^{\alpha\beta}\right) = \left(-\frac{\partial p'}{\partial t}, \frac{\partial p'}{\partial x^i}\right) \tag{2.82}$$

Then the zeroth component of (2.52) is

$$-\frac{\partial p'}{\partial t} + 4p\frac{\partial U'^i}{\partial x^i} + 4\frac{\partial p'}{\partial t} = 0 \tag{2.83}$$

because $\frac{\partial U'^0}{\partial t}$ may be neglected. Equation (2.83) is equivalent to (2.77). The ith equation is

$$\frac{\partial p'}{\partial x^i} + 4p\frac{\partial U'^i}{\partial t} = 0 \tag{2.84}$$

which is equivalent to (2.76). Using the same equations the result is, of course, the same.

2.3   General relativistic fluid dynamics

Determination of the space-time curvature from the energy momentum, using Einstein's field equations, is a topic that falls within the objectives of fluid dynamics. Indeed, the energy-momentum tensor is a four-dimensional generalization of the pressure tensor plus the heat conduction flux. The only difference now is that neither the pressure tensor nor the heat conduction flux are functions of the peculiar velocities. Peculiar velocities have not been mentioned in this chapter because they are identical to particle velocities. The definition $\vec{v} = \vec{v}_0 + \vec{V}$ is now replaced by a Lorentz transformation or any other transformation in general relativity.

Several examples, from the fluid dynamic point of view will be examined throughout this book. However, Einstein's field equations are considered to be known by the reader, and we merely note the equations, using them to present the standard nomenclature.

Einstein's field equations can be written in two forms:

$$R_{\mu\nu} - \frac{1}{2}g_{\mu\nu}R = -8\pi\tau_{\mu\nu}$$ (2.85)

$$R_{\mu\nu} = -8\pi\left(\tau_{\mu\nu} - \frac{1}{2}g_{\mu\nu}\tau^\lambda{}_\lambda\right)$$ (2.86)

where $g_{\mu\nu}$ is a metric tensor and $R_{\mu\nu}$ is the Ricci tensor

$$R_{\mu\nu} = R^\lambda{}_{\mu\lambda\nu}$$ (2.87)

that is, a contraction of $R^\lambda{}_{\mu k\nu}$, the Riemann curvature tensor

$$R^\lambda{}_{\mu k\nu} = \frac{\partial\Gamma^\lambda_{\mu k}}{\partial x^\nu} - \frac{\partial\Gamma^\lambda_{\mu\nu}}{\partial x^k} + \Gamma^\eta_{\mu k}\Gamma^\lambda_{\nu\eta} - \Gamma^\eta_{\mu\nu}\Gamma^\lambda_{k\eta}$$ (2.88)

and $\Gamma^\lambda_{\mu\nu}$ is the affine connection, usually calculated using

$$\Gamma^\lambda_{\mu\nu} = \frac{1}{2}g^{\lambda\epsilon}\left(\frac{\partial g_{\epsilon\mu}}{\partial x^\nu} + \frac{\partial g_{\epsilon\nu}}{\partial x^\mu} - \frac{\partial g_{\mu\nu}}{\partial x^\epsilon}\right)$$ (2.89)

Therefore (2.85) and (2.86) relate the energy-momentum at a given point with the curvature at this point.

2.3.1 Fluid dynamic equations

Boltzmann's equation has already been written in covariant form and is therefore easily generalized now. However, it is preferable to deduce the general relativistic fluid dynamic equations as a generalization of the special relativistic ones, which have also been written in covariant form.

In this way, the continuity equation (2.44) will now be

$$J^\alpha{}_{;\alpha} = 0$$ (2.90)

which can be interpreted either as

$$\frac{1}{\sqrt{g}}\frac{\partial}{\partial x^\mu}\left(\sqrt{g}J^\mu\right) = 0$$ (2.91)

where

$$g = -det\left(g_{\alpha\beta}\right)$$ (2.92)

or

$$\frac{\partial J^\alpha}{\partial x^\alpha} + \Gamma^\mu_{\mu\lambda}J^\lambda = 0$$ (2.93)

The energy-momentum tensor will now be written as

$$\tau^{\alpha\beta} = pg^{\alpha\beta} + (p+\epsilon)U^\alpha U^\beta$$ (2.94)

where

$$g_{\mu\nu}U^\mu U^\nu = -1$$ (2.95)

and the energy-momentum conservation equation will be written as

$$\tau^{\mu\alpha}{}_{;\alpha} = 0$$ (2.96)

which can be interpreted either as

$$\frac{\partial p}{\partial x^\mu} g^{\mu\alpha} + \left(\frac{\partial}{\partial x^\mu}(p+\epsilon)\right) U^\mu U^\alpha + \frac{\partial}{\partial x^\mu}(U^\mu U^\alpha)(p+\epsilon)$$
$$+ \left(\Gamma^\alpha_{\mu k} U^\mu U^k + \Gamma^\mu_{\mu k} U^k U^\alpha\right)(p+\epsilon) = 0 \tag{2.97}$$

or

$$\frac{\partial p}{\partial x^\mu} g^{\mu\alpha} + \frac{1}{\sqrt{g}}\frac{\partial}{\partial x^\mu}\left(\sqrt{g}(p+\epsilon)U^\mu U^\alpha\right) + \Gamma^\alpha_{\mu\lambda}(p+\epsilon)U^\mu U^\lambda = 0 \tag{2.98}$$

The simplest application to begin with is the case of hydrostatics, with $U^\mu \equiv (1,0,0,0)$ and $\frac{\partial}{\partial x^0} = 0$. Then, using (2.95):

$$U^0 = \frac{1}{\sqrt{-g_{00}}} \tag{2.99}$$

and from (2.98):

$$\frac{\partial p}{\partial x^i} g^{i\alpha} + \Gamma^\alpha_{00}(p+\epsilon)U^0 U^0 = 0 \tag{2.100}$$

where

$$\Gamma^\alpha_{00} = \frac{1}{2}g^{\alpha\lambda}\left(\frac{\partial g_{\lambda 0}}{\partial x^0} + \frac{\partial g_{\lambda 0}}{\partial x^0} - \frac{\partial g_{00}}{\partial x^\lambda}\right) = -\frac{1}{2}g^{i\alpha}\frac{\partial g_{00}}{\partial x^i} \tag{2.101}$$

Therefore,

$$\frac{\partial p}{\partial x^i} g^{i\alpha} + \frac{1}{2}g^{i\alpha}\frac{\partial g_{00}}{\partial x^i}(p+\epsilon)\frac{1}{g_{00}} = 0 \tag{2.102}$$

and multiplying by $g_{\alpha j}$:

$$\frac{\partial p}{\partial x^j} + (p+\epsilon)\frac{\partial}{\partial x^j}\left(\ln\sqrt{-g_{00}}\right) = 0 \tag{2.103}$$

For instance, for a hot gas with $\epsilon = 3p$ this equation has the solution

$$p = \text{constant}(-g_{00})^{-2} \tag{2.104}$$

These equations will be applied later to specific problems, but there are no really simple applications which we could include here as examples. In order to become familiar with the hydrostatic equation, let us consider a case in which general relativity is strictly unnecessary, but does provide a familiar classical result. It is always a relief to see how a general theory yields correct solutions to simple problems. Consider the altitude variation of the atmospheric pressure in a cold planetary stratified atmosphere. We know the answer in advance, as this question was put in Chapter 1, from the classical fluid point of view. The solution was given in equation (1.91).

As the atmosphere is very cold, $\epsilon \approx mn + \frac{3}{2}p \approx mn$. The gas is assumed to be ideal, $p = nT$, isothermal, $T = $ constant, and stationary. From (2.103):

$$T\frac{dn}{dz} + mn\frac{1}{2}\frac{1}{g_{00}}\frac{dg_{00}}{dz} = 0 \tag{2.105}$$

The gravitational potential being weak in a planet, the weak field approximation of general relativity can be adopted:

$$g_{00} = -1 - 2\phi = -1 - 2gz \qquad (2.106)$$

where ϕ is the classical Newtonian potential, $\phi = gz$, g being the acceleration of gravity. Then we obtain

$$n = n_0 \exp\left(-\frac{z}{H}\right) \qquad (2.107)$$

where n_0 is $n(z = 0)$ and H the scale height defined in (1.90); now $H = \frac{T}{mg}$ as $k \equiv 1$. Our result here (2.107) is clearly identical to that obtained in (1.93).

3 Photon fluids

Photon fluids are relativistic. The results obtained in the preceding chapter may now be used as a starting point. It is interesting to connect the covariant formalism of relativistic fluids with very simple applications such as atmospheric extinction.

For historical reasons and due to their exceptional observational interest, photon fluids are usually denoted by specific nomenclature. When the preceding chapter is applied to photon fluids, introducing them from the point of view of fluid dynamics, a glossary of radiation and transport terms needs to be provided.

A problem arises with the unit system adopted. Relativistic topics have been described using geometrized units. Most radiative transfer calculations use gaussian units, or other conventional units. As the connection between chapter 2 and the present chapter is of fundamental importance, we have used geometrized units here. Fortunately, nearly all frequently used formulae are the same in both systems. This is the case, for instance, for planetary and stellar atmospheres where the Eddington nomenclature is adopted.

3.1 The distribution function

The distribution function is defined as before, but it will now be termed f_ν. Therefore, $f_\nu \, d\tau_r \, d\tau_p$ gives the number of photons in a cell in the six-dimensional phase space of position and momentum. The difference is that, now, spherical coordinates in the momentum space will be used. The modulus of a photon's momentum three-vector is

$$p = h\nu \tag{3.1}$$

and its direction is given by two angles. These two angles and $h\nu$ are now the spherical coordinates in the momentum subspace. $d\tau_p$ is then specified by the value of the momentum between p and $p + dp$ and the direction contained in the solid angle $d\omega$. Photons having momentum between p and $p + dp$ are those contained in the spherical shell:

$$d\left(\frac{4}{3}\pi p^3\right) = 4\pi p^2 dp = 4\pi h^3 \nu^2 d\nu \tag{3.2}$$

Of these photons, we are interested solely in those contained in $d\omega$, that is, we must multiply by $d\omega/4\pi$:

$$d\tau_p = h^3 \nu^2 d\nu \, d\omega \qquad (3.3)$$

In the nomenclature of radiative transfer, the specific intensity I_ν is defined as the radiant energy per frequency element, with a direction within a solid angle element, traversing a surface element ds perpendicular to this direction, in a time element. Therefore, $\frac{1}{h\nu}(I_\nu d\omega \, d\nu \, ds \, dt)$ is the number of photons fulfilling the above specifications. This number of photons must be calculated by $f_\nu \, d\tau_r \, d\tau_p$ using the transport nomenclature. In this case, $d\tau_p$ is given by (3.3) and $d\tau_r$ is

$$d\tau_r = ds \, dt \qquad (3.4)$$

that is, the volume of a small cone with base ds and height $c \, dt$, where $c = 1$. If photons are further from the surface than $c \, dt$, they do not have time to reach ds in dt. Then

$$\frac{1}{h\nu} I_\nu \, d\omega \, d\nu \, ds \, dt = f_\nu h^3 \nu^2 d\nu \, d\omega \, ds \, dt \qquad (3.5)$$

and therefore

$$I_\nu = h^4 \nu^3 f_\nu \qquad (3.6)$$

which is the first entry of our transport–radiation glossary. The intensity I_ν and the distribution function are essentially a single concept, with different historical origins.

As is usual in optics, the omission of the subindex ν implies integration over all frequencies; for instance, let us define

$$I = \int_0^\infty I_\nu \, d\nu \qquad (3.7)$$

$$f = \int_0^\infty f_\nu \, d\nu \qquad (3.8)$$

Therefore,

$$I = h^4 \int_0^\infty f_\nu \nu^3 d\nu \qquad (3.9)$$

Note that I_ν, I, and f are not scalars.

3.1.1 Transport fluxes

The transport flux ϕ^α of a quantity G is again defined by (2.5). A photon current four-vector exists:

$$J^\alpha = \int_0^\infty \oint p^\alpha f_\nu \frac{d\tau_p}{h\nu} = \int_0^\infty \oint p^\alpha f_\nu h^3 \nu \, d\omega \, d\nu \qquad (3.10)$$

where

$$p^\alpha = h\nu(1, \sin\theta \cos\varphi, \sin\theta \sin\varphi, \cos\theta) \qquad (3.11)$$

θ and φ being spherical coordinates. We could write $p^0 = h\nu$, $p^i = h\nu u_i$, where u_i is a unitary three-vector denoting the direction of a photon. Note that the solid angle element with these coordinates is

$$d\omega = \sin\theta \, d\theta \, d\varphi \qquad (3.12)$$

Then the zeroth component of J^α becomes

$$J^0 = \int_0^\infty h^3 \nu^2 d\nu \oint f_\nu \, d\omega = \int_0^\infty \frac{d\nu}{h\nu} \oint I_\nu \, d\omega \tag{3.13}$$

and can still be interpreted as the photon number density, n.

As in (3.7) and (3.8), and as is usual in optics, we can calculate

$$J^0_\nu = \frac{dJ^0}{d\nu} = \frac{1}{h\nu} \oint I_\nu \, d\omega = n_\nu \tag{3.14}$$

n_ν is the number density of photons with frequency between ν and $\nu + d\nu$. On the other hand, astrophysicists define the mean intensity J_ν as

$$J_\nu = \frac{1}{4\pi} \oint I_\nu \, d\omega \tag{3.15}$$

which is the average of I_ν in all possible directions. It is unfortunate that the current four-vector and the mean intensity are denoted by the same symbol. The next entry in our glossary is (from (3.14) and (3.15))

$$J^0_\nu = \frac{4\pi J_\nu}{h\nu} \tag{3.16}$$

The spatial components of J^α are

$$J^i = \int_0^\infty \frac{d\nu}{h\nu} \oint I_\nu u_i \, d\omega \tag{3.17}$$

and therefore

$$J^i_\nu = \frac{dJ^i}{d\nu} = \frac{1}{h\nu} \oint I_\nu u_i \, d\omega \tag{3.18}$$

In the case of photons of a given frequency, the number of particles and the energy are essentially the same concept, $h\nu$ being a conversion factor. Therefore we can define ϵ_ν from J^0_ν as

$$\epsilon_\nu = h\nu J^0_\nu = h\nu n_\nu = \oint I_\nu \, d\omega \tag{3.19}$$

where we have deviated from the conventional way of defining ϵ_ν as the zero–zero component of the energy-momentum tensor. Then

$$\epsilon = \int_0^\infty \epsilon_\nu d\nu = 4\pi \int_0^\infty J_\nu \, d\nu = 4\pi J = \oint I \, d\omega \tag{3.20}$$

Equations (3.19) and (3.20) are also glossary entries. Instead of J^i_ν an energy flux can be defined as

$$q_{\nu i} = h\nu J^i_\nu = \oint I_\nu u_i \, d\omega \tag{3.21}$$

and therefore

$$q_i = \oint I u_i \, d\omega \tag{3.22}$$

One might have expected that the three-vector $\vec{q}$ would have been defined as a component of the energy momentum four-tensor, but (3.21) is based on a clear interpretation of what $q_{\nu i}$ and J^i_ν are. Let us, then, consider $\tau^{\alpha\beta}$, taking $G \equiv p^\alpha$ in (2.5):

$$\tau^{\alpha\beta} = \int_0^\infty \oint p^\alpha p^\beta \frac{I_\nu}{h^4 \nu^3} h^2 \nu \, d\nu \, d\omega = \int_0^\infty \oint p^\alpha p^\beta \frac{I_\nu}{h^2 \nu^2} \, d\nu \, d\omega \qquad (3.23)$$

which is a true contravariant tensor. As usual,

$$d\tau^{\alpha\beta} = \frac{d\nu}{h^2 \nu^2} \oint p^\alpha p^\beta I_\nu \, d\omega \qquad (3.24)$$

and it is possible to define

$$\tau_\nu^{\alpha\beta} = \frac{d\tau^{\alpha\beta}}{d\nu} \qquad (3.25)$$

but it is preferable to use $d\tau^{\alpha\beta}$ instead of $\tau_\nu^{\alpha\beta}$ because the former is a true four-tensor. Similarly, J^α and dJ^α are true four-vectors, but $J_\nu^\alpha = \frac{dJ^\alpha}{d\nu}$ is not.

The 00-component of $\tau^{\alpha\beta}$ is

$$\epsilon = \int_0^\infty d\nu \oint I_\nu \, d\omega = \oint I \, d\omega \qquad (3.26)$$

and

$$\epsilon_\nu = \frac{d\epsilon}{d\nu} = \oint I_\nu \, d\omega \qquad (3.27)$$

which results coincide with (3.20) and (3.19).

Let us, then, calculate τ^{0i}, which is equivalent to τ^{i0} due to the symmetry of $\tau^{\alpha\beta}$:

$$\tau^{0i} = \int_0^\infty \oint I_\nu u_i \, d\nu \, d\omega = \oint I u_i \, d\omega = q_i \qquad (3.28)$$

and

$$\tau_\nu^{0i} = \oint I_\nu u_i \, d\omega = q_{\nu i} \qquad (3.29)$$

obtaining (3.21) and (3.22) by an independent method.

Finally,

$$\tau^{ij} = \oint I u_i u_j \, d\omega \qquad (3.30)$$

and

$$\tau_\nu^{ij} = \oint I_\nu u_i u_j \, d\omega \qquad (3.31)$$

τ^{ij} corresponds to the radiative pressure three-tensor:

$$\tau^{ij} = \mathcal{P}_{ij} \qquad (3.32)$$

$$\tau_\nu^{ij} = \mathcal{P}_{\nu ij} \qquad (3.33)$$

The radiative hydrostatic pressure is

$$p_\nu = \frac{1}{3}\tau_\nu^{ii} = \frac{1}{3}\oint I_\nu u_i u_i \, d\omega = \frac{1}{3}\oint I_\nu \, d\omega = \frac{\epsilon_\nu}{3} \qquad (3.34)$$

and therefore

$$\epsilon_\nu = 3p_\nu \qquad (3.35)$$

and

$$\epsilon = 3p \tag{3.36}$$

which is the well-known energetic equation of state for photons.

Summarizing, we can write radiative fluxes in the form of the following mnemonic expressions:

$$J_\nu^\alpha = \frac{1}{h\nu}(\epsilon_\nu, (\vec{q}_\nu)) \tag{3.37}$$

$$\tau_\nu^{\alpha\beta} = \begin{bmatrix} \epsilon_\nu & (\vec{q}_\nu) \\ (\vec{q}_\nu) & (\mathcal{P}_\nu) \end{bmatrix} \tag{3.38}$$

We can schematize the meaning of some of the quantities defined above in a somewhat light-hearted fashion. Quantity I_ν says, for instance: two photons are going to the right, and one photon to the left. Quantity J_ν comments: I see three photons. Quantity q_ν states: net effect, one photon is going to the right.

3.1.2   Observational aspects

Some of the above-defined quantities are of observational interest, as they are commonly used to characterize the light coming from astrophysical objects. The specific intensity I_ν, also called the radiance, is one of them, and is useful when extended sources are studied.

The intensity I_ν is the radiative energy of photons with frequency ν ($\pm\frac{d\nu}{2}$) having direction OA in Figure 3.1 contained in a cone of solid angle $d\omega$ (per unit surface area, per unit time). These photons are coming from an area in the source that is determined by the solid angle $d\omega$. In this case, therefore, I_ν is the energy received from an angular area $d\omega$ in the source (per unit frequency, per unit surface area, per unit time), and characterizes the emission from different regions of the extended source. Of course, the number of photons emerging from the source coincides with the number of photons arriving at O if the medium between the source and the observer is perfectly transparent.

On the other hand, the specific intensity is, in principle, a quantity that is independent of the distance of the source and can therefore be interpreted directly in terms of physical processes in the observed source.

This property is due to the fact that light coming from a distant extended source, such as a galaxy, falls proportionally to r^{-2} (r being the distance to the galaxy) but the area covered by the same $d\omega$ increases proportionally to r^2; and the two effects compensate for one another. Of course, it is assumed in this argument that the area covered by $d\omega$ is sufficiently small in the galaxies observed and that true intensity variations in different regions do not modify the result. For instance, the central region may emit more than the peripheral regions. If the source is placed far enough away its angular diameter becomes smaller than the minimum resolved solid angle of observations $\Delta\omega$, which must take finite values in practice.

The flux is the appropriate quantity for describing point-like sources, such as stars or quasars, as well as for characterizing the total emission of an extended source. The flux

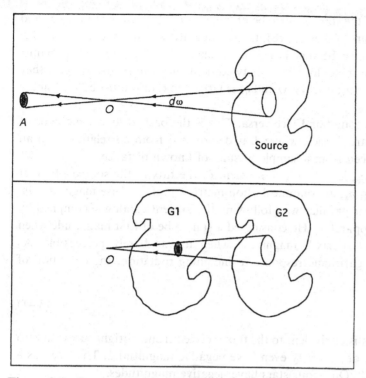

Figure 3.1 Intensity as an observational distance independent quantity.

follows an r^{-2} law in the absence of sources and opacity in the medium between the object and the observer. The flux at the source surface and the flux received at the Earth are quantitatively extremely different.

Let $q_\star$ be the flux at the source and q_e the flux at the Earth. $q_\star$ is a function of the internal processes in the source. It clearly depends on the temperature of the outer layers, and if stars were black bodies it would be related to the Planck function. The luminosity L of a star (or any source) is defined as

$$L = 4\pi R^2 q_\star \tag{3.39}$$

As $4\pi R^2$ is the star's surface area, L is the total energy emitted per unit time by the star. If the star is in equilibrium, the energy produced inside must be completely emitted into space. Therefore, L is related to the energy release in the star's interior, and so is $q_\star$. The flux at the Earth is calculable by measuring only a small fraction $(A/r^2)/4\pi$ of L arriving at a surface A in the telescope. A/r^2 is the solid angle subtended by the area A, and 4π the complete solid angle. Per unit surface, the flux at the Earth will then be given by

$$q_e = \frac{4\pi R^2 q_\star}{4\pi r^2} = \frac{R^2}{r^2} q_\star \tag{3.40}$$

q_e is measured directly, but R, r, and $q_\star$ are often unknown. $q_\star$ can be obtained from theoretical arguments using the star's spectrum or colour, as will be considered later

(Chapter 5). The radius of the star can then be obtained if r is known. For instance, red giants are luminous red stars; if they are red, their temperature is low and q_* is low; if L is high and q_* is low, R must be very large; this explains why they are called giants. White dwarfs are relatively blue, low luminosity stars; if they are relatively blue, they have high T and hence high q_*; if L is low and q_* is large, the radius must be very small; thus they are called dwarfs.

If L is known, r can be found and vice versa. This is the basis of many methods of measuring distances in astrophysics, where L is determined from correlation with an observable property deduced from a sample of stars of known distance.

The flux at the Earth, q_e, is in fact a component of the flux in the source–observer direction. Instead of using q_e, astronomical photometrists prefer to use magnitude m, which is defined in a rather peculiar way following the ancient catalogue compiled by the Greek astronomer Hipparchus. He considered a star to be of first magnitude when it was very brilliant, and of sixth magnitude when it was barely perceptible. As Hipparchus' eye had a logarithmic response to incoming radiation, the definition of m is

$$m = -2.5 \lg \frac{q_e}{q_0} \tag{3.41}$$

where q_0 is a constant that is equivalent to the flux received from a bright star with zero magnitude. Very brilliant objects may even have negative magnitudes. The Sun has a magnitude of about -26.8. Only four stars have negative magnitudes.

In practice, photometrists usually work with broad filters; therefore the flux is

$$q_e = \int_0^\infty q_\nu E_\nu \, d\nu \tag{3.42}$$

where E_ν is the transmittance of the filter at frequency ν. One of the most commonly adopted filter systems is the UBV system of Johnson and Morgan. It consists of filters U (ultraviolet), B (blue), and V (visible) centred at 365, 444, and 548 nm, the wavelength width being about 68 nm for U, 98 for B, and 89 for V. The magnitude symbol m is replaced by U, B, or V when this system is used, for instance

$$U = -2.5 \lg \frac{\int_0^\infty q_\nu E_{U\nu} \, d\nu}{q_{0\nu}} \tag{3.43}$$

The system has been extended into the near-infrared spectrum with filters R, I, J, H, K, L, M, N reaching 10 μ, and placed in selected windows of relatively high transparency of the atmosphere. The constant q_0 for U, B, and V filters, respectively, takes the values 1.9, 4.3, and 3.7×10^{-20} erg s^{-1} cm^{-2} Hz^{-1}.

A colour index, such as $U - B$ or $B - V$ is the difference in the magnitudes in two different filters. Blue stars have $U - V$ negative; red stars positive. (Note that $U - B$ is larger for a redder star!) A colour index provides a quick look at the temperature in the stellar atmosphere. The bluer the star, the higher is its temperature. For instance, if filters had a Dirac δ-function for E_ν and stars emitted as black bodies, then $U - B \propto T^{-1}$, a relationship which is easy to check.

As in the case of the flux, the magnitude of a star depends not only on how bright it is, but also on how far away it is. As a measure of the true luminosity, essentially independent of distance, the absolute magnitude M is defined as the magnitude a star would have if it were 10 pc away (1 pc $= 3.086 \times 10^{18}$ cm). It can then easily be seen that the relation between m and M is

$$m - M = 5 \lg \frac{r}{10} \tag{3.44}$$

if no extinction is produced along the star–observer path. The Sun has an absolute visual magnitude of 4.83. In (3.44), r must be expressed in pc (parsec). If the absolute magnitude is bolometric, that is, with $E_\nu = 1$ for $\nu = 0$ to $\nu = \infty$, then M_{bol} is related to L itself, and it can easily be shown that

$$M_{bol} = -2.5 \lg \frac{L}{4\pi 10^2 q_0} \tag{3.45}$$

M_{bol} is not directly obtainable from observations (except in very few cases) due to the difficulty of making observations over the whole range of frequencies.

For extended sources, we saw that I_ν denotes the emission coming from a small part of the source. The flux, however, comes from the whole surface area. Therefore,

$$q_{e\nu} = \int_{source} I_\nu \, d\omega \tag{3.46}$$

For an extended source, the concept of magnitude per arcsec^{-2}, μ is used:

$$\mu = -2.5 \lg \frac{I}{q_0} \tag{3.47}$$

instead of I_ν. Radioastronomy uses its own nomenclature. The Jansky is used as a unit for the flux which is 10^{-26} W m^{-2} Hz^{-1}. Instead of using the intensity, the brightness temperature $T_b(\nu)$ is preferred, that is, the temperature of a black-body which has the same intensity I_ν (in the Rayleigh–Jeans approximation):

$$T_b(\nu) = \frac{c^2 I_\nu}{2\nu^2 k} \tag{3.48}$$

in conventional units. As the source may not in fact be a black-body, T_b depends in general on the frequency.

3.1.3 Boltzmann's equation

Let us apply Boltzmann's equation (2.40) to the specific case of photons, taking as p^α the components given in (3.11), replacing f_ν by I_ν using (3.6), taking into account that $F^i = 0$ for photons, and recovering the three-dimensional description

$$\frac{\partial I_\nu}{\partial t} + \vec{u} \cdot \nabla I_\nu = \Gamma_\nu \tag{3.49}$$

where Γ_ν is not the same collision term indicated in (2.40), but is left as an unspecified function. However, photons interact with matter in such a way that neither the number of photons nor their energy-momentum is conserved, so that Γ_ν must be precisely determined in this case.

The different interactions between matter and radiation can be classified as absorption or scattering processes. In absorption, the photon is lost. In scattering, the photon just changes its direction. Typical absorption processes are bound-bound, bound-free and free-free absorptions, as well as photodissociation. Typical scattering processes are Thomson, Mie, and Rayleigh scatterings.

Following the argument of Section 1.1.4, in six-dimensional phase space the photons in cell 1 go to cell 2 in dt in the absence of interactions with matter. Some photons are lost on the way and do not reach cell 2 because they are either absorbed or scattered. Scattering also has the effect of taking away photons because the two angles indicating the direction are coordinates of phase space. Photons lost by absorption and scattering are given by

$$\Gamma_\nu^- = (k_\nu + \sigma_\nu)\rho I_\nu \tag{3.50}$$

that is, proportional to the number of photons flowing from cell 1 to cell 2, and proportional to the number of absorbing or scattering molecules. The constants of proportionality k_ν and σ_ν are called the coefficients of absorption and of scattering respectively. The opacity coefficient κ_ν is also defined as

$$\kappa_\nu = k_\nu + \sigma_\nu \tag{3.51}$$

Because of the quantized energy levels in an atom, k_ν may be a complicated function of ν, with many doppler-broadened peaks. On the other hand σ_ν is usually a smooth function of ν, and it may even be independent of ν, as is the case for Thomson scattering.

There is also a production function for photons, denoted by Γ^+. Two effects are usually responsible. One of them is the scattering itself, as photons which were lost in an initial direction must show up in another. We can write

$$\Gamma_{\nu_{scattering}}^+ = \oint \sigma_\nu \rho I_\nu P \, d\omega \tag{3.52}$$

where P is the probability of a given emerging angle when the initial direction is given. Not all photons, irrespective of their initial direction, have an equal probability of emerging in the direction corresponding to our cell 2. If they had the scattering would be said to be isotropic, for which P is a constant and can be taken outside the integral. In this case, the value of the constant P is determined, taking into account that

$$\oint P d\omega = P \oint d\omega = 4\pi P = 1 \tag{3.53}$$

since the probability of the photon emerging at some angle is unity. Then $P = 1/4\pi$, and (3.52) becomes

$$\Gamma_{\nu_{scattering}}^+ = \frac{1}{4\pi} \sigma_\nu \rho \oint I_\nu \, d\omega = \sigma_\nu \rho J_\nu \tag{3.54}$$

The isotropic scattering condition generally gives good results, as the scattering frequently behaves as statistically isotropic. We will assume this condition, noting, however, that it may not be satisfactory in some cases.

Photons are also produced when energy stored in atoms or molecules is transferred to the radiative field. The energy stored may be the excitation energy of electronic levels, chemical or ionization energy, or simply thermal kinetic energy which is always accompanied by radiative energy release, black-body radiation being an extreme example. All these effects can be considered as emission and we define the emission coefficient e_ν by

$$\Gamma^+_{\nu_{emission}} = \rho e_\nu \tag{3.55}$$

Let us give some examples. When photons are in thermodynamic equilibrium in a black-body system, e_ν is given by the well-known Kirchoff law

$$e_\nu = k_\nu B_\nu \tag{3.56}$$

where B_ν is the Planck function

$$B_\nu = 2h\nu^3 \frac{1}{\left(\exp\frac{h\nu}{T}\right) - 1} \tag{3.57}$$

Sometimes (3.56) is assumed, even if the system is not a black body. This assumption is called local thermodynamic equilibrium, LTE, but it differs from the LTE defined previously in Chapters 1 and 2 and will be applied to photons later. The assumption (3.56) is called Planckian emissivity here.

Suppose, as a second example, that photons are produced by the reaction $A + B \rightarrow C^*$, followed by $C^* \rightarrow C + h\nu$. Then e_ν will be proportional to the concentrations of A and B. A simple case of astrophysical interest is the recombination process $H^+ + e^- \rightarrow H^*$, followed by $H^* \rightarrow H + h\nu$. If the plasma is macroscopically neutral, that is, if the number density of H^+ ions is equal to the electron concentration n_e, then the number of recombinations per unit volume per unit time is Kn_e^2, where K is the recombination reaction rate. Then, $e_\nu \propto AKn_e^2$, where A is Einstein's coefficient of spontaneous emission.

If radiation in a galaxy is studied, e_ν is proportional to the density of stars. Many effects should then be included in the function e_ν, so that it will remain as an unspecified function, to be determined separately for each application.

Taking all production and loss processes, $\Gamma_\nu = \Gamma_\nu^+ - \Gamma_\nu^-$, into account Boltzmann's equation becomes

$$\frac{\partial I_\nu}{\partial t} + \vec{u} \cdot \nabla I_\nu = \rho(e_\nu + \sigma_\nu J_\nu - k_\nu I_\nu - \sigma_\nu I_\nu) \tag{3.58}$$

which is an integro-differential equation. By defining the source function S_ν by

$$S_\nu = \frac{e_\nu + \sigma_\nu J_\nu}{k_\nu + \sigma_\nu} \tag{3.59}$$

another alternative form of (3.58) is obtained:

$$\frac{\partial I_\nu}{\partial t} + \vec{u} \cdot \nabla I_\nu = \rho\kappa_\nu(S_\nu - I_\nu) \tag{3.60}$$

As a simple example, assume a steady beam of light for which $\partial/\partial t = 0$, and with no source present, $S_\nu = 0$. Then, after integrating (3.60):

$$I_\nu = I_{\nu 0} \exp(-\rho\kappa_\nu x) \tag{3.61}$$

In this well-known equation, x is the length coordinate along the direction in which I_ν is calculated and $I_{\nu 0}$ is the intensity for $x = 0$. The equation implies that

$$\lambda_\nu = \frac{1}{\rho\kappa_\nu} \tag{3.62}$$

is the mean-free-path of a photon.

3.2   Macroscopic equations

It is not possible to apply (2.44) and (2.52) directly because, in this case, neither the number of photons, nor their momentum, nor their energy are conserved quantities. Therefore, the macroscopic equations must be deduced from (3.58). The term 'macroscopic' is not very appropriate here, but it remains appropriate to emphasize the parallelism between this chapter and the preceding ones. As stated before, radiative number density and energy density are not different concepts; therefore only two macroscopic equations will be obtained: the energy flux equation and the momentum flux equation. Two alternative methods are available: either using the covariant Boltzmann equation to obtain a single four-dimensional energy-momentum flux equation, or starting with (3.58) and obtaining two three-dimensional equations. We will choose the second approach.

3.2.1   The energy flux equation

Equation (3.58) must be multiplied by $G\,d\tau_p$, where $G \equiv 1$. As $d\tau_p = h^3\nu^2\,d\nu\,d\omega$ (as found in (3.3)), we must then integrate over the solid angle and frequency. The frequency integration will be left to the end of this section, so that we start by keeping ν constant. Therefore, let us multiply (3.58) by $d\omega$ and perform $\oint$:

$$\oint \frac{\partial I_\nu}{\partial t}\,d\omega = \frac{\partial}{\partial t}\oint I_\nu\,d\omega = 4\pi\frac{\partial J_\nu}{\partial t} \tag{3.63}$$

$$\oint \vec{u}\cdot\nabla I_\nu\,d\omega = \nabla\cdot\oint \vec{u}I_\nu\,d\omega = \nabla\cdot\vec{q}_\nu \tag{3.64}$$

$$\oint (e_\nu + \sigma_\nu J_\nu)\rho\,d\omega = 4\pi\rho(e_\nu + \sigma_\nu J_\nu) \tag{3.65}$$

$$\oint (\sigma_\nu + k_\nu)I_\nu\,d\omega = (\sigma_\nu + k_\nu)4\pi J_\nu \tag{3.66}$$

We assume e_ν, σ_ν, and k_ν to be isotropic. This may be not true when the fluid is in motion. A doppler effect would, for instance, make k_ν anisotropic. Now we have

$$\frac{\partial J_\nu}{\partial t} + \frac{1}{4\pi}\nabla\cdot\vec{q}_\nu = \rho(e_\nu - k_\nu J_\nu) \tag{3.67}$$

Note that scattering has no net effect on the energy budget. It is now time to integrate in ν:

$$\frac{\partial J}{\partial t} + \frac{1}{4\pi}\nabla \cdot \vec{q} = \rho e - \rho \int_0^\infty k_\nu J_\nu \, d\nu \tag{3.68}$$

Under steady-state conditions and under the so-called radiative equilibrium condition, $\nabla \cdot \vec{q} = 0$, the equation simplifies to

$$e = \int_0^\infty k_\nu J_\nu \, d\nu \tag{3.69}$$

Many systems do in fact obey the radiative equilibrium condition. As will be shown at the end of this chapter, $\nabla \cdot \vec{q} = 0$ is the equation for energy equilibrium in the matter if energy is transported by radiation only. If the energy flux had a finite divergence it would produce heating in the matter. Note that $\nabla \cdot \vec{q} = 0$, without the subindex ν. The condition $\nabla \cdot \vec{q}_\nu = 0$ with subindex ν is more rarely met because the dominant frequencies at which heat flows may not remain constant.

Let us consider another example. Assuming equilibrium, with neither absorption nor emission in the region of interest, as well as spherical symmetry, (3.67) implies $\nabla \cdot \vec{q}_\nu = 0$, which is, in spherical coordinates,

$$\frac{dq_{\nu r}}{dr} + 2\frac{q_{\nu r}}{r} = 0 \tag{3.70}$$

When integrated, this differential equation gives $q_{\nu r} \propto r^{-2}$, a well-known formula, which has already been applied on the basis of intuitive simple interpretation.

3.2.2   The momentum flux equation

In this case, Boltzmann's equation should be multiplied by $h\nu \; \vec{u} \; h^3\nu^2 d\nu \, d\omega$, but again let us start by keeping ν constant. Then (3.58) must be multiplied only by $\vec{u} \, d\omega$ and then $\oint$ performed. The integrals of the different terms give

$$\oint \frac{\partial I_\nu}{\partial t}\vec{u} \, d\omega = \frac{\partial}{\partial t}\oint I_\nu \vec{u} \, d\omega = \frac{\partial \vec{q}_\nu}{\partial t} \tag{3.71}$$

$$\oint \vec{u}\vec{u} \cdot \nabla I_\nu \, d\omega = \nabla \cdot \oint I_\nu \vec{u}\vec{u} \, d\omega = \nabla \cdot \mathcal{P}_\nu \tag{3.72}$$

$$\oint \rho(e_\nu + \sigma_\nu J_\nu)\vec{u} \, d\omega = \rho(e_\nu + \sigma_\nu J_\nu)\oint \vec{u} \, d\omega = 0 \tag{3.73}$$

$$\oint \rho(\sigma_\nu + k_\nu)I_\nu \vec{u} \, d\omega = \rho(\sigma_\nu + k_\nu)\vec{q}_\nu \tag{3.74}$$

Therefore,

$$\frac{\partial \vec{q}_\nu}{\partial t} + \nabla \cdot \mathcal{P}_\nu = -\rho(\sigma_\nu + k_\nu)\vec{q}_\nu \tag{3.75}$$

Integrating in frequency:

$$\frac{\partial \vec{q}}{\partial t} + \nabla \cdot \mathcal{P} = -\rho \int_0^\infty \kappa_\nu \vec{q}_\nu \, d\nu \tag{3.76}$$

which will be needed later.

3.3   The black body

3.3.1   Boltzmann's solution

Consider, first, a perfect black body, that is, thermodynamic equilibrium. As in equilibrium there cannot be privileged directions; I_ν does not depend on the direction. By using the definition of J_ν (3.15) we obtain $J_\nu = I_\nu$. In perfect equilibrium no time and no space derivatives of the distribution function can exist. Therefore, $\frac{\partial I_\nu}{\partial t} = 0$ and $\nabla I_\nu = 0$. Boltzmann's equation then tells us that $e_\nu = k_\nu I_\nu$, which, when compared with (3.56), gives

$$I_\nu = B_\nu \tag{3.77}$$

which is the well-known intensity of a black-body system. Using this solution it is possible to calculate all representative quantities of the radiative field. We easily find that

$$S_\nu = J_\nu = I_\nu = B_\nu \tag{3.78}$$

The flux can now be calculated as

$$\vec{q}_\nu = \oint I_\nu \vec{u} \, d\omega = B_\nu \oint \vec{u} \, d\omega = 0 \tag{3.79}$$

which is an obvious conclusion: in the absence of privileged directions, no flux can be expected. A black body does not emit. This appears paradoxical, as we are used to hearing sentences such as 'assuming the star to emit as a black body. . .'. Black bodies are systems of photons in thermodynamic equilibrium, whose the best representations are boxes with small holes. The hole must be small enough so as not to perturb thermodynamic equilibrium. When we state that a star emits like a black body, we assume that its hole, in this case the whole star's surface, does not perturb the internal equilibrium. A black body star is a box with a hole, in which the hole is larger than the box! The pressure tensor becomes

$$\mathcal{P}_\nu = \oint I_\nu \vec{u}\vec{u} \, d\omega = B_\nu \oint \vec{u}\vec{u} \, d\omega = p_\nu \delta \tag{3.80}$$

where

$$p_\nu = \frac{1}{3} B_\nu \oint d\omega = \frac{4\pi}{3} B_\nu \tag{3.81}$$

It was shown in (3.34) that $\epsilon_\nu = 3p_\nu$; therefore

$$\epsilon_\nu = 4\pi B_\nu \tag{3.82}$$

and hence

$$n_\nu = 4\pi \frac{B_\nu}{h\nu} \tag{3.83}$$

It is well known, and can be checked, that

$$B = \int_0^\infty B_\nu \, d\nu = \frac{a}{4\pi} T^4 \tag{3.84}$$

where

$$a = \frac{8\pi^5}{15h^3} \tag{3.85}$$

is four times the so called Stefan–Boltzmann constant σ (in conventional units, however, $a = \frac{4\sigma}{c}$). Also,

$$\epsilon = 4\pi B = aT^4 \tag{3.86}$$

$$p = \frac{4\pi}{3} B = \frac{a}{3} T^4 \tag{3.87}$$

The number of photons per volume element is

$$n = 4\pi \oint_0^\infty \frac{B_\nu}{h\nu} d\nu = 3.7aT^3 \tag{3.88}$$

3.3.2 Entropy of black bodies

The second law of thermodynamics states that $T \, dS = dU + p \, dV$. For photons $U = \epsilon V$, where V is the volume of the black body. ϵ is given by (3.86), and p by (3.87). Then,

$$dS = 4aT^2 V dT + \frac{4}{3} aT^3 dV \tag{3.89}$$

which can be integrated as

$$S = \frac{4}{3} aT^3 V \tag{3.90}$$

with a choice of integration constant in agreement with the third law of thermodynamics. In astrophysics, these extensive quantities are not useful. Instead of S, it is preferable to work with σ, the entropy of photons per conserved material particle, or, frequently, the entropy of photons per baryon. There are $n_b V$ baryons in a volume V; therefore

$$\sigma = \frac{4}{3} \frac{aT^3}{n_b} \tag{3.91}$$

using only intensive quantities. From (3.88):

$$\sigma = 0.36 \frac{n_\gamma}{n_b} \tag{3.92}$$

where $n_\gamma = n$, with an added subindex γ to emphasize that the entropy of photons per baryon is proportional to the relative abundance of photons and baryons.

3.3.3   The black body in motion

It was shown in Chapter 2 that a perfect fluid has an isotropic momentum distribution for one observer only, called the wet observer. Another observer in motion with respect to the wet observer was called a dry observer, and for him the local microstate does not appear isotropic. A black body is a perfect photon fluid and there is therefore a wet observer who sees the isotropic properties accounted for by (3.79) and (3.80). If we now consider a dry observer with a relative motion with respect to the wet observer, the following question arises: what is the general expression of the energy-momentum tensor? We will now proceed to an analogous derivation of the deduction of (2.33) from (2.12), but for the particular case of photons.

The energy-momentum for the wet observer was given by (3.24), where I_ν was given by (3.77). If in (3.24) the integral $\oint$ were not performed, the differential energy-momentum would be

$$d\tau^{\lambda\mu} = \frac{2}{h} \frac{\nu}{\exp\left(\frac{h\nu}{T}\right) - 1} p^\lambda p^\mu \sin\theta \, d\theta \, d\varphi \, d\nu \tag{3.93}$$

Let us produce a Lorentz transformation for the new observer moving with respect to the wet one, along the OZ-axis. As p^α, given by (3.11), is a four-vector

$$h\nu' \begin{bmatrix} 1 \\ \sin\theta'\cos\varphi' \\ \sin\theta'\sin\varphi' \\ \cos\theta' \end{bmatrix} = h\nu \begin{bmatrix} \Gamma & 0 & 0 & \Gamma v \\ 0 & 1 & 0 & 0 \\ 0 & 0 & 1 & 0 \\ \Gamma v & 0 & 0 & \Gamma \end{bmatrix} \begin{bmatrix} 1 \\ \sin\theta\cos\varphi \\ \sin\theta\sin\varphi \\ \cos\theta \end{bmatrix} \tag{3.94}$$

where v is the speed of the dry observer and $\Gamma = (1 - v^2)^{-\frac{1}{2}}$. Then

$$\nu' = \nu\Gamma(1 + v\cos\theta) \tag{3.95}$$

$$\nu'\sin\theta'\cos\varphi' = \nu\sin\theta\cos\varphi \tag{3.96}$$

$$\nu'\sin\theta'\sin\varphi' = \nu\sin\theta\sin\varphi \tag{3.97}$$

$$\nu'\cos\theta' = \nu\Gamma(v + \cos\theta) \tag{3.98}$$

From this system of equations we obtain

$$\varphi = \varphi' \tag{3.99}$$

$$\cos\theta' = \frac{v + \cos\theta}{1 + v\cos\theta} \tag{3.100}$$

$$\nu' = \nu\Gamma(1 + v\cos\theta) \tag{3.101}$$

It can easily be shown that $\nu^2 d\omega = \nu^2 \sin\theta \, d\varphi \, d\theta$ is invariant. After derivation in (3.100) and taking (3.101) into account:

$$-\sin\theta' d\theta' = -\sin\theta \, d\theta \frac{\nu^2}{\nu'^2} \tag{3.102}$$

If $\nu^2 d\omega$ is invariant, $\nu \, d\nu \, d\omega$ is also invariant. Carrying out a Lorentz transformation on (3.93), which is a true four-tensor,

$$d\tau'^{\lambda\mu} = \Lambda_\sigma{}^\lambda \Lambda_\rho{}^\mu \frac{2}{h} p^\sigma p^\rho \frac{1}{\exp\frac{h\nu}{T} - 1} d\omega\, \nu\, d\nu$$

$$= \frac{2}{h} p'^\lambda p'^\mu \frac{1}{\exp\frac{h\nu'}{\Gamma(1+\nu\cos\theta)T} - 1} d\omega' \nu' d\nu' \qquad (3.103)$$

The spectrum $[d\tau'^{\lambda\mu}, \nu']$ is found to be that of a black body. The difference now is that the system is by no means isotropic, the equivalent temperature being

$$T' = T(1 + \nu\cos\theta)\Gamma = T\frac{1 + \nu\cos\theta}{\sqrt{1 - \nu^2}} \qquad (3.104)$$

Assume that the dry observer is embedded in the black body. By fitting the curve (T', θ) it is possible to deduce both the actual temperature and ν, the speed of the black body with respect to it. The line connecting the direction of the minimum and maximum equivalent temperatures indicates the direction of motion of the black body. This formula will be used when studying the relative motion of the Earth with respect to 2.7 K black-body radiation.

3.4 Solutions near thermodynamic equilibrium

3.4.1 Perfect photon fluids

A perfect photon fluid is a photon fluid in local thermodynamic equilibrium in the sense considered in preceding chapters, that is, an observer exists who sees its local microstate as isotropic. Black-body radiation is therefore a perfect photon fluid. But there are other systems that behave locally as black bodies. We can take $I_\nu = B_\nu$, as for a black body, but now we allow small time and space variations of B_ν, that is, variations of temperature. Also, e_ν is calculated taking Kirchoff's law (3.56) into account. Then we obtain from (3.58)

$$\frac{\partial B_\nu}{\partial t} + \vec{u} \cdot \nabla B_\nu = 0 \qquad (3.105)$$

This assumption is more restrictive than that of Planckian emissivity.

3.4.2 The diffusion approximation

The so-called diffusion approximation gives a solution for I_ν for systems which depart very slightly from equilibrium. Consider again the Boltzmann equation in the form of (3.60) under steady-state conditions and rewrite

$$I_\nu = S_\nu - \frac{\vec{u} \cdot \nabla I_\nu}{\kappa_\nu \rho} \qquad (3.106)$$

There is an iterative method for finding I_ν, consisting of using I_ν determined from the $(i-1)$th step, on the right-hand side, to determine a new value of I_ν for step i. When this iterative method has only one step, it is called the diffusion approximation. I_ν is calculated from (3.106), where the right-hand side is calculated using B_ν instead of S_ν and I_ν. Therefore, the solution of the diffusion approximation is

$$I_\nu = B_\nu - \frac{\vec{u} \cdot \nabla B_\nu}{\kappa_\nu \rho} \tag{3.107}$$

Once the distribution function is known, all radiative quantities can be calculated. For the mean intensity we have

$$4\pi J_\nu = \oint B_\nu \, d\omega - \frac{1}{\kappa_\nu \rho} \oint \vec{u} \cdot \nabla B_\nu \, d\omega = 4\pi B_\nu - 0 = 4\pi B_\nu \tag{3.108}$$

Therefore as in a black body,

$$J_\nu = B_\nu \tag{3.109}$$

Let us calculate the flux:

$$\begin{aligned}
\vec{q}_\nu &= \oint I_\nu \vec{u} \, d\omega = \oint B_\nu \vec{u} \, d\omega - \oint \frac{1}{\kappa_\nu \rho} (\vec{u}\vec{u}) \cdot \nabla B_\nu \, d\omega \\
&= 0 - \frac{1}{\kappa_\nu \rho} \nabla \cdot \oint B_\nu \vec{u}\vec{u} \, d\omega = -\frac{1}{\kappa_\nu \rho} \nabla \cdot (p_\nu \delta) \\
&= -\frac{\nabla p_\nu}{\kappa_\nu \rho} = -\frac{4\pi}{3} \frac{1}{\kappa_\nu \rho} \nabla B_\nu
\end{aligned} \tag{3.110}$$

B_ν is an increasing function of temperature. Thus (3.110) tells us that radiation flows from places at higher temperature to places at lower temperature. Similarly, the pressure tensor becomes diagonal:

$$\mathcal{P}_\nu = p_\nu \delta = \frac{4\pi}{3} B_\nu \delta \tag{3.111}$$

and the hydrostatic pressure is

$$p_\nu = \frac{4\pi}{3} B_\nu \tag{3.112}$$

Equation (3.112) also holds for a black body.

The diffusion approximation can be used when the system is so close to equilibrium that the second term on the right-hand side of (3.107) is much smaller than the first one, that is, it is a corrective term. Therefore, the condition under which the diffusion approximation is valid is

$$\lambda_\nu \ll L \tag{3.113}$$

where λ_ν is defined by (3.62) and L is a characteristic length over which significant variations in temperature take place.

3.4.3   The grey body

A grey body has k_ν and σ_ν independent of the frequency, so that we may write simply k and σ, without the ν subindex. This condition is obeyed exactly when interactions with matter are restricted to Thomson scattering, which is indeed independent of frequency.

The grey body is not an approximation near thermodynamic equilibrium, but it is included here as it lies behind an approximation method, the mean opacities method, which provides good results near equilibrium.

When k, σ, and therefore κ, have no ν subindex, our equations (3.60), (3.67), and (3.75) can be simplified considerably, and we can formulate them as

$$\frac{\partial I}{\partial t} + \vec{u} \cdot \nabla I = \rho\kappa(S - I) \tag{3.114}$$

$$\frac{\partial J}{\partial t} + \frac{1}{4\pi} \nabla \cdot \vec{q} = \rho e - \rho k J \tag{3.115}$$

$$\frac{\partial \vec{q}}{\partial t} + \nabla \cdot \mathcal{P} = -\rho\kappa\vec{q} \tag{3.116}$$

Remember that these equations are not independent. Equations (3.116) and (3.115) were obtained from (3.114). Suppose as an example that we are in a steady state, and that radiative equilibrium $\nabla \cdot \vec{q} = 0$ exists. Then (3.115) gives $e = kJ$, even if Planckian emissivity is not assumed. Then $S = J$ and (3.114) is completely determined after substitution of J for S.

3.4.4 Mean opacities

When the system is not far from equilibrium, an adequate definition of mean opacity permits the use of differential equations as simple as for a grey body. This definition depends on the equation under consideration, as it will be different if we are interested in Boltzmann's equation itself (3.58), the energy flux equation (3.68), or the momentum flux equation (3.76).

Consider, first, the case of Boltzmann's equation (3.58). Integration in frequency yields

$$\frac{\partial I}{\partial t} + \vec{u} \cdot \nabla I = \rho\left(e + \int_0^\infty \sigma_\nu J_\nu \, d\nu - \int_0^\infty \kappa_\nu I_\nu \, d\nu\right) \tag{3.117}$$

By defining a mean opacity κ as

$$\int_0^\infty \kappa_\nu I_\nu \, d\nu = \kappa \int_0^\infty I_\nu \, d\nu \tag{3.118}$$

the last term on the right-hand side is simplified. The problem is that to calculate κ we must know I_ν, but this requires the solution of equation (3.117). However, if the system is not far from equilibrium, so that $I_\nu \approx B_\nu$, we could define a Planck mean opacity κ_p instead of κ in (3.118) so that

$$\kappa_p = \frac{\int_0^\infty \kappa_\nu B_\nu \, d\nu}{\int_0^\infty B_\nu \, d\nu} \tag{3.119}$$

This Planck mean opacity is calculable without knowing the solution of (3.117). If $\int_0^\infty \kappa_\nu I_\nu \, d\nu$ is replaced by $\kappa_p I$ in this equation, the error will be small for a system near equilibrium. As in (3.119) we could define a Planck mean scattering coefficient σ_p and a Planck absorption coefficient so that $\kappa_p = k_p + \sigma_p$. With these concepts we would have

$$\frac{\partial I}{\partial t} + \vec{u} \cdot \nabla I = \rho(e + \sigma_p J - \kappa_p I) \tag{3.120}$$

close to equilibrium, which is a much simplified equation.

Suppose now that we are dealing with the energy flux equation (3.68). The use of the Planck mean absorption coefficient would give

$$\frac{\partial J}{\partial t} + \frac{1}{4\pi}\nabla \cdot \vec{q} = \rho(e - k_p J) \tag{3.121}$$

For instance, under steady-state and radiative equilibrium conditions, which can be considered as valid for many real systems, we would have $e = k_p J$, which greatly simplifies (3.121), in which e was unknown.

Suppose, lastly, that we are dealing with the momentum flux equation (3.76). We can define the Chandrasekhar opacity κ_c via

$$\int_0^\infty \kappa_\nu \vec{q}_\nu \, d\nu = \kappa_c \vec{q} \tag{3.122}$$

and have

$$\frac{\partial \vec{q}}{\partial t} + \nabla \cdot \mathcal{P} = -\rho \kappa_c \vec{q} \tag{3.123}$$

At this point we encounter the same problem. The determination of κ_c requires previous knowledge of $\vec{q}_\nu$, which requires the solution of (3.76). Here we cannot take $\vec{q}_\nu$ from the black-body solution as it vanishes in this case. A solution which yields a non-vanishing flux near equilibrium is the diffusion approximation. The Chandrasekhar mean opacity is then replaced by the Rosseland mean opacity, κ_R. Using (3.110), Rosseland's mean is defined by

$$\int_0^\infty \nabla B_\nu \, d\nu = \kappa_R \int_0^\infty \frac{1}{\kappa_\nu} \nabla B_\nu \, d\nu \tag{3.124}$$

which is equivalent to

$$\frac{1}{\kappa_R} = \frac{\int_0^\infty \frac{1}{\kappa_\nu}\frac{dB_\nu}{dT}\, d\nu}{\int_0^\infty \frac{dB_\nu}{dT}\, d\nu} \tag{3.125}$$

The flux for the diffusion approximation then becomes

$$\vec{q} = -\frac{4\pi}{3\rho}\int_0^\infty \frac{1}{\kappa_\nu}\nabla B_\nu \, d\nu = -\frac{4\pi}{3\rho}(\nabla T)\frac{1}{\kappa_R}\frac{d}{dT}\int_0^\infty B_\nu \, d\nu = -\frac{4a}{3\rho}\frac{1}{\kappa_R}T^3\nabla T \tag{3.126}$$

which is a simple formula with interesting astrophysical applications.

3.5   Basic astrophysical applications

3.5.1   Stellar and planetary atmospheres

Specific nomenclature

One of the simplest systems for which the above equations has direct application, and which is of great astrophysical interest, is the case of a stationary, plane, stratified atmosphere. The direction is specified by angles θ and φ, as in (3.11). Now θ is the angle between the direction of I_ν and the vertical, and φ is the azimuthal angle. Instead of θ, it is convenient to use

$$\mu = \cos\theta \tag{3.127}$$

$\theta = 0°$ that is, $\mu = 1$ defines the upward vertical; $\theta = 90°$, $\mu = 0$, is horizontal and $\theta = 180°$, $\mu = -1$ is vertically down. Using μ:

$$\vec{u} = (\sqrt{1-\mu^2}\cos\varphi, \sqrt{1-\mu^2}\sin\varphi, \mu) \tag{3.128}$$

and

$$d\omega = -d\mu\, d\varphi \tag{3.129}$$

Let us consider the flux $\vec{q}$. It is obvious that $q_{\nu x} = q_{\nu y} = 0$, due to the stratification condition. We also know that I_ν is not a function of φ. Then

$$q_{\nu z} = \int_0^{2\pi}\int_{-1}^1 I_\nu\mu\, d\mu\, d\varphi = 2\pi\int_{-1}^1 I_\nu\mu\, d\mu \tag{3.130}$$

Instead of $q_{\nu z}$, the quantities H_ν and F_ν are normally used, where

$$H_\nu = \frac{q_{\nu z}}{4\pi} = \frac{F_\nu}{4} \tag{3.131}$$

F_ν is also (unfortunately) called the flux, and H_ν the Eddington flux. From its definition, (3.33) with (3.31), we obtain for the pressure tensor

$$\mathcal{P}_\nu = \oint I_\nu \vec{u}\vec{u}\, d\omega = \pi \begin{bmatrix} PP\int_{-1}^1 I_\nu(1-\mu^2)\, d\mu & 0 & 0 \\ 0 & \int_{-1}^1 I_\nu(1-\mu^2)\, d\mu & 0 \\ 0 & 0 & 2\int_{-1}^1 I_\nu\mu^2 d\mu \end{bmatrix} \tag{3.132}$$

Eddington also introduced the function K_ν, which is widely used and is related to $\mathcal{P}_{\nu zz}$ by:

$$\mathcal{P}_{\nu zz} = 4\pi K_\nu \tag{3.133}$$

The mean intensity becomes

$$J_\nu = \frac{1}{4\pi}\oint I_\nu\, d\omega = \frac{1}{2}\int_{-1}^1 I_\nu\, d\mu \tag{3.134}$$

Eddington's nomenclature is justified because, in addition to (3.134),

$$H_\nu = \frac{1}{2}\int_{-1}^1 I_\nu\mu\, d\mu \tag{3.135}$$

$$K_\nu = \frac{1}{2}\int_{-1}^1 I_\nu\mu^2 d\mu \tag{3.136}$$

which are simply moments of intensity weighted by $\cos\theta$ and $\cos^2\theta$.

The hydrostatic pressure becomes

$$p_\nu = \frac{4\pi}{3}J_\nu \tag{3.137}$$

Instead of z, it is also customary to use the optical depth, defined as

$$d\tau_\nu = -\kappa_\nu\rho\, dz \tag{3.138}$$

taking $\tau_\nu = 0$ as the origin, for $z = \infty$. Note that τ_ν increases inwards. As ρ is an exponential-like function in the inward direction, the altitude separation of a constant interval in τ_ν, such as $\Delta\tau_\nu = 1$, decreases exponentially. This is clearly true when κ_ν is a constant and we may use (1.92). Then, integrating (3.138) from a given z to $z = \infty$, outside the atmosphere,

$$\tau_\nu = \kappa_\nu \rho(z) H \tag{3.139}$$

where H is the scale height. If the atmosphere is isothermal, H is a constant, and $\rho(z)$ is precisely exponential.

Conditions of flatness and stratification imply that Boltzmann's equation can be written in a simplified way since

$$\vec{u} \cdot \nabla I_\nu = \mu \frac{\partial I_\nu}{\partial z} \tag{3.140}$$

and, under stationarity, Boltzmann's equation reads

$$\mu \frac{\partial I_\nu}{\partial z} = \kappa_\nu \rho (S_\nu - I_\nu) \tag{3.141}$$

With the change of variable (3.138):

$$\mu \frac{\partial I_\nu}{\partial \tau_z} = I_\nu - S_\nu \tag{3.142}$$

although S_ν remains unknown.

The formal solution

Even if S_ν remains unknown, equation (3.142) can be formally integrated, because $e^{-\tau_\nu/\mu}$ is an integrating factor. When (3.142) is multiplied by this, it can be written in the form

$$\mu \frac{\partial}{\partial \tau_\nu} \left(I_\nu e^{-\tau_\nu/\mu} \right) = -S_\nu e^{-\tau_\nu/\mu} \tag{3.143}$$

as can easily be proved. If we integrate from $\tau_\nu = 0$ to a given τ_ν,

$$\mu \left[I_\nu e^{-t/\mu} \right]_0^{\tau_\nu} = - \int_0^{\tau_\nu} S_\nu e^{-t/\mu} dt \tag{3.144}$$

where t is an integration variable which replaces τ_ν. This equation can be used to obtain $I_{\nu\infty} = I_\nu(\tau_\nu = 0)$, the intensity outside the atmosphere, which is of obvious astrophysical interest; this is the intensity or radiance measured at the Earth. If we make $\tau_\nu = \infty$, we obtain

$$I_{\nu\infty} = \frac{1}{\mu} \int_0^\infty S_\nu e^{-t/\mu} dt \tag{3.145}$$

For a better interpretation of this result, let us first take $\mu = 1$, that is, the vertical direction. Equation (3.145) tells us that $I_{\nu\infty}$ is the sum of the contribution of the sources in each layer, each contribution being reduced by a transmission factor $e^{-t/\mu}$, which is larger for deeper layers. Observe that in a spherical star τ_ν cannot be ∞, but a flat star is being assumed here. However, for very deep layers, $e^{-t/\mu}$ is so small that they do not contribute at all to the radiation observed outside.

Besides being a direction coordinate at any point inside an atmosphere, μ is also an observational parameter. When the telescope points to the centre of the Sun, we are observing $I_\nu(\mu = 1)$. When the telescope points to the limb (i.e. the edge) of the Sun, we are observing $I_\nu(\mu = 0)$. Halfway between these two points, $\theta = 60°$, therefore, we are observing $I_\nu(\mu = 0.5)$. In the visible we find limb darkening, I_ν decreases as μ decreases. A detailed study of the limb darkening effect provides valuable information about the structure of the solar atmosphere. To obtain some insight into this problem, let us consider an oversimplified picture. Suppose, as Eddington and Barbier did, that the source function is a linear function of τ_ν:

$$S_\nu = a\tau_\nu + b \qquad (3.146)$$

where a and b are unknown constants. This is the simplest assumption for the relation (S_ν, τ_ν); a must be positive, because the temperature is higher in the deeper layers; b must be positive or zero, as the outer layers, with $\tau_\nu \approx 0$, have a positive source function. Applying (3.145) it can easily be seen that

$$I_\nu = a\mu + b \qquad (3.147)$$

Fitting the observed curve $[I_\nu, \mu]$ to (3.147) yields a and b, and thus $[S_\nu, \tau_\nu]$, the source function, as a function of optical depth, which is the result of physical processes in the atmosphere. Observations give $a/b \approx 3/2$, so that $I_\nu(0)/I_\nu(1) = 2/5$. This is substantial limb darkening. However, note that using this value of a/b we obtain $I_\nu(0.5)/I_\nu(1) = 7/10$. The limb darkening only affects the outer parts of the observed solar disc considerably. To zero order I_ν is constant; considered in more detail it is not, and this departure is very important for the physical interpretation of the Sun's atmosphere.

Atmospheric extinction

Having left the surface of a star, and after a long journey, light arrives at the top of the Earth's atmosphere. Let us assume that there are no sources and no extinction in the interstellar medium. Assume that $I_{\nu\infty}$ is the intensity for $\mu = 1$ at the star's surface. The same value, $I_{\nu\infty}$, will characterize the light coming from the centre of the observed star's surface (if it is observable without spatial resolution problems), and at the top of our atmosphere. But, now, we must change the coordinate system. If the star has local horizontal coordinates A (azimuth) and χ (zenith distance), I_ν is the intensity in the direction $\varphi = -A$, $\theta = \chi + \pi$. The value of μ is then $\mu = \cos\theta = -\cos\chi$. Let us use equation (3.144) for $\tau_\nu = \tau_{\nu s}$, the optical depth of our atmosphere at the Earth's surface, and, assuming that no light sources exist in the atmosphere, $S_\nu = 0$. Then

$$I_{\nu s} = I_{\nu\infty} e^{\frac{\tau_{\nu s}}{\mu}} = I_{\nu\infty} e^{-\tau_{\nu s}\sec\chi} = I_{\nu\infty} e^{-\sec\chi \int_0^\infty \kappa_\nu \rho \, dz} \qquad (3.148)$$

where $I_{\nu s}$ is the intensity at the ground. What we observe from a star is, however, its flux, and not the intensity at its disc centre. We have also shown that I_ν is nearly constant across the star's disc. Then, using (3.46),

$$q_\nu = I_\nu \omega \qquad (3.149)$$

where q_ν is the component of $\vec{q}_\nu$ along the direction $(-A, \chi + \pi)$ and ω is the solid angle from the Earth formed by the star. Observe that q_ν and ω follow a r^{-2} law, whilst I_ν remains independent of r. As ω does not vary appreciably from the top of the atmosphere to the ground, equation (3.149) can be rewritten for the flux:

$$q_{\nu s} = q_{\nu\infty} e^{-\sec\chi \int_0^\infty \kappa_\nu \rho \, dz} \tag{3.150}$$

or, in magnitudes,

$$m_s = m_\infty + K \sec\chi \tag{3.151}$$

where

$$K = \lg e \int_0^\infty \kappa_\nu \rho \, dz \tag{3.152}$$

is more or less constant. κ depends on weather conditions and therefore changes from night to night, but can normally be measured and taken into account during observation. Of course, our interest is m_∞, but m_s is what we measure. K is determined by observing a standard star for which we know m_∞ to be time independent. This star is observed from time to time during the observations, for different values of χ, which varies due to the rotation of the Earth. Fitting of the curve $[m_s, \sec\chi]$ for the standard star provides the value of K during the observing period, and this value is used in (3.151) to calculate m_∞ of the star under study.

Although I_ν has been assumed to be constant over the star's observed disc, this condition is not essential to deduce (3.152), as ω could be an equivalent solid angle.

Our atmosphere not only produces extinction, it also alters colours. Using, for instance, B and V magnitudes:

$$B_s = B_\infty + K_B \sec\chi \tag{3.153}$$

$$V_s = V_\infty + K_V \sec\chi \tag{3.154}$$

Therefore,

$$(B - V)_s = (B - V)_\infty + (K_B - K_V) \sec\chi \tag{3.155}$$

K_B and K_V are not equal since K is a function of κ_ν. In our atmosphere, the most important effect that produces extinction is Rayleigh scattering, which is proportional to λ^{-4}. Blue light is therefore more affected, $K_B > K_V$, and thus $(B - V)$ is higher after traversing the atmosphere, that is, the star becomes redder.

The atmosphere has been assumed flat, which is a good approximation other than for $\chi \geq 75°$. Then a function which includes Earth's sphericity, which is not difficult to deduce, must replace the secant function.

Temperature profile

The upper layers of any stellar atmosphere cannot be considered to be in equilibrium, because, just outside the star, departure from equilibrium becomes complete. The mean free path of photons increases upwards, as it depends on ρ^{-1}, and it must eventually become very large. Indeed, a rough definition of stellar atmosphere as a distinct region with respect to the stellar interior would be that part of the star that can be seen, that is, from which a non-infinitesimal flux of photons can escape. Quantitatively this can be expressed by stating that the depth of the atmosphere corresponds to τ_ν of order unity. Equation (3.139) then indicates that this occurs approximately where the mean free path of the photons is of the order of the scale height. Roughly speaking, the atmosphere is that region of the star which is not close to thermodynamic equilibrium. However, an adequate approximation is to apply some formulae obtained under conditions near equilibrium in order to obtain an estimate of the relation $[T, \tau]$.

Under the conditions assumed for an atmosphere equation (3.123) gives

$$\frac{d\mathcal{P}_{zz}}{dz} = -\rho \kappa_c q_z \tag{3.156}$$

or, using the specific atmospheric nomenclature,

$$\frac{dK}{d\tau} = \frac{F}{4} \tag{3.157}$$

κ_c being the Chandrasekhar mean opacity. Let us assume radiative equilibrium, F being a constant, so that (3.157) can be integrated.

Near equilibrium we could assume that $J = 3K$, called the Eddington condition. It can be checked that this holds for the black-body and the diffusion approximations. Then

$$J = \frac{3}{4}F(\tau + Q) \tag{3.158}$$

where Q is an integration constant. From (3.146), taking the value $a/b \approx 3/2$ (and $J \approx S$), Q must be $\approx 2/3$. Considering the diffusion approximation result: $J = B = (a/4\pi)T^4$, and defining the effective temperature, with $F = \frac{a}{4\pi}T_{eff}^4$, we obtain

$$T^4 = \frac{3}{4}T_{eff}^4\left(\tau + \frac{2}{3}\right) \tag{3.159}$$

The definition of effective temperature tells us that this temperature would correspond to a black body emitting a total flux F. As F was considered to be a constant, T_{eff} is, in turn, a constant. Then (3.159) shows that T_{eff} is approximately the temperature at the depth, where $\tau = \frac{2}{3}$.

3.5.2 Energy flux in stellar interiors

The nuclear energy released in the stellar interior must escape, and much of this energy is transported to the surface as radiative flux. Even if the mean free path is very short, a flux must exist, so that thermodynamic equilibrium cannot be assumed. However, we may use the diffusion approximation to quantify this energy transport. Following (3.126), the vertical component of the flux is

$$q = -\frac{4a}{3\rho}\frac{1}{\kappa_R}T^3\frac{dT}{dr} \tag{3.160}$$

This is one of the fundamental equations for modelling stellar interiors.

In atmospheres the diffusion approximation is not fulfilled, as was shown, but it will now be used to justify why a star's radiation is similar to that of a black body, even

though we know that a star is not a black body. A rough and ready application of
(3.160) consists of calculating $\frac{dT}{dr}$ for two points only, one of them fairly deep in the
atmosphere, at $\tau = 1$, and the other outside at $\tau = 0$:

$$q = \frac{4a}{3} T^3 \frac{\Delta T}{\Delta \tau} = \frac{4a}{3} T^3 \frac{T - 0}{1 - 0} = \frac{4a}{3} T^4 \qquad (3.161)$$

in agreement with the T^4 dependence for black-body radiation. Here T would correspond to an optical depth slightly less than unity. The calculation can be performed a
little more precisely using a better approximation to $\frac{dT}{dr}$.

3.5.3 Interstellar extinction

In the absence of sources, interstellar dust produces extinction, as does our atmosphere.
The intensity decreases by an exponential factor, so that the magnitude increases by an
additive term. If the observed star lay in the galactic plane, we would write

$$m = M + 5 \lg \frac{r}{10} + ar \qquad (3.162)$$

instead of (3.44), where r must be measured in parsecs. The function a depends closely
on the frequency and density of dust, which is distributed with a fair degree of homogeneity. For example, at visible wavelengths, a is of the order of 1 magnitude per
kiloparsec. If we are observing a bright star with $M = 0$, at approximately 10 kpc
(about the distance to the galactic centre), we have $m = 15 + 10 = 25$, practically
unobservable with the best telescopes; the effect of extinction produces an increase of 10
magnitudes. With the naked eye we cannot observe even intrinsically bright stars
further than 1 kpc if they are close to the galactic plane. At near infrared wavelengths
the visibility is better, a being of the order of 0.1 mag/kpc, and red giants can be
observed in the neighbourhood of the galactic centre.

Visibility is also much improved when looking in directions away from the galactic
plane. As the Sun lies near the symmetry plane of the flat thin galactic disc, the situation
is similar to the case of a flat atmosphere. If b is the galactic latitude, the angle between
the direction of the observed object and the galactic plane, and if extragalactic objects
are observed, there will be an increase in magnitude of

$$\Delta m = \frac{\Delta m_0}{\sin b} \qquad (3.163)$$

where $(\Delta m)_0$ is the increment corresponding to observations perpendicular to the
galactic plane.

Finally, consider the observation of a galaxy similar to ours. There is also an internal
extinction in the observed galaxy. Let us make a rough calculation to estimate it. In this
case we cannot ignore sources along the line of sight as stars are emitting in the volume
of space where dust is absorbing light. S is now due to starlight and depends on the star
number density. We now have

$$\mu \frac{\partial I}{\partial z} = \rho_e e - k\rho I \qquad (3.164)$$

where $\mu = \cos i$, where i is the inclination of the galaxy (90° for an edge-on galaxy, 0° for a face-on galaxy), z is perpendicular to the galactic plane, and ρ_e is the density of stars. If the star distribution in the z-direction can be treated as gaussian:

$$\rho_e e = \frac{I_0}{\sqrt{2\pi}H_S} e^{-\frac{z^2}{2H_S^2}} \tag{3.165}$$

where I_0 is the intensity that would be observed in the absence of extinction, H_S is the scale thickness of the stellar disc, and ρ is the gas density.

Extinction is not produced by gas, but by dust. However, the densities of gas and dust are found to be proportional, so that $\kappa \rho_{dust}$ can be replaced by $K\rho$, where K is an equivalent opacity for visible wavelengths of the order of $5 \times 10^{-2} M_\odot^{-1} pc^2$. The use of the density of gas, instead of the density of dust, is practical because the column of hydrogen through a galactic disc is not optically thick, and radio observations enable us to directly infer the column density $\int_0^\infty \rho \, dz = \sigma$, which is termed the disc surface density. However, far infrared observations made with satellites are the only direct method of observing dust, apart from extinction effects themselves. We assume that the gas density z-distribution is another gaussian:

$$\rho = \frac{\sigma}{\sqrt{2\pi}H_g} e^{-\frac{z^2}{2H_g}} \tag{3.166}$$

where H_g indicates the thickness of the gas z-distribution.

Equation (3.164) can now be integrated. Let us express the result so as to give A, the effective increase in magnitude due to the presence of dust:

$$A = -2.5 \lg \left[\frac{1}{\sqrt{2\pi}H_S\mu} \int_{-\infty}^{\infty} e^{-\left[\frac{z^2}{2H_S^2} + \frac{K\sigma}{2\mu}\left(1 - erf\left(\frac{z}{2H_g^2}\right)\right)\right]} dz \right] \tag{3.167}$$

Here σ, A, H_S, and H_g are functions of the galactocentric radius. An implicit assumption in the above calculation is that i be small enough to make the approximation that all volume elements in the line-of-sight, are at the same galactocentric radius. Of course, this formula cannot be used for $i = 90°$.

3.6 Classical fluids in a radiation field

Photons may interchange momentum and energy with particles, hence they may have an influence on the particles' motion and temperature. One clear example is provided by the Universe at early epochs and is to be considered in chapters 8, 9, and 10. Here we will deal with classical fluids embedded in a relatively moderate radiative field, such as those constituting a stellar atmosphere. The classical fluid dynamic equations must be solved together with the radiative Boltzmann equation. The simplified case of a stationary, flat, and stratified atmosphere will be considered.

Diffusion properties are modified by radiation. For instance, a species may be a good absorber, absorb momentum from photons, and become overabundant in the upper layers. Indeed, some stars present peculiar abundances which are attributable to this effect. The modifications required in the molecular diffusion equation are not difficult to obtain, but we will concentrate on the most important aspects, on the fluid dynamic equations of a mixture. The continuity equation obviously remains unmodified. Under these conditions, the equation of motion would be

$$\nabla \cdot \mathcal{P} = mn\vec{g} \tag{3.168}$$

where $\mathcal{P}$ is not solely due to the hydrostatic pressure; a contribution due to photons must be included:

$$\mathcal{P} = \begin{pmatrix} p + \mathcal{P}_{xx}^R & 0 & 0 \\ 0 & p + \mathcal{P}_{yy}^R & 0 \\ 0 & 0 & p + \mathcal{P}_{zz}^R \end{pmatrix} \tag{3.169}$$

where $\mathcal{P}_{ij}^R$ is the ij-component of the radiation pressure tensor. Only the z-equation from the vector equation (3.168) is of interest:

$$\frac{dp}{dz} + \frac{d}{dz}\mathcal{P}_{zz}^R = -\rho g \tag{3.170}$$

Using equations (3.133) and (3.157): [1]

$$\frac{1}{\rho}\frac{dp}{dz} = \pi F \kappa_c - g \tag{3.171}$$

or

$$\frac{1}{p}\frac{dp}{dz} = -\frac{1}{H} + \pi F \kappa_c \frac{m}{T} \tag{3.172}$$

where H is the scale height and m the mass of a molecule. Equation (3.171) clearly indicates that the flux term has a direction opposite to that of gravity. When photons are coming from inside, they push matter outwards, apparently resulting in a diminished effect of gravity. In other words, the radiative pressure in a stellar atmosphere partially compensates the action of gravity.

The new term is negligible for normal stars. For instance, for the Sun, with $\kappa_c \approx 3.5 \text{ g}^{-1}\text{cm}^2$, $g \approx 3 \times 10^4 \text{cm s}^{-2}$ and $T_{eff} \approx 5 \times 10^3$ K (to calculate F), we quickly find that the effect of radiation is negligible. It may however be larger for very hot stars of type O, for which F is very large, or for red giants for which g is very low.

Let us take the energy conservation equation into account. Under the conditions specified, if the energy flux due to conduction and convection is negligible, it can be obtained directly from (1.79):

$$\nabla \cdot \vec{q} = 0 \tag{3.173}$$

or $H = $ constant, for a plane atmosphere. This result has already been used in preceding sections, where it was referred to as the radiative equilibrium condition.

---

1 When gaussian or conventional units are used, F must be replaced by F/c, where c is the speed of light.

4 Plasma fluids

4.1 Introduction

Plasmas are fluids in which ionization is significant. Motion and thermal energy are affected by electromagnetic fields, which are in turn affected by motion and thermal energy. As a consequence, Maxwell's equations must be merged with the fluid dynamic equations to form a closed system. On the other hand, the fluid dynamic equations must be modified to take magnetic forces into account. These forces depend on the particle's momentum, and this possibility was excluded in Chapter 1.

Most astrophysical plasmas are macroscopically neutral. This means that in a volume element containing a statistically important collection of ions and electrons, the net charge is zero. For instance, if atoms are singly ionized, this implies that the ion number density n_i, and the electron number density n_e are equal. This condition is usually met because if some effect suddenly separates the positive and negative charges, the electric field generated quickly restores neutrality. In the following, only macroscopically neutral plasmas will be considered.

Recombination may destroy a plasma. If a plasma is stable, either there is a permanent source of ionization – as in interstellar HII regions – or recombination is negligible. A typical distance between an ion and an electron is $n_e^{-1/3}$; therefore the coulomb potential energy is of the order of $e^2 n_e^{1/3}$, where e is the electron charge. When the thermal energy of a typical electron kT is much higher than $e^2 n_e^{1/3}$, recombination may be neglected. This criterion is frequently expressed as recombination and is negligible when

$$\lambda_L \ll n_e^{-1/3} \tag{4.1}$$

where λ_L is the Landau length

$$\lambda_L = \frac{e^2}{kT} \tag{4.2}$$

Strict criteria are often defined to specify when a fluid is a plasma, but they are not very useful in practice.

4.1.1 Cyclotron and collision frequencies

Apart from the usual thermodynamic parameters, the degree of ionization and the magnetic field strength determine most of the properties of a plasma. A set of new characteristic parameters may be defined in order to infer, *a priori*, which effects will dominate.

The gyrofrequency, or cyclotron frequency, of an electron is defined as

$$\Omega_e = \frac{eB}{m_e c} \tag{4.3}$$

where B is the magnetic field strength and m_e is the mass of an electron. This is the frequency with which an electron gyrates about $\vec{B}$. This frequency is to be compared with the electron collision frequency, which was calculated in (1.139) and in (1.112). If the gas is weakly ionized, electrons mainly collide with neutrals. Then, from (1.112) and (1.111), the collision frequency can be calculated from

$$\nu_e = \frac{16}{3} n_n \sigma_{en}^2 \sqrt{\frac{\pi kT}{2m_e}} \tag{4.4}$$

where n_n is the number density of neutrals. σ_{en} was defined in (1.111), and is the sum of the equivalent radii of neutrals and electrons, so that $\Sigma = \pi \sigma_{en}^2$ is the collision cross-section. For an equivalent air molecule σ_{en} is about 10^{-8} cm; for atomic hydrogen, σ_{en} is 0.5×10^{-8} cm, the first Bohr orbit radius. Equation (1.139) provides similar orders of magnitude:

$$\nu_e = n_n \Sigma \sqrt{\frac{3kT}{m_e}} \tag{4.5}$$

If the plasma is highly ionized, electrons mainly collide with ions. The collision is then of the coulomb type, the cross-section being proportional to T^{-2}. The mean-free-path can be estimated as $10^4 T^2 n_e^{-1}$ (cgs units) and the collision frequency as $50\, n_e T^{-\frac{3}{2}}$.

When Ω_e is much higher than ν_e, an electron has time to gyrate about $\vec{B}$ many times before colliding. The magnetic field is then comparatively important and the anisotropy it introduces is macroscopically apparent. The anisotropy associated with the magnetic field is due to the fact that no work is needed to displace a charged particle along the magnetic field lines. When the magnetic field strength is very high, the magnetic field lines become natural paths for particles, rendering all transport coefficients anisotropic. The degree of anisotropy x_e is then defined as

$$x_e = \frac{\Omega_e}{\nu_e} = \frac{eB}{cm_e \nu_e} \tag{4.6}$$

4.1.2   Maxwell's equations

Maxwell's equations have a complementary interpretation from the fluid dynamic point of view, illustrated here by a particular case. Let us assume a fluid in which all particles are identical and have the same velocity $\vec{v}_0$, no restriction being made about the density. (Note that this is not a macroscopically neutral plasma.) As a transport question assume that the microscopic behaviour is known, so that we know the Coulomb and

Lorentz forces, and that the macroscopic electromagnetic behaviour is to be found. We also start from the knowledge that there is a relation between the $\vec{D}$ and $\vec{H}$ fields created by a given charged particle at a given point. It is confirmed that this relation is

$$\vec{H} = \frac{\vec{v}_0}{c} \times \vec{D} \tag{4.7}$$

But $\vec{H}$ and $\vec{D}$ produced by all particles at a given point are additive so that (4.7) is also valid for $\vec{H}$ and $\vec{D}$ created by the fluid at a given point.

The standard introduction to Maxwell's equations provides a clear macroscopic interpretation of the two formulae:

$$\nabla \cdot \vec{D} = 4\pi\rho_c \tag{4.8}$$

$$\nabla \cdot \vec{B} = 0 \tag{4.9}$$

where ρ_c is the charge density.

Let us calculate the curl of (4.7), taking the tensor relation for any two vectors $\vec{A}$ and $\vec{B}$ into account:

$$\nabla \times (\vec{A} \times \vec{B}) = \vec{B} \cdot \nabla\vec{A} - \vec{A} \cdot \nabla\vec{B} + \vec{A}\nabla \cdot \vec{B} - \vec{B}\nabla \cdot \vec{A} \tag{4.10}$$

As $\vec{v}_0$ is a constant,

$$\nabla \times \vec{H} = -\frac{\vec{v}_0}{c} \cdot \nabla\vec{D} + \frac{\vec{v}_0}{c}\nabla \cdot \vec{D} \tag{4.11}$$

Applying (1.54) to $\vec{D}$:

$$\frac{d\vec{D}}{dt} = \frac{\partial\vec{D}}{\partial t} + \vec{v}_0 \cdot \nabla\vec{D} \tag{4.12}$$

The wet observer detects no time variation as all particles have the same velocity. Inserting (4.12) with $\frac{d\vec{D}}{dt} = 0$ and (4.8) in (4.11), and introducing the charge current density, which in this case is $\vec{j} = \rho_c\vec{v}_0$,

$$\nabla \times \vec{H} = \frac{1}{c}\frac{\partial\vec{D}}{\partial t} + \frac{4\pi}{c}\vec{j} \tag{4.13}$$

which is one of the fundamental Maxwell equations.

Another transport argument may provide an interpretation of the other fundamental Maxwell equation. The wet observer appreciates only an electrical field $\vec{E}'$, but no magnetic field as all particles are at rest for him. The dry observer, however, observes $\vec{E}$ and $\vec{B}$. The force on any of the particles must be the same for both observers, as we are not considering a relativistic system. Therefore,

$$q\vec{E}' = q(\vec{E} + \frac{\vec{v}_0}{c} \times \vec{B}) \tag{4.14}$$

From (1.54), and noting that $\frac{d\vec{B}}{dt}$ vanishes for the wet observer:

$$0 = \frac{d\vec{B}}{dt} = \frac{\partial\vec{B}}{\partial t} + \vec{v}_0 \cdot \nabla\vec{B} \tag{4.15}$$

Taking the curl in (4.14), using (4.10), taking (4.9) and (4.15) into account, and noting that the wet observer sees an electrostatic system, $\nabla \times \vec{E}' = 0$, we obtain

$$\nabla \times \vec{E} + \frac{1}{c}\frac{\partial \vec{B}}{\partial t} = 0 \tag{4.16}$$

the other fundamental Maxwell equation. Obviously, these equations need not be deduced in this way, but the above arguments provide a transport interpretation of them which gives an interesting physical insight. Of course, (4.8), (4.9), (4.13), and (4.16) are valid for many fluids, and in particular for the macroscopically neutral plasmas to which this chapter is devoted.

4.2   Macroscopic equations

Boltzmann's equation in the form given in Chapter 1 is perfectly valid in a plasma context. In order to deduce the fluid dynamic equations from it we assumed that forces acting on a particle do not depend on the particle's momentum, which is no longer valid for Lorentz forces.

We will assume a plasma to have only three components: electrons, ions, and neutrals. When not otherwise specified, we will assume that the charge on an ion is the same as that on the electron, e, for singly ionized atoms. This is the most common situation in astrophysics, and for multiply ionized atoms the derivation of the corrected formulae is straightforward. The mass of the neutrals and the mass of the ions will also be considered to be the same, m_i.

The forces will now be

$$\vec{F}_j \rightarrow \vec{F}_j + q_j\vec{E} + \frac{q_j}{m_j}\left(\frac{\vec{p}}{c} \times \vec{B}\right) \tag{4.17}$$

where $\vec{F}_j$ is the non-electromagnetic force. Obviously, q_j is equal to e (for ions), equal to $-e$ (for electrons), or zero (for neutrals). Note that in order to obtain the macroscopic equations it was necessary to calculate $\sum_j \int_p [\textit{Boltzmann's equation}]_j G_j d\tau_p$, where G is any one of the collisional invariants. The nomenclature is that used in Chapter 1.

4.2.1   Continuity equations

In this case the sum $\sum_j$ is not necessary, as the individual mass of any of the three components is conserved if ionization and recombination are negligible. Let $G \equiv m_j$ and we will restrict ourselves to calculating those terms involving only the Lorentz force. Therefore, the derivation carried out in (1.48) must be completed, with

$$\int_p m_j \frac{q_j}{m_j}\left(\frac{\vec{p}}{c} \times \vec{B}\right) \cdot \nabla_p f_j \, d\tau_p = \frac{q_j}{c}\int_p (\vec{p} \cdot \hat{B}) \cdot \nabla_p f_j \, d\tau_p = \frac{q_j}{c}\int_p p_\alpha \hat{B}_{\alpha\beta}\frac{\partial f_j}{\partial p_\beta} \, d\tau_p$$

$$= \frac{q_j}{c}\int_p \hat{B}_{\alpha\beta}\frac{\partial}{\partial p_\beta}(p_\alpha f_j) \, d\tau_p - \frac{q_j}{c}\int_p \hat{B}_{\alpha\beta}f_j\frac{\partial p_\alpha}{\partial p_\beta} \, d\tau_p$$

$$= 0 - \frac{q_j \hat{B}_{\alpha\beta}}{c}\int_p f_j\delta_{\alpha\beta} \, d\tau_p = 0 \tag{4.18}$$

because $\hat{B}_{\alpha\alpha} = 0$, since $\hat{B}_{\alpha\beta}$ is antisymmetric. (It is so defined that, for any two vectors, $\vec{A} \times \vec{B} = \vec{A} \cdot \hat{B}$.) The presence of magnetic forces does not alter the individual continuity equations, as magnetic forces do not modify the conservation of mass. We then have

$$\text{Electrons} \quad \frac{\partial n_e}{\partial t} + \nabla \cdot (n_e(\vec{v}_0 + \langle \vec{V}_e \rangle)) = 0 \tag{4.19}$$

$$\text{Ions} \quad \frac{\partial n_e}{\partial t} + \nabla \cdot (n_e(\vec{v}_0 + \langle \vec{V}_i \rangle)) = 0 \tag{4.20}$$

$$\text{Neutrals} \quad \frac{\partial n_n}{\partial t} + \nabla \cdot (n_n(\vec{v}_0 + \langle \vec{V}_n \rangle)) = 0 \tag{4.21}$$

When (4.19), (4.20) and (4.21) are summed (after previously being multiplied by m_e, m_i, and m_i respectively) the continuity equation for the mixture is seen to be unmodified:

$$\frac{\partial \rho}{\partial t} + \nabla \cdot (\rho \vec{v}_0) = 0 \tag{4.22}$$

using (1.31). By subtracting (4.19) from (4.20), we obtain

$$\nabla \cdot (n_e(\langle \vec{V}_i \rangle - \langle \vec{V}_e \rangle)) = 0 \tag{4.23}$$

The charge current density is

$$\vec{j} = en_e(\langle \vec{V}_i \rangle - \langle \vec{V}_e \rangle) \tag{4.24}$$

Hence (4.23) is simply

$$\nabla \cdot \vec{j} = 0 \tag{4.25}$$

which is the charge continuity equation; note that the time derivative of the charge density is zero because the charge density was set equal to zero as a consequence of macroscopic neutrality.

4.2.2 The equation of motion

Let us now repeat the derivation of (1.65), taking the presence of Lorentz forces into account. Their contribution is taken into consideration via

$$\int_p \vec{p} \frac{q_j}{m_j} \left(\frac{\vec{p}}{c} \times \vec{B} \right) \cdot \nabla_p f_j \, d\tau_p = -q_j n_j \frac{\vec{v}_0}{c} \times \vec{B} - q_j n_j \frac{\langle \vec{V}_j \rangle}{c} \times \vec{B} \tag{4.26}$$

Obtaining this result may require some effort, but similar integrals have been used frequently throughout the book, and they should have become familiar. Now (1.66) must be replaced by

$$\vec{A}_j = m_j \frac{\partial}{\partial t} \left(n_j(\vec{v}_0 + \langle \vec{V}_j \rangle) \right) + \nabla \cdot (P_j + m_j n_j \vec{v}_0 \vec{v}_0)$$
$$n_j \vec{F}_j - q_j n_j \vec{E} - q_j n_j \left(\frac{\vec{v}_0 + \langle \vec{V}_j \rangle}{c} \times \vec{B} \right) \tag{4.27}$$

These are the individual equations of motion. Summing these equations, and using (1.58), an equation similar to (1.67) is obtained. But, again, a form of equation similar to (1.68) is preferred. The equation of motion then becomes

$$\rho \frac{\partial \vec{v}_0}{\partial t} + \rho \vec{v}_0 \cdot \nabla \vec{v}_0 + \nabla \cdot \mathcal{P} = \sum_j n_j \vec{F}_j + \frac{\vec{j}}{c} \times \vec{B} \tag{4.28}$$

From this we can infer that electric fields will have no direct influence on the motion of the plasma, as they act in opposite directions for ions and electrons. They may, however, have an indirect influence. For example, they produce heating, therefore temperature gradients, and therefore pressure gradients which explicitly produce accelerations in the plasma. Or they simply produce a charge current density.

4.2.3   The energy balance equation

A similar derivation to that employed to obtain (1.72) must be performed. Plasmas exist in which the temperature of electrons and ions differs. This is due to the fact that the electron does not interchange very much energy when colliding with an ion or a neutral, both of which are much heavier. Therefore, it is sometimes difficult for electrons and ions to reach a common thermodynamic equilibrium. However, let us consider the simplest case in which a single temperature characterizes the system, that is, where $T_i \approx T_e$. Again, we must concentrate our attention on the Lorentz forces and calculate

$$\int_p \frac{p^2}{2m_j} \frac{q_j}{m_j} \left(\frac{\vec{p}}{c} \times \vec{B} \right) \cdot \nabla_p f_j \, d\tau_p = 0 \tag{4.29}$$

The details of this derivation are again omitted, as it is similar to others encountered throughout the book. It can be seen that magnetic fields have no direct influence on thermal balance. An intuitive explanation for this is that, although charged particles gyrate about $\vec{B}$, their speed is not modified by the presence of $\vec{B}$.

   The effect of electrical fields is calculated, as is that of any other force, using (1.72):

$$-\sum_j (n_j q_j \vec{E} \cdot \langle \vec{V}_j \rangle) = (-n_e e \langle \vec{V}_i \rangle + n_e e \langle \vec{V}_e \rangle) \cdot \vec{E} = -\vec{j} \cdot \vec{E} \tag{4.30}$$

The energy balance equation for monatomic gases is therefore, as in (1.72),

$$\frac{3}{2} nk \left(\frac{\partial T}{\partial t} + \vec{v}_0 \cdot \nabla T \right) + \nabla \cdot \vec{q} + \mathcal{P}_{ij} \frac{\partial v_{0i}}{\partial x_j} = \sum_j n_j \vec{F}_j \cdot \langle \vec{V}_j \rangle + \vec{j} \cdot \vec{E} \tag{4.31}$$

The term $\vec{j} \cdot \vec{E}$ is clearly identified as Joule heating.

4.3   Diffusion

The diffusion velocities of the three different components of these plasmas $\langle \vec{V}_i \rangle$, $\langle \vec{V}_e \rangle$, and $\langle \vec{V}_n \rangle$ must satisfy two basic formulae. One of them is (1.31):

$$m_i n_e \langle \vec{V}_i \rangle + m_e n_e \langle \vec{V}_e \rangle + m_i n_n \langle \vec{V}_n \rangle = 0 \tag{4.32}$$

and the other is (4.24). From these two formulae it is possible to obtain $\langle \vec{V}_i \rangle$ and $\langle \vec{V}_e \rangle$ as functions of $\vec{j}$ and $\langle \vec{V}_n \rangle$:

$$\langle \vec{V}_i \rangle = -\frac{n_n}{n_e} \langle \vec{V}_n \rangle + \frac{1}{en_e} \frac{m_e}{m_i} \vec{j} \tag{4.33}$$

$$\langle \vec{V}_e \rangle = -\frac{n_n}{n_e} \langle \vec{V}_n \rangle - \frac{1}{e n_e} \vec{j} \tag{4.34}$$

which are equivalent to

$$\vec{\phi}_i = \vec{\phi} + \frac{m_e \vec{j}}{m_i e} \tag{4.35}$$

$$\vec{\phi}_e = \vec{\phi} - \frac{\vec{j}}{e} \tag{4.36}$$

where, obviously, $\vec{\phi}_i = n_e \langle \vec{V}_i \rangle$ and $\vec{\phi}_e = n_e \langle \vec{V}_e \rangle$ are the diffusion fluxes of ions and electrons, whilst

$$\vec{\phi} = -n_n \langle \vec{V}_n \rangle \tag{4.37}$$

is called the ambipolar diffusion flux. The interpretation provided by (4.35) and (4.36) is valuable. The diffusion of the ions and electrons has been decomposed into two fluxes, one associated with the charge current density, different for ions and electrons, and the other called ambipolar diffusion, which is the same for both charges. Electrons and ions diffuse together through neutrals. The diffusion of each may be different, but this difference is interpreted as a charge current density. No ambipolar diffusion will take place in fully ionized plasmas.

The interpretation of plasma equations is somewhat complicated if all the different terms corresponding to different effects are considered together. A progressive understanding of these effects is gained by isolating them. Thus we first assume that no ambipolar diffusion takes place, and study the charge current density; then ambipolar diffusion will be studied alone. We must, however, be aware that the actual description is more complicated, unless one or other of the two effects is negligible.

Our analysis must start from the individual equations of motion. It is convenient to write the collision term $\vec{A}_j$ in (4.27) more explicitly for ions and electrons. From (1.109) and (1.110):

$$\vec{A}_i = \vec{A}_{in} + \vec{A}_{ie} = s_{in} n_e n_n (\langle \vec{V}_n \rangle - \langle \vec{V}_i \rangle) + s_{ie} n_e^2 (\langle \vec{V}_e \rangle - \langle \vec{V}_i \rangle) \tag{4.38}$$

$$\vec{A}_e = \vec{A}_{en} + \vec{A}_{ei} = s_{en} n_e n_n (\langle \vec{V}_n \rangle - \langle \vec{V}_e \rangle) + s_{ei} n_e^2 (\langle \vec{V}_i \rangle - \langle \vec{V}_e \rangle) \tag{4.39}$$

4.3.1 Ohm's law

Consider, first, the absence of ambipolar diffusion, $\vec{\phi} = 0$; then, clearly, $n_n \langle \vec{V}_n \rangle = 0$. Using (4.38) and (4.39) let us calculate:

$$e \left(\frac{\vec{A}_i}{m_i} - \frac{\vec{A}_e}{m_e} \right) \approx e s_{en} n_e n_n \frac{\langle \vec{V}_e \rangle}{m_e} - s_{ei} n_e \frac{\vec{j}}{m_e}$$

$$\approx -s_{en} n_n \frac{\vec{j}}{m_e} - s_{ei} n_e \frac{\vec{j}}{m_e} = -\nu_e \vec{j} \tag{4.40}$$

in which several approximations have been used. Collisions of ions and electrons with neutrals are not of the coulomb type, so that the rigid spheres model (1.111) can be adopted, and $\frac{s_{in}}{m_i} \propto \frac{1}{\sqrt{m_i}}$ and $\frac{s_{en}}{m_e} \propto \frac{1}{\sqrt{m_e}}$. Also, (4.33) and (4.34) tell us that $|\langle \vec{V}_e \rangle| \gg |\langle \vec{V}_i \rangle|$. Hence the term containing s_{in} is much lower than the term containing s_{en}. Similarly, the term containing s_{ie} is negligible when compared with the term containing s_{ei}, because we know that $s_{ie} = s_{ei}$ and $m_e \ll m_i$. We know that all terms in (4.38) were neglected, which means that the small electrons are more affected by collisions, although their contribution to the charge current density is important.

Let us now perform the same calculation, with the right-hand side of (4.27). Assume that the external forces are unimportant in producing a charge current density, and that $\mathcal{P}_j = p_j \delta$, with the temperature also being constant:

$$-\nu_e \vec{j} = \frac{\partial \vec{j}}{\partial t} - \frac{ekT}{m_e} \nabla n_e - \frac{e^2 n_e}{m_e} \vec{E}$$
$$- \frac{e^2 n_e}{m_e} \left(\frac{\vec{v}_0}{c} \times \vec{B} \right) + \frac{e}{m_e c} (\vec{j} \times \vec{B}) \tag{4.41}$$

Very often, $\frac{\partial \vec{j}}{\partial t}$ is negligible compared with $\nu_e \vec{j}$, as collisions are quite frequent, with $\nu_e^{-1} \ll \tau$ where τ is a characteristic time for the system. This condition is familiar from domestic electrical currents; the current intensity reaches a constant value shortly after switching on. When the characteristic time of evolution of the cosmic fluid under study is much longer than ν_e^{-1}, the system reaches a constant value for the charge current density after a transient regime time of the order of ν_e^{-1}. This constant $\vec{j}$ may, however, have time variations over longer periods. This assumption is usually satisfied in astrophysical plasmas and will be adopted here. As will be shown later, gradients in electron concentration are responsible for ambipolar diffusion. If this is assumed to be negligible, the term containing ∇n_e may also be considered to be negligible. By defining the scalar electrical conductivity:

$$\sigma_0 = \frac{e^2 n_e}{m_e \nu_e} \tag{4.42}$$

Ohm's law is obtained from (4.41):

$$\vec{j} = \sigma_0 \vec{E} + \sigma_0 \left(\frac{\vec{v}_0}{c} \times \vec{B} \right) - x_e (\vec{j} \times \vec{u}_B) \tag{4.43}$$

where x_e was defined in (4.6) and $\vec{u}_B$ is a unit vector with direction $\vec{B}$.

A simple application of (4.43) is the case of a fluid at rest, with $\vec{v}_0 = 0$, for which

$$\vec{j} = \sigma_0 \vec{E} - x_e (\vec{j} \times \vec{u}_B) \tag{4.44}$$

We then find:

$$\vec{j} = \sigma_0 \vec{E} + x_e \frac{\hat{B}}{B} \cdot \vec{j} \tag{4.45}$$

$$\left(\delta - x_e \frac{\hat{B}}{B} \right) \cdot \vec{j} = \sigma_0 \vec{E} \tag{4.46}$$

$$\vec{j} = \sigma_0 \left(\delta - x_e \frac{\hat{B}}{B} \right)^{-1} \cdot \vec{E} \tag{4.47}$$

or defining the electrical conductivity tensor

$$\sigma = \sigma_0 \left(\delta - x_e \frac{\hat{B}}{B} \right)^{-1} \tag{4.48}$$

we obtain

$$\vec{j} = \sigma \cdot \vec{E} \tag{4.49}$$

It can be checked that

$$\sigma = \sigma_0 \left(\delta - x_e \frac{\hat{B}}{B} \right)^{-1} = \frac{\sigma_0}{1 + x_e^2} \left(\delta + \frac{x_e^2}{B^2} \vec{B}\vec{B} + \frac{x_e}{B} \hat{B} \right)$$

$$= \sigma_P \left(\delta + \frac{x_e^2}{B^2} \vec{B}\vec{B} \right) + \sigma_H \frac{\hat{B}}{B} \tag{4.50}$$

where

$$\sigma_P = \frac{\sigma_0}{1 + x_e^2} \tag{4.51}$$

is called the Pedersen conductivity, and

$$\sigma_H = \frac{x_e \sigma_0}{1 + x_e^2} \tag{4.52}$$

is called the Hall conductivity. There will therefore be a 'Hall charge current density'

$$\vec{j}_H = \sigma_H \frac{1}{B} \hat{B} \cdot \vec{E} = \sigma_H \frac{1}{B} \vec{B} \times \vec{E} \tag{4.53}$$

associated with the $\vec{E} \times \vec{B}$-drift of charged particles, taking the effect of collisions into account. In the absence of collisions the $\vec{E} \times \vec{B}$-drift would not produce any charge current density, as both electrons and ions would move in the same direction. When x_e is very high, the Hall conductivity is unimportant, as shown in (4.52). When $x_e = 0$, in the absence of magnetic fields, it is also negligible. It becomes important for intermediate degrees of anisotropy, taking its maximum value for $x_e = 1$.

When x_e is very low, $\sigma_H \approx 0$, $\sigma_P \approx \sigma_0$, and the conductivity tensor becomes $\sigma_0 \delta$, which means a perfectly isotropic electrical conductivity. When the magnetic field becomes very high, that is, x_e is very high, equation (4.50) tells us that

$$\vec{j} = \sigma_0 \frac{x_e^2}{B^2} (\vec{B}\vec{B}) \cdot \vec{E} = \sigma_0 \frac{x_e^2}{B^2} (\vec{B} \cdot \vec{E}) \vec{B} \tag{4.54}$$

that is, the charge current density vector lies in the direction of $\vec{B}$, and only the projection of $\vec{E}$ along the direction of $\vec{B}$ is effective in producing it. Therefore, the electrical conductivity becomes highly anisotropic.

For any degree of anisotropy, the component of the charge current density along $\vec{B}$ is

$$j_B = \vec{j} \cdot \frac{\vec{B}}{B} = \sigma_0 E_B \tag{4.55}$$

where E_B is the component of $\vec{E}$ along $\vec{B}$. Equation (4.55) is Thonk's theorem and states that the relation between j and E in the direction of $\vec{B}$ is the same as if the system were isotropic. It can be also shown that if the OX-axis is chosen to coincide with the direction of $\vec{B}$ the electrical conductivity tensor (4.50) is:

$$\sigma = \begin{pmatrix} \sigma_0 & 0 & 0 \\ 0 & \sigma_P & -\sigma_H \\ 0 & \sigma_H & \sigma_P \end{pmatrix} \tag{4.56}$$

Using this expression, the above properties can easily be verified.

4.3.2 Ambipolar diffusion

Ambipolar diffusion cannot exist in a fully ionized plasma as there is no neutral component through which ions and electrons can diffuse. We will therefore consider a weakly ionized medium, with $n_n \gg n_e$. Equation (4.27) is again our starting point, together with (4.38) and (4.39). For ions:

$$
\begin{aligned}
\vec{A}_i &= s_{in} n_e n_n (\langle \vec{V}_n \rangle - \langle \vec{V}_i \rangle) + s_{ie} n_e^2 (\langle \vec{V}_e \rangle - \langle \vec{V}_i \rangle) \\
&= -s_{in} n_e \vec{\phi} - s_{in} n_n \left(\vec{\phi} + \frac{m_e}{m_i} \frac{\vec{j}}{e} \right) - s_{ie} \frac{n_e}{e} \vec{j} \\
&= -(s_{in} n_e + s_{in} n_n) \vec{\phi} - \left(s_{in} \frac{m_e}{m_i} \frac{n_n}{e} + s_{ie} \frac{n_e}{e} \right) \vec{j} \\
&\approx -m_i \nu_i \vec{\phi} - m_e \nu_i \frac{\vec{j}}{e}
\end{aligned}
\tag{4.57}
$$

Suppose again that the plasma is at rest, $\vec{v}_0 = 0$, that $\mathcal{P}_j = p_j \delta$, that ∇T is negligible, and that no non-electromagnetic forces act on the particles. Then (4.27) and (4.57) give

$$-m_i \nu_i \vec{\phi} - m_e \nu_i \frac{\vec{j}}{e} = kT \nabla n_e - e n_e \vec{E} - e \frac{\vec{\phi}}{c} \times \vec{B} - \frac{1}{c} \frac{m_e}{m_i} \vec{j} \times \vec{B} \tag{4.58}$$

$\frac{\partial \langle \vec{V}_i \rangle}{\partial t}$ is ignored for the same reasons as $\frac{\partial \vec{j}}{\partial t}$ is neglected in (4.41). Similarly for electrons:

$$-m_e \nu_e \vec{\phi} + m_e \nu_e \frac{\vec{j}}{e} = kT \nabla n_e + e n_e \vec{E} + e \frac{\vec{\phi}}{c} \times \vec{B} - \frac{1}{c} \vec{j} \times \vec{B} \tag{4.59}$$

Let us consider the case where $\vec{B} = 0$, that is, ambipolar diffusion in the absence of magnetic fields.

Let us also assume naïvely $\vec{j} = 0$, $\vec{E} = 0$, and try to solve (4.58) and (4.59):

$$\vec{\phi} = -D_i \nabla n_e \tag{4.60}$$

$$\vec{\phi} = -D_e \nabla n_e \tag{4.61}$$

where

$$D_i = \frac{kT}{m_i \nu_i} \tag{4.62}$$

$$D_e = \frac{kT}{m_e \nu_e} \tag{4.63}$$

As $D_i \neq D_e$, $\vec{\phi}$ cannot be calculated simultaneously using (4.60) and (4.61). Therefore, too many simplifying hypotheses have been made. Intuitively, electrons ought to diffuse from a region of high electron density at a higher speed than the heavier ions. But electrons and ions will not be separable, so that an electric field must be generated, acting in a direction which impedes their separation. Therefore, this problem cannot be solved without accepting the existence of a coupling connective, or binding, electric field that maintains macroscopic neutrality. This means that we should not set $\vec{E} = 0$. Hence

$$-m_i\nu_i\vec{\phi} = kT\nabla n_e - en_e\vec{E} \tag{4.64}$$

$$-m_e\nu_e\vec{\phi} = kT\nabla n_e + en_e\vec{E} \tag{4.65}$$

which, taking the fact that $m_i\nu_i \gg m_e\nu_e$ into account, yields

$$\vec{\phi} = -2D_i\nabla n_e \tag{4.66}$$

Therefore, the ambipolar diffusion coefficient is twice the ion diffusion coefficient, but is much lower than D_e. The binding electric field is

$$\vec{E} = \frac{m_i\nu_i}{2en_e}\vec{\phi} \tag{4.67}$$

In the absence of magnetic fields, ambipolar diffusion is isotropic.

4.3.3   The electric field of an atmosphere

Once more, the case of a static, stationary, flat, and stratified atmosphere provides a simple application. Under such conditions neither a charge current density nor ambipolar diffusion are expected. However, molecular diffusive equilibrium would distribute the heavier ions in the lower layers in the main, and the light electrons mainly in the upper layers. Again, a binding electric field prevents this gravitational separation of charges.

In (4.27), we assume that $\langle \vec{V}_j \rangle = 0$, $\vec{v}_0 = 0$, $T = \text{constant}$, $\vec{F}_j = m_j\vec{g}$, and obtain for ions

$$kT\nabla n_e - m_i n_e \vec{g} - en_e\vec{E} = 0 \tag{4.68}$$

and for electrons

$$kT\nabla n_e - m_e n_e \vec{g} + en_e\vec{E} = 0 \tag{4.69}$$

Summing the two equations:

$$\frac{\nabla n_e}{n_e} = \frac{m_i}{2kT}\vec{g} \tag{4.70}$$

which is the equilibrium vertical profile for a constituent of mass $m_i/2$. This is not surprising as the equivalent mass of the ensemble of ions and electrons is deduced from (1.26):

$$m_e n_e + m_i n_e = m_{eq}(n_e + n_e) \tag{4.71}$$

which tells us that this equivalent mass is half the ion mass.

The binding electric field is obtained by subtracting (4.68) and (4.69):

$$\vec{E} = -\frac{m_i}{2e}\vec{g} \tag{4.72}$$

directed outwards. Other electric fields are present in atmospheres, such as those of tidal origin in our ionosphere, but they are associated with winds.

4.4   **Magnetohydrodynamics**

Magnetohydrodynamics – abbreviated to MHD – is the macroscopic description of a plasma under several specified assumptions: MHD equations are the fluid dynamic equations plus Maxwell's equations under these assumptions, which will now be justified for most cosmic systems.

No distinction between either $\vec{H}$ and $\vec{B}$ or $\vec{E}$ and $\vec{D}$ will be considered to be important. On the other hand, $\frac{1}{c}\frac{\partial \vec{D}}{\partial t}$ will be considered negligible in (4.13). This is justified as follows. Equation (4.16) indicates that E is of the order of $\frac{B}{c}\frac{L}{\tau}$, where τ is the characteristic time of the system, and the continuity equation always implies that $\frac{L}{\tau} \approx v_0$; therefore E is of the order of $\frac{v_0}{c} B$. Classical MHD does not deal with relativistic systems, therefore $E \ll B$. Then $\nabla \times \vec{H}$ in (4.13), of order $\frac{B}{L}$, is much higher than $\frac{E}{c\tau}$, which has order of magnitude $\frac{1}{c}\frac{\partial \vec{D}}{\partial t}$.

The conditions leading to the simple expression for Ohm's law (4.43) will now be outlined. In particular it is important to note that $\tau \gg \nu_e^{-1}$. There are three electromagnetic equations: (4.43), (4.16), and (4.13), with $\frac{1}{c}\frac{\partial \vec{D}}{\partial t} \sim 0$. There are also three electromagnetic quantities, $\vec{j}$, $\vec{E}$, and $\vec{B}$. Considerable agility in the mathematical treatment of this large number of parameters is achieved when these electromagnetic equations are used to obtain $\vec{j}$ and $\vec{E}$ as functions of $\vec{B}$ and insert them into the single equation (4.16), called the induction equation, with a single electromagnetic unknown, the magnetic field strength $\vec{B}$. Of course, these substitutions must also be used in the remaining hydrodynamic equations. From (4.13) $\vec{j}$ is easily obtained as a function of $\vec{B}$:

$$\vec{j} = \frac{c}{4\pi} \nabla \times \vec{B} \tag{4.73}$$

Ohm's law (4.43) then gives $\vec{E}$ as a function of $\vec{B}$

$$\vec{E} = \frac{c}{4\pi\sigma_0} \nabla \times \vec{B} + \frac{c}{\sigma_0} \frac{x_e}{4\pi B} ((\nabla \times \vec{B}) \times \vec{B}) - \frac{\vec{v}_0}{c} \times \vec{B} \tag{4.74}$$

The induction equation then becomes

$$\frac{\partial \vec{B}}{\partial t} = -\frac{c^2}{\sigma_0} \frac{x_e}{4\pi B} \nabla \times ((\nabla \times \vec{B}) \times \vec{B}) + \frac{c^2}{4\pi\sigma_0} \nabla^2 \vec{B} + \nabla \times (\vec{v}_0 \times \vec{B}) \tag{4.75}$$

which is a complicated expression. By analogy with the simplifying hypothesis of infinite conductivity for metals, we can assume here that n_e is sufficiently high to render σ_0 infinite. As we will see, this condition is satisfactory for most cosmic plasmas. A plasma with $\sigma_0 = \infty$ is called an ideal plasma, or a plasma with frozen-in magnetic field lines. Under this condition the induction equation (4.75) is simply

$$\frac{\partial \vec{B}}{\partial t} = \nabla \times (\vec{v}_0 \times \vec{B}) \tag{4.76}$$

and (4.74) becomes

$$\vec{E} = -\frac{\vec{v}_0}{c} \times \vec{B} \tag{4.77}$$

We will now establish an order of magnitude criterion for an ideal plasma. Dividing the dynamic term containing $\vec{v}_0$ by the electrical conductivity terms containing σ_0, a dimensionless number is obtained:

$$\frac{4\pi\sigma_0 v_0 L}{c^2(1 + x_e)} = \frac{\mathcal{R}_m}{1 + x_e} \tag{4.78}$$

where $\mathcal{R}_m$ is called the magnetic Reynolds number. When the left-hand side of (4.78) is much greater than unity, conductivity is high enough to ensure the condition of an ideal plasma. For isotropic plasmas with $x_e \ll 1$, the dimensionless number to be compared with unity is the magnetic Reynolds number itself, but for highly anisotropic plasma the value of $\frac{\mathcal{R}_m}{x_e}$ determines whether or not dynamic processes are more effective in controlling the magnetic field.

The effect of finite electrical conductivity is in general to dissipate the magnetic field. With appropriate boundary conditions the magnetic field would eventually disappear. For instance, an isotropic plasma with finite electrical conductivity and zero velocity would obey the simplified induction equation

$$\frac{\partial \vec{B}}{\partial t} = \frac{c^2}{4\pi\sigma_0} \nabla^2 \vec{B} \tag{4.79}$$

which is the differential equation for any diffusive process. The magnetic field would then diffuse out in a characteristic time of $\frac{4\pi\sigma_0 L^2}{c^2}$. Cosmic systems are in general characterized by the very large value of the characteristic length L, so that $\mathcal{R}_m$ and the magnetic diffusion time become very large. For this reason $\sigma_0 = \infty$ and, therefore, equation (4.76) can be adopted in many cosmic plasmas. Nevertheless, this condition will be justified for each particular case treated in this book.

A plasma with infinite conductivity is said to have frozen-in magnetic field lines. In these systems the magnetic field has no tendency to dissipate; the magnetic field lines seem to move as if held in the fluid. This picture is not exact but does give an intuitive approximate insight into the behaviour of the magnetic field. From (4.76), (4.10), and (4.22) the time derivative of $\vec{B}$ for the wet observer becomes

$$\frac{d\vec{B}}{dt} = \vec{B} \cdot \nabla \vec{v}_0 - \vec{B}\nabla \cdot \vec{v}_0 = \vec{B} \cdot \nabla \vec{v}_0 + \frac{\vec{B}\,d\rho}{\rho\,dt} \tag{4.80}$$

If the spatial variations of the velocity are negligible, then $\frac{d\vec{B}}{dt} = 0$ and the wet observer would notice no variation in $\vec{B}$. The second equality implies that if $\nabla\vec{v}_0$ were negligible $\frac{d}{dt}\left(\frac{\vec{B}}{\rho}\right)$ would vanish, and $\vec{B} \propto \rho$. In an expansion of the fluid, B would decrease as a consequence of the separation of the different magnetic field lines. In contraction of the fluid, magnetic field lines would approach each other, raising the modulus of $\vec{B}$, while the direction of $\vec{B}$ would not be modified. Although this is not always true, it justifies the name of frozen-in magnetic field lines.

As an example, consider a radial expression of the type $\vec{v}_0 = a\vec{r}$, where a is a constant. Equation (4.80) gives

$$\frac{d\vec{B}}{dt} = -2a\vec{B} \tag{4.81}$$

The time variation of $\vec{B}$ is parallel to $\vec{B}$, therefore the direction of $\vec{B}$ is conserved. If $\vec{B}$ was initially along the z-axis $(0,0,B)$, with B homogeneous, this property would be maintained, with B being a function of time. We also have

$$\frac{d\rho}{dt} = -3a\rho \tag{4.82}$$

Therefore $B\rho^{-\frac{2}{3}}$ remains constant. If the radial expansion is limited and affects only a sphere of radius R, we obtain $BR^2 = $ constant as the mass contained in this sphere can be considered to be a constant, that is, the magnetic flux remains constant. This is depicted in Figure 4.1. The problem is really not so simple because, firstly, if $\vec{B}$ is not modified beyond R, the continuity of $\vec{B}$, that is, $\nabla \cdot \vec{B} = 0$, implies that the direction of $\vec{B}$ must change near the surface of the contracting sphere. And, secondly, gradients of $\vec{B}$ will eventually affect the motion, so that $\vec{v}_0 = a\vec{r}$ cannot always be valid. In practice, contraction can take place along $\vec{B}$, and is prevented in directions perpendicular to $\vec{B}$. The problem requires the full system of MHD equations. As this is important in star formation it will be reconsidered later.

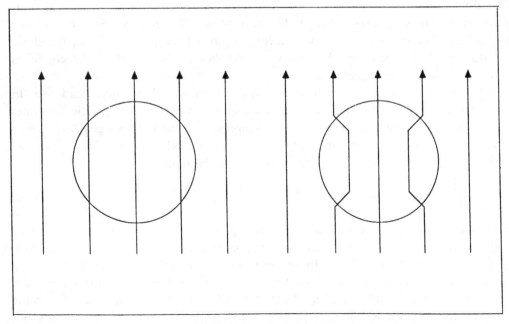

Figure 4.1   Evolution of magnetic field lines in the collapse of a magnetized cloud.

The remaining MHD equations must also be modified so that $\vec{B}$ is the only electromagnetic quantity present. Hence, in the equation of motion the force $\frac{\vec{j}}{c} \times \vec{B}$ must be calculated. It may take several alternative expressions

$$\frac{\vec{j}}{c} \times \vec{B} = \frac{1}{4\pi}(\nabla \times \vec{B}) \times \vec{B} = \frac{1}{4\pi}\vec{B} \cdot \nabla \vec{B} - \nabla \frac{B^2}{8\pi} = \nabla \cdot \mathcal{M} \qquad (4.83)$$

where $\mathcal{M}$ is the Maxwell tensor

$$\mathcal{M} = \frac{1}{4\pi}\left(\vec{B}\vec{B} - \frac{B^2}{2}\delta\right) \qquad (4.84)$$

The second expression in (4.83) is the most commonly used. The last one shows, yet again, that a force can be expressed as the divergence of a momentum transport tensor.

Joule heating can be rewritten in several alternative ways

$$\begin{aligned}
\vec{j} \cdot \vec{E} &= -\frac{1}{4\pi}(\nabla \times \vec{B}) \cdot (\vec{v}_0 \times \vec{B}) = \frac{1}{4\pi}\vec{v}_0 \cdot ((\nabla \times \vec{B}) \times \vec{B}) \\
&= \vec{v}_0 \cdot \left(\frac{\vec{j}}{c} \times \vec{B}\right) = \vec{v}_0 \cdot (\nabla \cdot \mathcal{M}) \\
&= \frac{1}{4\pi}\vec{v}_0 \cdot (\vec{B} \cdot \nabla\vec{B}) - \vec{v}_0 \cdot \nabla\frac{B^2}{8\pi}
\end{aligned} \qquad (4.85)$$

The final set of MHD equations adopted for most plasmas considered in this book is:

$$\frac{\partial \rho}{\partial t} + \nabla \cdot (\rho\vec{v}_0) = 0 \qquad (4.86)$$

$$\rho\frac{\partial \vec{v}_0}{\partial t} + \rho\vec{v}_0 \cdot \nabla\vec{v}_0 + \nabla\left(p + \frac{B^2}{8\pi}\right) = n\vec{F} + \frac{1}{4\pi}\vec{B} \cdot \nabla\vec{B} \qquad (4.87)$$

$$\frac{3}{2}nk\left(\frac{\partial T}{\partial t} + \vec{v}_0 \cdot \nabla T\right) + \nabla \cdot \vec{q} + \mathcal{P}_{ij}\frac{\partial v_{0i}}{\partial x_j} = \frac{1}{4\pi}\vec{v}_0 \cdot (\vec{B} \cdot \nabla\vec{B}) - \vec{v}_0 \cdot \nabla\frac{B^2}{8\pi} \qquad (4.88)$$

$$\frac{\partial \vec{B}}{\partial t} = \nabla \times (\vec{v}_0 \times \vec{B}) \qquad (4.89)$$

which obviously retain the names of continuity, motion, energy balance, and induction. Viscosity terms have been not included. They may be important, but their inclusion when necessary does not present any problem.

4.4.1 Magnetohydrostatics

Consider once more a static ($\vec{v}_0 = 0$) and stationary ($\frac{\partial}{\partial t} = 0$) system. Equation (4.87) can be written as

$$\nabla p = -\nabla\left(\frac{B^2}{8\pi}\right) + \frac{1}{4\pi}\vec{B} \cdot \nabla\vec{B} + n\vec{F} \qquad (4.90)$$

or alternatively as

$$\nabla p = \frac{1}{c}(\vec{j} \times \vec{B}) + n\vec{F} \tag{4.91}$$

These equations are only apparently simple. It is, in general, not very difficult to find equilibrium configurations, but they frequently correspond to unstable equilibria. Magnetohydrostatics thus becomes a very complex matter, presenting a very large variety of instabilities, difficult even to classify, which lie beyond the scope of this introductory work.

Magnetic energy density also acts as a pressure in the dynamic case (4.87) and is therefore called magnetic pressure. Its gradient even has the capacity to confine the plasma. If the magnetic pressure is created by the plasma itself, autoconfinement of the plasma is possible.

We see in (4.91), in the absence of non-electromagnetic forces, that $\vec{j}$ and $\vec{B}$ are aligned with the isobar surfaces. Suppose that $\vec{B} \cdot \nabla \vec{B}$ is negligible or null, as would actually be the case, for example, for $\vec{B} \equiv (0, 0, B)$, $\vec{B}$ being in the z-direction and with $\partial B/\partial z = 0$ ($\partial B/\partial x$ and $\partial B/\partial y$ non-vanishing). It is then found using (4.90) that

$$p + \frac{B^2}{8\pi} = constant \tag{4.92}$$

This equation represents a distribution of plasmas in equilibrium which is often obeyed in static systems.

Let us return to the case of an atmosphere, with $\vec{B}$ in one of the horizontal directions. From (4.90) we find

$$\frac{d}{dz}\left(p + \frac{B^2}{8\pi}\right) = -\rho g \tag{4.93}$$

and for an isothermal ideal gas:

$$\frac{d\rho}{dz} + \frac{1}{Hg}\frac{d}{dz}\left(\frac{B^2}{8\pi}\right) = -\frac{\rho}{H} \tag{4.94}$$

There is even a theoretical possibility that at some point $\rho = 0$, and above this level it can increase with altitude if $\frac{d}{dz}B^2 < 0$. This equilibrium is unstable but it is important for the formation of solar prominences (Chapter 7).

4.4.2   The Virial theorem for continuous and magnetized systems

Let us consider a finite volume, τ, in the fluid, multiply the equation of motion (4.87) by $\vec{r}\, d\tau$ ($\vec{r}$ being the position vector), and integrate this over a finite volume under equilibrium conditions. The pressure gradient term will give

$$\int_{\tau}(\nabla p) \cdot \vec{r}\, d\tau = \int_{\tau} \nabla \cdot (p\vec{r})\, d\tau - \int_{\tau} 3p\, d\tau = \iint_{s} p\vec{r} \cdot d\vec{S} - 3\Pi \tag{4.95}$$

where $d\vec{S}$ is a surface element vector of the surface around τ, the integral $\int\int_s$ is performed over this surface, and Π is defined as

$$\Pi = \int_\tau p \, d\tau \tag{4.96}$$

This is proportional to the total thermal energy in the volume τ.

Similarly, we obtain from the magnetic pressure gradient

$$\int_\tau \nabla\left(\frac{B^2}{8\pi}\right) \cdot \vec{r} \, d\tau = \int\int_s \frac{B^2}{8\pi} \vec{r} \cdot d\vec{S} - 3\mathcal{B} \tag{4.97}$$

where

$$\mathcal{B} = \int_\tau \frac{B^2}{8\pi} \, d\tau \tag{4.98}$$

This is the total magnetic energy in the volume τ.

From the inertial term:

$$\int_\tau \rho(\vec{v}_0 \cdot \nabla\vec{v}_0) \cdot \vec{r} \, d\tau = \int\int_s (\rho(\vec{v}_0\vec{v}_0) \cdot \vec{r}) \cdot d\vec{S} - 2\mathcal{Q} \tag{4.99}$$

where

$$\mathcal{Q} = \frac{1}{2}\int_\tau \rho v_0^2 \, d\tau \tag{4.100}$$

This is the total macroscopic kinetic energy in the volume τ.

Similarly,

$$\frac{1}{4\pi}\int_\tau (\vec{B} \cdot \nabla\vec{B}) \cdot \vec{r} \, d\tau = \frac{1}{4\pi}\int\int_s ((\vec{B}\vec{B}) \cdot \vec{r}) \cdot d\vec{S} - 2\mathcal{B} \tag{4.101}$$

If the non-electromagnetic force is gravitational, we obtain

$$W = -\int_\tau \rho(\nabla\mathcal{F}) \cdot \vec{r} \, d\tau \tag{4.102}$$

where $\mathcal{F}$ is the gravitational potential. This is in general more difficult to calculate and interpret. Let us calculate (4.102) for some particular cases. If the gravitational potential $\mathcal{F}$ is due to gravitation produced by the fluid itself, and if the fluid has spherical symmetry,

$$\nabla\mathcal{F} = \frac{GM(r)m}{r^3}\vec{r} \tag{4.103}$$

where $M(r)$ is the mass contained in the sphere of radius r. Then W is clearly identified with the gravitational potential energy. We obtain

$$W = -\frac{G M^2}{2 R_*} \tag{4.104}$$

where M is the total mass contained in τ, which in this case must be a spherical volume. R_* is the equivalent radius, which is related to the radius R of the sphere, this relation being dependent on the density distribution. R_* is of the order of R. For instance, it is easy to calculate for a sphere of constant density that

$$W = -\frac{3}{5} G \frac{M^2}{R} \tag{4.105}$$

and hence $R_\star = 5/6 \ R$. A similar result is obtained for a disc of radius R and thickness H. Again, W is the potential energy. For a constant density distribution within the disc it can be calculated that $R_\star = 3/4 \, R$. For more realistic density distributions it is always found that $R_\star$ is of the order of the typical radius of the system.

Sometimes $\mathcal{F}$ may not be produced by the system itself. A case of astrophysical interest would be the calculation of W for the gas of a galaxy, assuming that $\mathcal{F}$ is not produced by the gas but by stars, which contribute the major fraction of the mass of the galaxy. Then $\nabla \mathcal{F} = \frac{GM(r)}{r^3} \vec{r}$, where $M(r)$ is now the mass of all stars within a radius less than r, but ρ is now the density of the gas. In this case W is still the potential energy of the gas and can be written as

$$W = -\frac{G \, M_s M_g}{2 \ R_\star} \tag{4.106}$$

where M_s is the total mass of stars and M_g the total mass of gas. $R_\star$ is now the equivalent radius of the gas volume, again of the order of the typical radius of the volume containing the gas. For most practical cases, W can be taken as the potential energy of the system.

The Virial theorem can be expressed as

$$2\mathcal{Q} + 3\Pi + \mathcal{B} + W = \iint_s (\rho \vec{v}_0 \vec{v}_0 \cdot \vec{r}) \cdot d\vec{S} + \iint_s \frac{B^2}{8\pi} \vec{r} \cdot d\vec{S}$$
$$+ \iint_s p\vec{r} \cdot d\vec{S} - \frac{1}{4\pi} \iint_s ((\vec{B}\vec{B}) \cdot \vec{r}) \cdot d\vec{S} \tag{4.107}$$

which is considerably reduced by using the Maxwell tensor defined in (4.84) and $\mathcal{R}$, the momentum flux tensor defined in (1.14), which in the absence of viscosity would be

$$\mathcal{R} = p\delta + \rho \vec{v}_0 \vec{v}_0 \tag{4.108}$$

Then

$$2\mathcal{Q} + 3\Pi + \mathcal{B} + W + \iint_s ((\mathcal{M} - \mathcal{R}) \cdot \vec{r}) \cdot d\vec{S} = 0 \tag{4.109}$$

It is sometimes possible to adopt a value of τ so large that the whole system under study can lie within it, so that ρ and $\vec{B}$ can be considered to vanish at the surface, and therefore the surface integral provides negligible values. Equation (4.109) is indeed a generalization of the commonly used Virial theorem. If there are no magnetic fields, so that $\mathcal{B} = 0$, and there are no macroscopic motions in the system, then

$$3\Pi + W = 0 \tag{4.110}$$

which corresponds to a very classical expression. When a discrete system is considered, for instance stars in a globular cluster or galaxies in a cluster, the distinction between macroscopic and microscopic motions becomes meaningless and the Virial theorem can be expressed as

$$2\mathcal{Q} + W = 0 \tag{4.111}$$

In this case

$$\frac{1}{2}\sum_{j} m_{j}v_{j}^{2} = \frac{1}{2}M\langle V^{2}\rangle \tag{4.112}$$

where m_{j} is (for instance) the mass of a star, $\langle V^{2}\rangle$ is the mean squared velocity, and M is the mass of the whole system. Using (4.104):

$$M = \frac{2R_{\star}\langle V^{2}\rangle}{G} \tag{4.113}$$

As $R_{\star}$ and $\langle V^{2}\rangle$ are quantities for which observational information is available, (4.113) provides the possibility of determining the mass of a cluster of either stars or galaxies if the system is virialized, that is, if it fulfils our simplifying hypotheses, in particular, that it is in equilibrium.

As a simple application we can infer that, from (4.113) for stars in a galaxy,

$$R_{\star_{stars}} = \frac{GM_{stars}}{2\langle V_{stars}^{2}\rangle} \tag{4.114}$$

while for a gas, from (4.106), (4.110), and (4.113),

$$R_{\star_{gas}} = \frac{GM_{stars}}{2\langle V_{gas}^{2}\rangle} \tag{4.115}$$

It is observed that $\langle V^{2}\rangle$ is larger for stars (~ 20 km s^{-1}) than for atomic hydrogen gas ($T \sim 100$ K), which is consistent with the observation that the atomic hydrogen in a galaxy typically extends to a larger radius than the stars.

4.4.3 Alfvén waves

A large variety of waves are capable of propagating through a plasma, such as ion-acoustic waves, Langmuir waves, electromagnetic waves, and Alfvén waves. Ion-acoustic waves are similar to normal acoustic waves, but, due to the coupling of ions and electrons in a situation of macroscopic neutrality, they have a speed slightly below that of sound, $\sqrt{\frac{kT}{m_{i}}}$ without the factor $\gamma = c_{p}/c_{v}$. Ion-acoustic waves are isothermal because characteristic thermal velocities are greater than the speed of sound. The speed of ion-acoustic waves has the value of the first, erroneous, theoretical determination of the speed of sound by Newton, who was unaware of the adiabatic nature of acoustic waves.

It has been stated that the separation of charges creates an electrical field which restores equilibrium. This is, however, attained by means of oscillations, which may propagate. This is the mechanism of Langmuir waves, of frequency $(\frac{4\pi n_{e}e^{2}}{m_{e}})^{\frac{1}{2}}$, which is also termed the plasma frequency.

Electromagnetic waves in plasmas exhibit peculiar properties, as electromagnetic fields oscillating in waves influence the free charges of the plasma directly. We will, however, concentrate our attention on Alfvén waves, whose properties are defined next.

Let us start with the MHD-equations in very simple form in order to emphasize the real nature of Alfvén waves. A homogeneous, static, and stationary plasma is assumed to be the medium in which waves propagate. This plasma will be considered ideal

($\sigma_0 = \infty$) and incompressible ($\nabla \cdot \vec{v}_0 = 0$). It will further be assumed that no temperature and no pressure changes are produced as a consequence of the passage of a wave. Our analysis will be standard. Any quantity, such as $\vec{B}$, is decomposed into two elements: $\vec{B}_0$, the mean magnetic field strength characterizing the field in the fluid without wave propagation, propagating in it, and $\vec{B}'$, which represents the oscillatory perturbation due to the wave motion. The velocity of the fluid will merely be the perturbation velocity $\vec{v}'$, as the fluid is considered to be at rest bodily. Terms containing the product of two perturbed quantities are considered to be negligible. No non-electromagnetic forces are present.

From (4.87), using (4.83),

$$\rho \frac{\partial \vec{v}'}{\partial t} = \frac{1}{4\pi} (\nabla \times \vec{B}') \times \vec{B}_0 \tag{4.116}$$

and (4.89) takes the form

$$\frac{\partial \vec{B}'}{\partial t} = \nabla \times (\vec{v}' \times \vec{B}_0) \tag{4.117}$$

which for an incompressible fluid yields

$$\frac{\partial \vec{B}'}{\partial t} = \vec{B}_0 \cdot \nabla \vec{v}' \tag{4.118}$$

The time derivative of (4.116), using (4.118), is

$$4\pi \rho \frac{\partial^2 \vec{v}'}{\partial t^2} = \left(\left(\nabla \times (\vec{B}_0 \cdot \nabla \vec{v}') \right) \times \vec{B}_0 \right) \tag{4.119}$$

Let OZ be the direction of the mean magnetic field. Developing this equation it is easily found that

$$B_0^2 \begin{pmatrix} \frac{\partial^2 v_x'}{\partial z^2} - \frac{\partial^2 v_z'}{\partial x \partial z} \\ \frac{\partial^2 v_y'}{\partial z^2} - \frac{\partial^2 v_z'}{\partial y \partial z} \\ 0 \end{pmatrix} = 4\pi \rho \begin{pmatrix} \frac{\partial^2 v_x'}{\partial t^2} \\ \frac{\partial^2 v_y'}{\partial t^2} \\ \frac{\partial^2 v_z'}{\partial t^2} \end{pmatrix} \tag{4.120}$$

The third component yields $\partial^2 v_z'/\partial t^2 = 0$, therefore there is no oscillation along the direction of $\vec{B}_0$.

If as initial conditions we assume that $v_z' = 0$, and $\partial v_z'/\partial t = 0$, we will always have $v_z' = 0$ at any point. Then (4.120) becomes

$$\frac{B_0^2}{4\pi \rho} \begin{pmatrix} \frac{\partial^2 v_x'}{\partial z^2} \\ \frac{\partial^2 v_y'}{\partial z^2} \\ 0 \end{pmatrix} = \begin{pmatrix} \frac{\partial^2 v_x'}{\partial t^2} \\ \frac{\partial^2 v_y'}{\partial t^2} \\ 0 \end{pmatrix} \tag{4.121}$$

which is a wave equation for v_x' and v_y' with speed

$$v_A = \frac{B_0}{\sqrt{4\pi \rho}} \tag{4.122}$$

which is called the Alfvén speed. Taking the time derivative of (4.118), together with (4.116),

$$\frac{B_0^2}{4\pi\rho} \begin{pmatrix} \frac{\partial^2 B_x{}'}{\partial z^2} - \frac{\partial^2 B_z'}{\partial x \partial z} \\ \frac{\partial^2 B_y{}'}{\partial z^2} - \frac{\partial^2 B_z'}{\partial y \partial z} \\ 0 \end{pmatrix} = \begin{pmatrix} \frac{\partial^2 B_x'}{\partial t^2} \\ \frac{\partial^2 B_y'}{\partial t^2} \\ \frac{\partial^2 B_z'}{\partial t^2} \end{pmatrix} \tag{4.123}$$

Again, we see that there is no oscillation of B_z', and, taking appropriate boundary conditions, we may set $B_z' = 0$. Then

$$\frac{B_0^2}{4\pi\rho} \begin{pmatrix} \frac{\partial^2 B_x'}{\partial z^2} \\ \frac{\partial^2 B_y'}{\partial z^2} \\ 0 \end{pmatrix} = \begin{pmatrix} \frac{\partial^2 B_x'}{\partial t^2} \\ \frac{\partial^2 B_y'}{\partial t^2} \\ 0 \end{pmatrix} \tag{4.124}$$

obtaining wave equations for B_x' and B_y', with (4.122) indicating the speed of these waves. As shown in (4.121) and (4.124), Alfvén waves are transverse waves propagating along the direction of the mean magnetic field.

Alfvén waves become more complicated when other representative terms of effects taking place in the fluid are included in the analysis. In particular, there are dissipative effects due to a finite conductivity – which is important in stars – and due to viscosity and thermal conduction, which are important in the interstellar medium.

4.5 Synchrotron radiation

A very large fraction of the radiative energy reaching our radiotelescopes has been produced by the synchrotron emission mechanism. Since first suggested by Alfvén and Herlofson (1950), this has been the accepted interpretation of the radio continuum of supernova remnants, galaxies, quasars, and extragalactic radio sources. Together with thermal and inverse compton emissions, it is responsible for the continuum spectra of most cosmic objects, mainly in the radio, but also at other wavelengths. Synchrotron emission is produced by the acceleration of relativistic charged particles (mainly electrons) when they follow helical trajectories around magnetic field lines, so that observing it can provide information about magnetic fields as well as about the density and the energy spectrum of the relativistic electrons, either within the source itself or in the space between the source and the observer. It even provides the only direct way of obtaining information about intergalactic magnetic fields.

A single electron emits in the direction of motion, nearly perpendicular to $\vec{B}$ and with a strong concentration of the emitted radiation in a narrow beam. This is the relativistic effect which was considered in Chapter 3. In an analogous inference to that in the derivation of equation (3.104), we now deduce that if we have a particle which emits isotropically at rest with intensity I', the radiation field $I(\varphi)$ becomes highly anisotropic when the particle is relativistic. φ is the angle between the direction of motion and any given direction:

$$I(\varphi) = I' \left[\left(1 + \frac{v}{c} \cos\varphi \right) \gamma \right]^4 \tag{4.125}$$

where v is the particle speed and $\gamma = 1/(1 - v^2/c^2)^{\frac{1}{2}}$. For relativistic electrons v is close to c and γ is very large. Therefore, $I(\varphi)$ will emerge as a very narrow beam having the same direction as the direction of motion and nearly perpendicular to $\vec{B}$. When many electrons per unit volume describe circular orbits around the same magnetic field lines, their overall emission will be perpendicular to the magnetic field.

The general mathematical treatment is rather complicated (see, for instance, the books by Harwit, Spitzer, Lang, etc.). Some approximate formulae are given here without derivation in order to illustrate problems and possibilities. In many astrophysical objects the energy spectrum of the electrons can be fitted by a power law:

$$N = N_0 E^{-\gamma} \tag{4.126}$$

where NdE is the number density of electrons with energy between E and $E + dE$, and N_0 and γ are constants. For sufficiently high frequencies, an optically thin emitting region produces an intensity in the direction of the observer given by

$$I_\nu = c_5 (2c_1)^{\frac{\gamma-1}{2}} S N_0 \nu^{\frac{1-\gamma}{2}} (B \sin \theta)^{\frac{1+\gamma}{2}} \tag{4.127}$$

where $c_5 = 8.38 \times 10^{-24}$ (c.g.s. units) when $\gamma = 2.8$; $c_1 = 6.27 \times 10^{18}$. S is the source depth, ν is the frequency and θ is the angle between the line of sight and the magnetic field. We see in this formula that

$$I_\nu \propto \nu^{\frac{1-\gamma}{2}} = \nu^{-\alpha} \tag{4.128}$$

$$\alpha = \frac{\gamma - 1}{2} \tag{4.129}$$

another power law that can be checked by observations. α is termed the spectral index of the source, and its determination provides information about the energy spectrum of the electrons. In practice, most extragalactic sources have $0.2 \leq \alpha \leq 1.2$. The disc of our galaxy in the solar neighbourhood has $\alpha \approx 0.8$, in agreement with $\gamma \approx 2.6$ measured by satellites. In the solar neighbourhood N_0 is of order $4.6 \times 10^{17} \text{erg cm}^{-3}$.

Equation (4.127) can in principle be used to determine the magnetic field strength; the problem is that N_0 is unknown in general, and some assumptions must be made. One of them is the condition of energy equipartition, which states that the magnetic energy density is equal to the energy density of the relativistic electrons. Another possibility for galactic discs is the assumption that N_0 is proportional to the gas pressure. This is based on the fact that the most important sources of relativistic electrons, the type II Supernovae, are probably concentrated in regions where there is more gas, since relativistic electrons and type II Supernovae are shortlived and do not travel far from their birthplace. Another possibility is to assume that $N_0 \propto \rho/B$, which takes the energy loss of cosmic electrons due to the synchrotron emission itself into account. More elaborate parametric expressions have also been proposed.

But the determination of the magnetic field from synchrotron radiation observations is particularly powerful when one is able to determine the polarization of the emission. The high synchrotron polarization is due to the fact that the helical orbits of the electrons around $\vec{B}$ do not entail accelerations in the direction of $\vec{B}$. Therefore no $\vec{E}$-field time variations are produced in the $\vec{B}$-direction, so the plane of polarization is

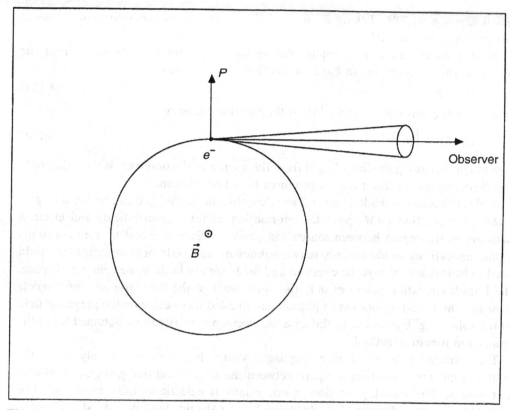

Figure 4.2 Synchrotron emission by a single electron. The direction of emission, the plane of polarization, and the magnetic field are perpendicular to each other.

perpendicular to $\vec{B}$. We have shown that only the $\vec{B}$-component perpendicular to the line of sight can be observed. This is illustrated in Figure 4.2 and equation (4.127).

When many electrons with different energies are emitting, the polarization must persist in regions where $\vec{B}$ remains homogeneous. An intrinsic polarization then arises as a result of the emission mechanism itself. The degree of polarization is

$$p = p_0 \left(\frac{B_u}{B_{tot}} \right)^2 \tag{4.130}$$

where

$$p_0 = \frac{3\gamma + 3}{3\gamma + 7} = \frac{3\alpha + 3}{3\alpha + 5} \tag{4.131}$$

and

$$B_{tot}^2 = B_u^2 + B_{ran}^2 \tag{4.132}$$

B_u denotes the uniform field, B_{ran} the random field, and B_{tot} the total field. If the random field is negligible, the intrinsic polarization is very high. For $B_{ran} = 0$,

$\alpha = 0.8$, $p = p_0$ is 73%. Taking $B_u/B_{tot} \approx 1/2$, $\alpha = 0.8$ as representative conditions in our galaxy, we obtain 20%.

But the polarization plane rotates during the propagation of the wave from the source to the observer due to Faraday rotation by an amount

$$\beta = (RM)\lambda^2 \tag{4.133}$$

where λ is the wavelength and (RM) is the rotation measure

$$(RM) \propto \int n_e \vec{B} \cdot \vec{dl} \tag{4.134}$$

The integral is along the line of sight from the source to the observer. We see that only the $\vec{B}$-component in this direction produces Faraday rotation.

If observations are made at several wavelengths, the curve $[\beta, \lambda]$ can be fitted to give (RM). We see that (RM) provides information about magnetic fields and electron densities in the region between source and observer. Even if $\vec{B}$ and n_e were constants along this path we would again have two unknowns, and additional assumptions would need to be invoked. Where the electrons and fields are not in the same volumes of space, the Faraday rotation produces only the lower limit of the field strength and particle density. The Faraday rotation of pulsars has enabled us to estimate the magnetic field in the solar neighbourhood. In this case the electron density can be obtained using the dispersion measure method.

The Faraday rotation of an extragalactic source is produced not only inside the source itself, but also along the path between the source and our galaxy, and also in our galaxy. The Faraday rotation in our galaxy is galactic latitude dependent. For instance, in the direction of M31 with a galactic latitude of about $-20°$, $(RM) = -88$ rad m^2. For a wavelength of 11 cm, this produces $\beta = -62°$.

5 The fluid in a star

This chapter is organized as follows: first, three basic types of star are studied in some detail: main-sequence stars, white dwarfs, and black holes; other transient and related types are then discussed more descriptively, classified according to the parameters mass and time.

5.1 Main-sequence stars

Main-sequence stars or normal stars are self-gravitating bodies, in hydrostatic equilibrium, consisting of hydrogen, helium, and other minor constituents, in a physical state that can reasonably be described as that of an ideal gas, releasing nuclear energy from the fusion of hydrogen to helium.

The mass loss in the fusion process $4H \rightarrow He$ is very large; some 0.7% of the rest mass is converted to radiative and kinetic energy. If the star were pure hydrogen, the fusion of hydrogen could supply energy of $0.007 \mathcal{M} c^2$, where $\mathcal{M}$ is the star's mass. Other fusion processes are much less effective. Atoms heavier than iron cannot be produced by exothermic fusion reactions. A hypothetical process $14 He \rightarrow Fe$ would have a mass loss of only 0.1%, supplying only approximately $0.001 \mathcal{M} c^2$. Hydrogen is thus the major fuel constituent and the most energetic nuclear process is $4H \rightarrow He$. After leaving the main sequence other heavier nuclei are produced, but the energy available is less and the post-main-sequence period is short. Before nuclear burning, a star radiates for a short time due to gravitational contraction. As the largest fraction of the energy is released during the main-sequence epoch, this will be the longest lasting stellar phase. Therefore, main-sequence stars are often called normal stars. Pre- and post-main-sequence stages are shortlived and are therefore observed less frequently.

Once all nuclear reactions cease in the interior of stars, they become dead or 'fossil' objects: white dwarfs, neutron stars, or black holes. Of course, dead stars are also very long-lived. If they still shine, they will eventually be quenched.

If the Sun maintains its present brightness, its total lifetime can be estimated as $0.007 \mathcal{M} c^2 / L$. For $\mathcal{M} = 2 \times 10^{33}$ g and $L = 3.8 \times 10^{33}$ erg s^{-1}, the lifetime of the Sun is of the order of 10^{11} years. This is greater than the age of the Universe, which need not surprise us as the Sun is not yet dead.

The condition of hydrostatic equilibrium is a very reasonable one for a main-sequence star, as we can easily show. Hydrostatic equilibrium is restored at the speed of sound in a characteristic time of $R\sqrt{m/kT}$. We still do not know the temperature in the interior of a star, but it must be higher, much higher, than the effective temperature of the Sun at 5770 K. Taking m to be of the order of the proton mass, and for R the solar radius 7×10^{10} cm, the characteristic time for equilibrium to be restored is less than 10^5 s, which is very short compared with the solar lifetime.

The physics of these normal stars is rather well understood. Other types of star, pre- and post-main-sequence and dead stars, are still subject to much active research to understand their structure and evolution.

5.1.1 Hydrostatics of a normal star

Consider a non-rotating spherical star. Under steady-state and stationarity conditions the continuity equations simply give the trivial result zero = zero. Another expression, which is the definition of the variable $M(r)$:

$$dM = \rho 4\pi r^2 dr \tag{5.1}$$

is given (incorrectly) the name of the continuity equation. Here, ρ, as usual, is the density, and r is the radial coordinate from the centre. As part of the definition of the radial variable M, we also state that $M(r = 0) = 0$. It is then clear that $M(r)$ is the mass in a sphere of radius r, and that $M(r = R) = \mathcal{M}$, with R being the stellar radius.

From the equation of motion (1.78) we find under these conditions that

$$\frac{1}{\rho}\frac{dp}{dr} + g = 0 \tag{5.2}$$

where g is clearly

$$g = \frac{GM}{r^2} \tag{5.3}$$

therefore (5.2) is rewritten as

$$\frac{dp}{dr} = -\rho\frac{GM}{r^2} \tag{5.4}$$

which is the hydrostatic equation.

It is not possible to use (1.79) directly as an energy balance equation, because energy is not conserved in a star. It is obvious that a production term, taking the nuclear energy input into account, must be included. We may write this production of energy as $\rho\epsilon$, where ϵ is the energy gained by the fluid per unit time and unit mass. Then (1.79) gives

$$\nabla \cdot \vec{q} = \rho\epsilon \tag{5.5}$$

Here, $\vec{q}$ may be due to conduction, convection, or radiation. The latter is usually the most important in normal stars, and in this preliminary study it is considered to be dominant. Nevertheless, convection is also important in some layers, particularly of low mass stars, and practical computations cannot ignore this contribution. Equation (5.5) is, in spherical coordinates,

$$\frac{1}{r^2}\frac{d}{dr}(r^2 q) = \rho\epsilon \tag{5.6}$$

Therefore,

$$r^2 q = \int_0^r \rho \epsilon r^2 \, dr \qquad (5.7)$$

Introducing the radial variable

$$l = 4\pi r^2 q \qquad (5.8)$$

equation (5.7) becomes

$$l = \int_0^r 4\pi r^2 \rho \epsilon \, dr \qquad (5.9)$$

Physical interpretation is not difficult here. As q is the energy per unit time passing through unit surface in the sphere of radius r, equation (5.8) tells us that l is the total energy passing through the sphere of radius r in one second. In particular, $l(r = R) = L$, the star's luminosity. Equation (5.9) informs us that all the nuclear energy released inside a sphere of radius r traverses its surface, which must hold for an equilibrium situation.

We have shown that equation (3.160) is suitable for stellar interiors, and it will be adopted here. We rewrite this equation using l as

$$l = -\frac{4^3 \pi}{3} \frac{\sigma}{\kappa \rho} r^2 T^3 \frac{dT}{dr} \qquad (5.10)$$

where κ is the Rosseland opacity. Obviously, l should be positive, which implies that $dT/dr < 0$, as expected.

The complete set of equations required has now been found. These are (5.1), (5.4), (5.9), and (5.10). As usual the system is closed by including the equation of state. The full system of equations is then

Definition of M:

$$dM = 4\pi \rho r^2 dr \qquad (5.11)$$

Motion:

$$\frac{dp}{dr} = -\rho \frac{GM}{r^2} \qquad (5.12)$$

Energy balance:

$$dl = 4\pi r^2 \rho \epsilon \, dr \qquad (5.13)$$

Radiative transport:

$$l = -\frac{4^3 \pi}{3} \frac{\sigma}{\kappa \rho} r^2 T^3 \frac{dT}{dr} \qquad (5.14)$$

State:

$$pm = \rho k T \qquad (5.15)$$

The chemical composition has not been introduced. It plays a role, however, as it determines the values of m, κ and, mainly, ϵ. In order to solve the system it is clearly necessary to know m, κ, and ϵ as functions of the other variables. Then the system can be solved, yielding ρ, M, p, l, and T as functions of r.

5.1.2   A rough solution

A numerical integration is most precise when the step Δr used is as short as possible. However, the most important characteristics of the behaviour of a star are easily shown using the simplest numerical model. This simplest model is equivalent to an order-of-magnitude calculation and consists of adopting $\Delta r = R$, with a grid with only two points: one at the centre of the star and the other at its surface. Values of quantities in this interval will be considered to be of the order of magnitude of the average of the central and surface values. Of course, a more realistic model must be constructed in practice, but our aim here is purely didactic.

Let us first consider (5.11). Here dM will be approximated by $(\mathcal{M} - 0)$ and dr by R. For ρ, we must use $(0 + \rho_c)/2$, with ρ_c being the unknown central density. Instead of r, $(R + 0)/2 = R/2$ will be used, so that r^2 will be replaced by $R^2/4$. Then, in order of magnitude, (5.11) can be approximated by

$$\mathcal{M} = \frac{\pi}{2}\rho_c R^3 \tag{5.16}$$

The other equations give

$$p_c = \rho_c G \frac{\mathcal{M}}{R} \tag{5.17}$$

$$L = \frac{\pi}{4}\rho_c \epsilon_c R^3 \tag{5.18}$$

$$L = \frac{4}{3}\pi \frac{\sigma T_c^4 R}{\kappa \rho_c} \tag{5.19}$$

$$p_c = \frac{\rho_c}{m} k T_c \tag{5.20}$$

where (5.20) is exact, as (5.15) is not a differential equation. We need to know m, κ, and ϵ_c as functions of the remaining quantities. In a first approximation m and κ may be considered as constants, but we will keep ϵ_c as an unknown function. All unknowns can then be expressed as functions of two parameters, such as $\mathcal{M}$ and R, which are well known for the Sun. Then, for instance,

$$\rho_c = \frac{2}{\pi}\mathcal{M}R^{-3} \tag{5.21}$$

$$p_c = \frac{2}{\pi}G\mathcal{M}^2 R^{-4} \tag{5.22}$$

$$T_c = \frac{G\mathcal{M}m}{kR} \tag{5.23}$$

provide p_c, ρ_c, and T_c in the centre of the star. For the Sun, $\mathcal{M} = 2 \times 10^{33}$g, $R = 7 \times 10^{10}$cm, and we obtain $T_c \approx 2 \times 10^7$ K, $\rho_c \approx 3.7$g cm^{-3}, $p_c \approx 7 \times 10^{15}$dyn cm^{-2}. These may be compared with more precise values obtained from more realistic models: $T_c \approx 1.5 \times 10^7$ K, $\rho_c \approx 160$ g cm^{-3}. Our rough values may therefore be used to explore the physical conditions in the interior quickly. Substituting these values in (5.19) a luminosity of the order of 5×10^{33}erg s^{-1} is obtained, close to the observed value of 3.8×10^{33}erg s^{-1} (with $\kappa \approx 2$g^{-1}cm^2). But the interest of the above solution mainly lies in quantitative discussions.

5.1.3 Some properties of normal stars

The nuclear reaction 4H $\rightarrow$ He can be produced either directly (the proton–proton, or *pp* chain) or through the CNO cycle, carbon acting as a catalyst. In the first case ϵ depends on T^4 and in the second case it depends on T^{20} where the indices are close approximations. At about 2×10^7 K both processes are equally important, and the temperature at which they come into action is slightly lower, at about 1.5×10^7 K. This is also the temperature that we found for the centre of a main-sequence star. This coincidence is not fortuitous. Initially the star contracts and the temperature increases until nuclear reactions begin. Then the temperature undergoes a sudden increase, producing a higher pressure gradient which prevents further contraction. At this stage the temperature becomes more or less constant because a stable equilibrium is established. If there is a small contraction, T increases and ϵ increases enhancing heating and producing a restoring expansion. A small expansion is followed by a decrease in temperature, then ϵ decreases and the star gravitationally contracts, restoring the equilibrium state. Therefore, the central temperature of a normal star is a constant and equals the temperature at which nuclear reaction begins. The central temperature must also be the same for all types of normal star.

If T_c is a constant, then equation (5.20) informs us that p_c/ρ_c has the same value for all normal stars. Therefore, (5.17) tells us that R is proportional to $\mathcal{M}$. It is found that

$$R = \frac{Gm}{kT_c}\mathcal{M} \tag{5.24}$$

More massive stars are longer, but (5.16) indicates that ρ_c is proportional to $\mathcal{M}^{-2}$, so low mass stars are denser.

Equation (5.19) is the so-called mass–luminosity relation

$$L \approx \left[\frac{4\pi}{3}\frac{\sigma}{\kappa}\left(\frac{Gm}{k}\right)^4\right]\mathcal{M}^3 \tag{5.25}$$

If κ were constant, then the mass–luminosity relation would be of the type $L \propto \mathcal{M}^3$. Observations indicate that $L \propto \mathcal{M}^a$, where a lies in the range 3–4. The exponent 3 is valid for low mass stars, becoming closer to 4, or even larger, for brighter stars. In fact κ is not a constant. For instance, Kramers' opacity is of the form $K\rho T^{-7/2}$, with K a constant. In this case a relation of the type $L \propto \mathcal{M}^5$ would be obtained. Therefore, our rough model is unable to predict a precise mass–luminosity relation, but it provides a reasonable approximation.

It was shown in (3.39) how the values of L and R determine the value of q, the radiative flux through the stellar surface:

$$q = \frac{L}{4\pi R^2} \propto \frac{\mathcal{M}^a}{\mathcal{R}^2} \propto \mathcal{M}^{a-2} \propto L^{\frac{a-2}{a}} \tag{5.26}$$

The surface flux was related to the effective temperature, T_{eff}, by the formula (3.159). Therefore, a luminosity-effective temperature relation exists, of the form

$$L \propto T_{eff}^{\frac{4a}{a-2}} \tag{5.27}$$

This is an interesting relation because it can be tested observationally. For $a = 3$, in agreement with (5.25), we have a relation of the form $L \propto T_{eff}^{12}$. This exponent is too high when compared with observations which are fitted approximately by

$$L \propto T_{eff}^{7.5} \tag{5.28}$$

For a Kramers opacity, $L \propto T_{eff}^{6.7}$. A plot $[L, T_{eff}]$ is called the Hertzsprung–Russell diagram. Instead of L, the absolute magnitude M defined in (3.44) is preferred by astronomers. Note that the effective temperature is representative of the atmosphere, and may be related to other observational parameters. Clearly, the atmospheric temperature has a considerable influence on its spectrum. When T_{eff} is high, spectral lines of ionized atoms predominate. When T_{eff} is low, molecular bands are present. It is well known that the spectra of normal stars may be arranged in a sequence. From high to low, T_{eff} stars are said to be of types . . . O, B, A, F, G, K, M, ... Each class is divided into ten groups designated by a digit 0, 1, 2, . . . , 9. For instance, the Sun is a typical G2 star. The spectral class is denoted by S. This classification is called Henry Draper (HD) or, in its improved version, the Yerkes classification. Also T_{eff} is related to the colour of a star, as was argued in Section 3.1.2, which is given directly by photometric observations. More precisely, the main sequence of stars occupies a rather broad straight strip in a $[\log L, \log T_{eff}]$ diagram that can be roughly fitted by

$$\log L = 7.5 \log T_{eff} + 5.44 \tag{5.29}$$

where L is measured in erg s^{-1} and T_{eff} in Kelvins. Again we see that the simplest numerical model provides an introductory insight.

The timelife $t_\star$ of a main-sequence star has been shown to be of the order of 0.007 $\mathcal{M}c^2/L$, the nuclear available energy divided by the present luminosity; therefore

$$t_\star \propto \frac{\mathcal{M}c^2}{L} \propto \frac{\mathcal{M}}{\mathcal{M}^a} = \mathcal{M}^{1-a} \tag{5.30}$$

A massive star is a bright star and consumes more energy per unit time. On the other hand, it possesses more fuel. We see now that the effects are not compensated for, as $a \geq 3$. Massive stars exhaust their fuel earlier and are short-lived. Low mass stars have characteristic lifetimes greater than the age of the Universe.

Luminosity is fairly constant throughout the life of a normal star (equation (5.25)). We have seen how, initially, the star contracts until the ignition temperature of nuclear reactions is reached. In the simple model when part of the hydrogen has been consumed, the star should contract a little to ensure the same temperature and therefore the same rate of energy release. During its complete history the radius and the mass should decrease slightly, by only a small fraction of the initial mass and radius. This is not true in practice; L is not constant but increases a little; T_{eff} decreases; and R increases a little throughout the whole period on the main-sequence. In fact, fuel is not used at a uniform rate, but the increase is very small.

Many of the above formulae are expressed as proportionality relations. Instead of calculating the proportionality constant, it becomes more convenient to use solar values to determine them. Solar values are usually used as units. Some typical approximate solar values are

> Mass: 2×10^{33} g
> Radius: 7×10^{10} cm
> Mean density: 1.4 g cm^{-3}
> Surface gravity: 2.7×10^4 cm s^{-2}
> Mean distance from Earth: $1 AU = 1.5 \times 10^{13}$ cm
> Solar constant: 0.137 watts cm^{-2}
> M_V absolute magnitude: 4.83
> m_V, magnitude: -26.74
> Bolometric magnitude: 4.75
> Effective temperature: 5770 K
> Luminosity: 3.8×10^{33} erg s^{-1}
> Spectral type: G2
> Central density: 150 g cm^{-3}
> Central temperature: 1.5×10^7 K
> Characteristic lifetime: 10^{11} years

For instance, the lifetime of a star with 10 $M_\odot$, taking $a = 5$, would be 10^7 years. A star with 0.1 $M_\odot$ would have a lifetime of the order of 10^{15} years.

5.2 White dwarfs

When the nuclear fuel has been exhausted, the star cools and collapses. The density increases until electrons become degenerate. Then, the so-called Fermi pressure may stop the collapse. A white dwarf is a dead star in which electrons constitute a degenerate Fermion system. It is also in equilibrium, and a dead star does not produce nuclear energy. White dwarfs originate in main-sequence stars.

There are other types of degenerate star. We have seen that in a normal star $\rho_c \propto \mathcal{M}^{-2}$. Therefore, small stars may reach such high densities that they may become degenerate before reaching the main sequence. Then, the Fermi pressure prevents the contraction from reaching the ignition temperature of hydrogen burning and the main sequence cannot be attained. This kind of star is called a brown dwarf, although none have been discovered. Giant planets are also partially degenerate. Within degenerate stars there are also neutron stars – for which neutrons constitute the degenerate Fermion system. We will concentrate, however, on white dwarfs, and reserve an overview of stellar evolution for the conclusion.

5.2.1   Statistical mechanics of degenerate stars

Let us remember some basic properties of a Fermion system. It is well known that the number n_i of Fermions with energy ϵ_i is calculated from the Fermi–Dirac distribution

$$n_i = \frac{1}{e^{(\epsilon_i - \mu)/kT} + 1} \tag{5.31}$$

where μ is the chemical potential. Classical systems have a large, negative value of μ; therefore the 1 in the denominator of (5.31) is negligible, and the Fermi–Dirac distribution becomes a Maxwell–Boltzmann distribution.

A first question of interest is when the Fermion system is degenerate. As we are looking for an order-of-magnitude criterion, we may use approximate formulae, so that we will take it that a Maxwell–Boltzmann distribution may be applied. If we then calculate the total number of particles in the star:

$$N = \Sigma_i n_i \approx Z e^{\mu/kT} \tag{5.32}$$

where Z is the partition function for a classical particle:

$$Z = \Sigma_i e^{-\epsilon_i/kT} \tag{5.33}$$

Let us suppose that we are dealing with an ideal system of electrons, that is, the energy of the system is the sum of the energies of each particle, ignoring the interaction energy (except in particle–particle collisions). We should note that this condition is not always met in real stars. Suppose that ϵ_i is due exclusively to the kinetic energy of electrons, and let us first consider a non-relativistic case in which $\epsilon_i \sim \frac{1}{2} m_e v_e^2$. The Σ_i may be replaced by an integral. To calculate Σ_i it is necessary to sum over all energy levels, which are of the order of $d\tau_p \, d\tau_r / h^3$ in a volume element, where $d\tau_p = 8\pi p^2 dp$ (instead of $4\pi p^2 dp$ because of the two possibilities for the electron spin) and with $\int d\tau_v = V$

$$Z = \frac{1}{h^3} \int d\tau_r \, d\tau_p \, e^{-\frac{p^2}{2m_e kT}} \tag{5.34}$$

The integral gives

$$Z = \left(\frac{2\pi m_e kT}{h^2}\right)^{\frac{3}{2}} V = \frac{V}{\Lambda^3} \tag{5.35}$$

where

$$\Lambda = \frac{h}{(2\pi m_e kT)^{\frac{1}{2}}} \tag{5.36}$$

is the de Broglie wavelength or the wavelength associated with the thermal mean momentum. Then the number density of electrons is

$$n_e = \frac{e^{\mu/kT}}{\Lambda^3} \tag{5.37}$$

and the chemical potential may be expressed in terms of more familiar quantities:

$$\mu = kT \ln(n_e \Lambda^3) \tag{5.38}$$

Classically, $\mu \ll 0$, which means that $n_e \Lambda^3 \ll 1$. Therefore, if

$$n_e^{-1/3} \gg \Lambda \tag{5.39}$$

the system can be considered to be classical and no quantum effects need to be considered. The interpretation of (5.39) is clear. As $n_e^{-1/3}$ is a typical distance between electrons, when this interparticle distance is very large compared to the de Broglie wavelength, then fermions behave like classical particles. If, on the other hand, $\Lambda \gg n_e^{-1/3}$, the fermion system is said to be completely degenerate. Complete degeneracy will pertain for $\Lambda \to \infty$, that is, for $T \to 0$. The usual approach to studying complete degeneration is to assume that $T = 0$ K, which is equivalent to having a high electron number density. In a plot $[\log n_e, \log T]$ the transition between non-degenerate/degenerate states would be a straight line of slope 3/2:

$$n_e^{2/3} T^{-1} = \text{constant} \tag{5.40}$$

In general, Σ_i can be replaced by an integral when the interval between energy levels is much lower than kT. Values for n_e will be calculated in general using

$$n_e = \frac{1}{h^3} \int \frac{d\tau_p}{e^{(\epsilon - \mu)/kT} + 1} \tag{5.41}$$

If the relativistic case is not excluded, ϵ will no longer be $\frac{1}{2} m_e v^2$ but

$$\epsilon^2 = \epsilon_0^2 + p^2 c^2 \tag{5.42}$$

where ϵ_0 is the rest energy. When $\epsilon_0 \gg pc$ the system is non-relativistic, whether it is degenerate or not. Then it can be assumed that $\epsilon = p^2/2m_e$, as is usually assumed in statistical mechanics. Chandrasekhar realized that some degenerate stars could be relativistic, with $pc \gg \epsilon_0$, so that $\epsilon \approx pc$, and that this could have serious consequences on stellar evolution. For a very high density, all the low energy levels are occupied and because of Pauli's principle, very high energy levels must also be occupied, rendering the star relativistic. Four extreme situations can be presented: non-degenerate and non-relativistic, non-degenerate and relativistic, degenerate and non-relativistic, degenerate and relativistic. Of course, intermediate situations may occur, which are in general more difficult to study than are the extreme asymptotic cases.

Non-relativistic stars

Let us then assume, instead of (5.42), that

$$\epsilon = \frac{p^2}{2m_e} \tag{5.43}$$

which should be inserted in (5.41):

$$n_e = \frac{4\pi}{h^3} (2m_e)^{3/2} \int_0^\infty \frac{\epsilon^{1/2} d\epsilon}{e^{(\epsilon - \mu)/kT} + 1} \tag{5.44}$$

The energy density will be calculated from $\Sigma_k n_k \epsilon_k$, or by the continuous description

$$u = \frac{4\pi}{h^3} (2m_e)^{3/2} \int_0^\infty \frac{\epsilon^{3/2} d\epsilon}{e^{(\epsilon-\mu)/kT} + 1} \tag{5.45}$$

Statistical mechanics provides a standard procedure to obtain the equation of state. This takes the form

$$p_e V = kT \ln \zeta \tag{5.46}$$

where ζ is the grand partition function, which for ideal systems is

$$\zeta = \Pi_i \zeta_i \tag{5.47}$$

where

$$\zeta_i = 1 + e^{(\mu-\epsilon_i)/kT} \tag{5.48}$$

The equation of state is

$$p_e V = kT \Sigma_i \ln \zeta_i \tag{5.49}$$

Again Σ_i may be replaced by an integral:

$$p_e V = kT \frac{4\pi V}{h^3} (2m_e)^{3/2} \int_0^\infty \epsilon^{1/2} \ln\left(1 + e^{(\mu-\epsilon)/kT}\right) d\epsilon \tag{5.50}$$

which can be integrated by parts (with $\epsilon^{1/2} d\epsilon = dv$), giving the result

$$p_e = \frac{4\pi}{h^3} (2m_e)^{3/2} \int_0^\infty \frac{2}{3} \epsilon^{\frac{3}{2}} \frac{1}{1 + e^{(\epsilon-\mu)/kT}} d\epsilon \tag{5.51}$$

Comparing with (5.45) we obtain

$$p_e = \frac{2}{3} u \tag{5.52}$$

The above equations are valid for any degree of degeneration. They are, then, valid for Newtonian stars, that is, non-degenerate and non-relativistic stars. Main-sequence stars are Newtonian stars; other stages shortly after the main sequence, including red giants, are also Newtonian.

To find the equation of state, we need to eliminate μ by equations (5.44) of the type $n_e = n_e(\mu, T)$, and (5.51), of the type $p_e = p_e(\mu, T)$. To obtain the thermal equation of state, we need to eliminate μ from equations (5.44) and (5.45).

For classical systems, with $1 \ll e^{(\epsilon-\mu)/kT}$, it is simple to obtain the obvious and well-known result

$$p_e = n_e kT \tag{5.53}$$

$$u = \frac{3}{2} n_e kT \tag{5.54}$$

In the other extreme situation, the completely degenerate star also gives easy, simple results. Here, we may assume that $T = 0$ and $\mu = \infty$, but the integrals to be solved become undetermined. To avoid this, let us again consider the Fermi–Dirac distribution (5.31), where $\mu \to +\infty$. If ϵ_k is small compared with μ, we find $n_k = 1$. If ϵ_k is very large, then $n_k = 0$. Values of n_k between 0 and 1 will be obtained for moderate values of ϵ_k. The curve $n_k(\epsilon_k)$ becomes closer to a step function for $T \to 0$. This step function is typical of zero temperature Fermionic systems. For $T = 0$, $n_k = 1$ for $\epsilon_k < \mu$, and

$n_k = 0$ for $\epsilon_k > \mu$. The chemical potential μ is also called the Fermi energy, and the associated momentum

$$p_F = \sqrt{2\mu m_e} \tag{5.55}$$

is called the Fermi momentum. Then the integral of (5.44) may easily be calculated:

$$n_e = \frac{8\pi}{3h^3}(2m_e)^{3/2}\mu^{3/2} = \frac{8\pi}{3h^3}p_F^3 \tag{5.56}$$

that is, the integration from μ to ∞ gives zero as no electron has an energy larger than μ. Then the number density of electrons and the Fermi momentum are related quantities. If the electron density is very high, p_F and therefore the Fermi energy μ is very large, and electrons have a high energy on average. The specific internal energy u is also a function of p_F. The adoption of a step function for $(1 + e^{(\mu-\epsilon)/kT})$, (5.45), gives

$$u = \frac{8\pi}{5h^3}(2m_e)^{3/2}\mu^{5/2} = \frac{8\pi}{5h^3}\frac{1}{2m_e}p_F^5 \tag{5.57}$$

The thermal equation of state is obtained by eliminating p_F using (5.56) and (5.57):

$$u = \frac{3^{5/3}}{5 \cdot 8^{2/3}}\frac{h^2}{\pi^{2/3}}\frac{1}{2m_e}n_e^{5/3} = constant \times n_e^{5/3} \tag{5.58}$$

which is obviously independent of the temperature as this was set equal to zero. Similarly, for the equation of state we obtain

$$p_e = \frac{h^2}{20m_e}\left(\frac{3}{\pi}\right)^{2/3}n_e^{5/3} = constant \times n_e^{5/3} \tag{5.59}$$

p_e does not depend on the first power of n_e (as in classical ideal gases), but, rather, on $n_e^{5/3}$, and it obviously does not depend on the temperature. Partial degeneracy is more difficult to handle. Equations (5.56), (5.57), (5.58), and the equation of state (5.59) are valid for a non-relativistic, completely degenerate white dwarf.

Relativistic stars

Let us carry out a similar calculation for extreme relativistic stars, for which we may assume in (5.42) that

$$\epsilon \approx pc \tag{5.60}$$

In this case

$$n_e = \frac{8\pi}{c^3h^3}\int_0^\infty \frac{\epsilon^2 d\epsilon}{e^{(\epsilon-\mu)/kT} + 1} \tag{5.61}$$

$$u = \frac{8\pi}{c^3h^3}\int_0^\infty \frac{\epsilon^3 d\epsilon}{e^{(\epsilon-\mu)/kT} + 1} \tag{5.62}$$

$$p_e = kT\frac{8\pi}{c^3h^3}\int_0^\infty \epsilon^2 d\epsilon \ln\left(1 + e^{(\mu-\epsilon)/kT}\right) \tag{5.63}$$

which can again be integrated by parts ($\epsilon^2 d\epsilon = dv$), giving

$$p_e = \frac{1}{3}\frac{8\pi}{c^3h^3}\int_0^\infty \frac{\epsilon^3 d\epsilon}{e^{(\epsilon-\mu)/kT} + 1} = \frac{u}{3} \tag{5.64}$$

which is a very familiar result (2.24).

Let us first consider a relativistic white dwarf (an extreme relativistic, completely degenerate electron star); adopting $T = 0$ K:

$$n_e = \frac{8\pi}{c^3 h^3} \int_0^\mu \epsilon^2 d\epsilon = \frac{1}{3} \mu^3 \frac{8\pi}{c^3 h^3} \tag{5.65}$$

$$u = \frac{2\pi}{c^3 h^3} \mu^4 = \frac{1}{8} \left(\frac{3^4}{\pi} \right)^{1/3} chn_e^{4/3} \tag{5.66}$$

and the equation of state

$$p_e = \frac{1}{8} \left(\frac{3}{\pi} \right)^{1/3} chn_e^{4/3} \tag{5.67}$$

now p_e depends on $4/3$ power of the electron number density.

We can also perform the calculations for relativistic non-degenerate systems, which are, however, not identified with any particular type of star. The results are

$$u = 3n_e kT \tag{5.68}$$

$$p_e = n_e kT \tag{5.69}$$

Equations of state

Let us summarize the equation of state as previously deduced and establish criteria to determine *a priori* which equation of state holds for any type of star. It has already been found that the transition non-degeneracy–degeneracy is characterized by (5.40), that is,

$$n_e = \frac{(2\pi m_e kT)^{3/2}}{h^3} = 2.41 \times 10^{15} T^{3/2} \tag{5.70}$$

This last equality requires the use of gaussian units. We now need a criterion to state *a priori* whether the star is relativistic or not. The transition will take place for $\epsilon_0 = pc$, where p is a characteristic momentum of the electron. Therefore, $p = p_F$ may be adopted as a characteristic momentum. Thus the transition takes place for $m_e c^2 = p_F c$, or, taking (5.56) into account, the required criterion is

$$n_e = \frac{8\pi}{3h^3} m_e^3 c^3 \tag{5.71}$$

which is a constant value. If n_e is much lower than this value no correction will be needed. The critical electron density is $n_e = 5.87 \times 10^{29} \mathrm{cm}^{-3}$.

It is desirable to know the critical density corresponding to this critical electron number density. There are two basic formulae for calculating the equivalent mass of an atom of any star and the relation between ρ and n_e. These are

$$m = \frac{2}{3X + \frac{Y}{2} + 1} m_H \tag{5.72}$$

and

$$\rho = \frac{2m_H n_e}{1 + X} \tag{5.73}$$

where m_H is the proton mass. X is the proportion of hydrogen by mass, and Y the proportion of helium by mass. These formulae are easily deduced for the case in which all nuclei have lost all their electrons, assuming the approximate relation $m_i \approx 2Z_i m_H$ for metals (adopting the astrophysical concept of a metal as being any element heavier than helium). Let us take a hypothetical pure hydrogen star, with $X = 1$, $Y = 0$. The equivalent mass would be $\frac{m_H}{2}$, which is a familiar result. When considering plasmas in Chapter 4, characterized by atoms that have been ionized only once, this result was assumed valid; it is the average mass of an ion and an electron (in this case, the average of the proton and the electron masses). If the star consists of helium, then $m = \frac{4}{3}m_H$. If the star has exhausted all He and H, with $X = 0$, $Y = 0$, we obtain $m = 2m_H$. The formula (5.73) tells us that a pure hydrogen star, with $X = 1$, will have a density of $\rho = m_H n_e$, which is also a familiar result. But, if hydrogen has been exhausted, with $X = 0$, the relation is $\rho = 2m_H n_e$. When considering white dwarfs $X = 0$, $Y = 0$, and the relation between density ρ and electron number density n_e is simply

$$\rho = 2m_H n_e \tag{5.74}$$

Therefore, the critical n_e value corresponds to a density of $2 \times 10^6 \mathrm{g\ cm}^{-3}$. Low mass white dwarfs have $Y \approx 1$ in the core, and in more external layers neither X nor Y is zero. But (5.74) is a reasonable approximation.

Another important transition separating two states of matter with very different physical properties is that between regions for which the electron pressure or the photon pressure is dominant. The two pressures are equal when

$$n_e kT = \frac{4}{3c} \sigma T^4 \tag{5.75}$$

that is, when

$$T = \left(\frac{3ck}{4\sigma}\right)^{1/3} n_e^{1/3} \approx 0.38 n_e^{1/3} \tag{5.76}$$

where σ is the Stefan–Boltzmann constant. The latter equality is valid for gaussian units only. When the temperature exceeds this value the equation of state clearly also changes, becoming the equation of state for photons of the type $p \propto T^4$.

The total pressure will be, in general,

$$p = p_i + p_e + p_r \tag{5.77}$$

where p_i, p_e, and p_r are the ion, electron, and radiation pressures. Under stellar conditions, however, there are many more electrons than ions, with $p_i \ll p_e$. Except for the upper part of Figure 5.1 the radiation pressure may also be ignored in most stars.

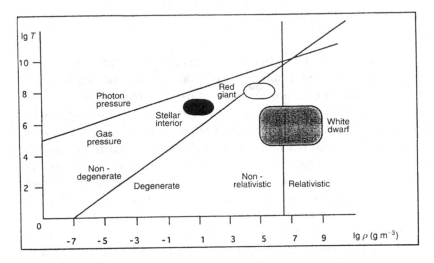

Figure 5.1   log T versus log ρ showing the transitions: photon dominated pressure, gas dominated pressure; non-degenerate, degenerate systems; non-relativistic, relativistic systems. The position of some characteristic stellar systems is plotted.

The following table summarizes the different equations of state:

p_e	*Non-degenerate*	*Degenerate*	
Non-relativistic	$n_e kT$	$\dfrac{h^2}{5m_e}\left(\dfrac{3}{8\pi}\right)^{2/3} n_e^{5/3}$	(5.78)
Relativistic	$n_e kT$	$\dfrac{1}{8}\left(\dfrac{3}{\pi}\right)^{1/3} chn_e^{4/3}$	(5.79)

The transition regions are more difficult to handle mathematically. Therefore, the above plot indicates that a system with $\rho \approx 10^7 \text{g cm}^{-3}$, $T \approx 5 \times 10^9$ K, where the three transition lines meet, will be very complicated.

5.2.2   Hydrostatics of white dwarfs

Equations (5.11) and (5.12) are also valid for white dwarfs. On the other hand, (5.13) is no longer valid as it was deduced under the assumption that radiation was the only important factor in transporting heat. Conduction is in fact dominant in a degenerate star. The mean-free-path is very large and the temperature becomes uniform. Although the extreme condition $T = 0$ K is a good approximation, the actual effective temperature may be rather high, of the order of 10^4 K. On the other hand, it is clear that $\epsilon = 0$. If there are no temperature gradients, then $l = 0$. If the star shines, a thin, non-degenerate atmosphere will be the source of the surface emission. Equation (5.15) must be replaced by (5.78) if it is a non-relativistic star or by (5.79) if is relativistic.

Let us first consider a non-relativistic white dwarf. The hydrostatic equations describing its behaviour are

$$dM = 4\pi\rho r^2 dr \tag{5.80}$$

$$\frac{dp}{dr} = -\rho\frac{GM}{r^2} \tag{5.81}$$

$$p = C_{NR}\rho^{5/3} \tag{5.82}$$

where

$$C_{NR} = \frac{h^2}{5m_e}\left(\frac{3}{8\pi}\right)^{2/3}\frac{1}{(2m_H)^{5/3}} \tag{5.83}$$

Let us carry out the simplest numerical model, which is very similar to the one developed for main-sequence stars. From (5.80) and (5.81), equations (5.16) and (5.17) are obtained again. Equation (5.82) is exact and does not need any approximation. By solving the resulting system we obtain

$$R \approx \left(\frac{2}{\pi}\right)^{2/3}\frac{C_{NR}}{2G}\mathcal{M}^{-1/3} \tag{5.84}$$

$$\rho_c \approx \frac{\pi}{2}(2G)^3 C_{NR}^{-3}\mathcal{M}^2 \tag{5.85}$$

The mechanical properties of a non-relativistic white dwarf are completely different from those of a Newtonian star. The larger the mass, the smaller the radius, and as an obvious consequence the density is an increasing function of the mass. Let us assume, then, that the mass of the white dwarf increases either because several white dwarfs of different mass are being analysed, or because the mass actually increases as a result of accretion. The density will increase up to the relativistic limit, when the equation of state will suddenly change. The new equations to be solved are (5.80), (5.81) again, and the equation of state:

$$p = C_R\rho^{4/3} \tag{5.86}$$

where

$$C_R = \frac{1}{8}\left(\frac{3}{\pi}\right)^{1/3}(2m_H)^{-4/3}ch \tag{5.87}$$

The result obtained using the simplest rough numerical model is now

$$\mathcal{M} \approx \left(\frac{C_R}{G}\right)^{3/2}\left(\frac{2}{\pi}\right)^{1/2} \tag{5.88}$$

which is a constant value, independent of the star's radius. The value of this constant is $0.25\ M_\odot$, but this is a very crude estimate. When precise calculations are carried out, we obtain for this limiting value of the star's mass:

$$\mathcal{M}_{CH} = 1.46\ M_\odot \tag{5.89}$$

for a white dwarf where all hydrogen and helium have been exhausted. The subindex *CH* stands for Chandrasekhar, who first found this important value. Equation (5.89) gives the Chandrasekhar mass for a degenerate electron star. A white dwarf cannot have a higher mass than the Chandrasekhar mass. Observations confirm this important theoretical result. No white dwarf has been found with a mass higher than 1.46 M$_\odot$.

A lower limit has also been observed for the mass of a white dwarf, of about 0.3 M$_\odot$, which can be justified as follows. White dwarfs are the last evolutionary stage of moderately low mass main-sequence stars. If the mass is too low, the star cannot reach the main sequence, and therefore never becomes a white dwarf. Such a star cannot reach the temperature of 10^7 K needed to generate nuclear reactions, because degeneration prevents collapse. Thus, the density must be that at which degeneration becomes important, that is, for an electron concentration given by (5.70). These objects have X close to unity; therefore $m \approx m_H/2$:

$$\rho = 2.41 \times 10^{15} \frac{m_H}{2} T^{3/2} \tag{5.90}$$

The largest of these objects reaches a temperature just below that required for nuclear burning. For $T \approx 10^7$, a density of 63 g cm^{-3} is obtained. For this density, equation (5.24) gives a value of $\mathcal{M} \approx 0.25$ M$_\odot$. This is the mass of the largest brown dwarf and the mass of the smallest white dwarf.

For a typical white dwarf of about 1 M$_\odot$, as the extreme relativistic conditions are not reached, equation (5.84) yields a radius of about 2000 km, similar to the order of magnitude of the Earth's radius. The density is of the order of the value given by (5.74). These high density properties of white dwarfs have long been known from observations.

5.3   Polytropic stars

The simplest numerical model provided a quick insight into the physics of stellar interiors. More precise numerical procedures will not be discussed here, but there is an approximation which can be made in a large variety of idealized stars and which is brief and instructive. The hypothesis will be that the star is a polytrope, that is, that the density and the pressure are related by a relation of the type

$$p = K\rho^{\frac{n+1}{n}} \tag{5.91}$$

where K and n are constants. In particular, n is called the polytropic index. For such stars, only the hydrostatic equation need be integrated. Some real stars obey the polytropic condition fairly well, so that, for these, equation (5.91) may be considered a realistic approximation. Four different types of star will be considered: (a) adiabatic stars, (b) stars where the radiative pressure is not negligible, (c) completely degenerate non-relativistic stars, (d) completely degenerate extremely relativistic stars.

(a) Adiabatic stars. It was shown in Chapter 1 that when the convective heat flux is dominant the radial profiles of p, ρ, and T are related by the adiabatic law (1.104). Therefore, when the profile is adiabatic,

$$p = K\rho^\gamma \tag{5.92}$$

the star can be considered as a polytrope. If the star is made up of monatomic gases $\gamma = c_p/c_v = 5/3$, and the polytropic index is 1.5. Convective or adiabatic stars were not considered in our previous rough description, but real stars, especially stars of inter-mediate or low mass, including our Sun, have extensive convective layers.

(b) Stars with radiative pressure. It was shown in Chapter 3, Section 3.6 that in the outer layers of blue giants with a high radiative flux, and those of red giants with a very low surface gravity, radiation pressure could not be neglected in the hydrostatic equa-tions. This may also be the case in the interiors. Assume, first, that β is a constant coefficient giving the relative importance of the gas pressure with respect to the total pressure:

$$\frac{\rho}{m} kT = \beta p \tag{5.93}$$

$$\frac{1}{3} aT^4 = (1 - \beta)p \tag{5.94}$$

β must lie in the range 0–1. For $\beta = 0$, the pressure would be exclusively radiative. For $\beta = 1$, the radiation pressure would be negligible.

Then

$$T^3 = 3 \frac{\rho}{m} \frac{k}{a} \frac{1 - \beta}{\beta} \tag{5.95}$$

and

$$p = \left[\frac{3k^4}{am^4} \frac{1 - \beta}{\beta^4} \right]^{1/3} \rho^{4/3} \tag{5.96}$$

Therefore, this type of star is a polytrope with index 3. The problem is that, in practice, β is a function of the radial coordinate. If $\beta = 0$ (radiative pressure only), $p = \infty$ and the polytropic condition cannot be used; this is reasonable since such a star would be completely unstable. If $\beta = 1$ (zero radiation pressure) $p = 0$ and again the polytropic approximation cannot be used. Rather curiously, the extreme cases $\beta = 0$ and $\beta = 1$ cannot be treated, but the intermediate cases can.

(c) Degenerate non-relativistic star. The equation of state (5.78) shows that this type of star is a polytrope with polytropic index $n = 1.5$, that is, the same index as that of a convective star.

(d) Degenerate relativistic star. The equation of state (5.79) shows that this type of star is a polytrope with polytropic index 3, that is, the same index as that of a star dominated by radiation pressure.

Let us introduce a change of variable, defining λ and ϕ by

$$\rho = \lambda \phi^n \tag{5.97}$$

where λ is the central density and ϕ is a function of r, where $\phi(r = 0) = 1$. From (5.1) and (5.4), which are valid for all the stars studied,

$$\frac{d}{dr} \left(\frac{r^2}{\rho} \frac{dp}{dr} \right) = -4\pi r^2 \rho G \tag{5.98}$$

From (5.91) and (5.97),

$$\frac{(n+1)\lambda^{\frac{1-n}{n}}K}{4\pi G}\frac{d}{dr}\left(r^2\frac{d\phi}{dr}\right) = -\phi^n r^2 \tag{5.99}$$

Defining a constant a by

$$a = \left(\frac{(n+1)\lambda^{\frac{1-n}{n}}K}{4\pi G}\right)^{1/2} \tag{5.100}$$

and introducing the radial variable ξ:

$$r = a\xi \tag{5.101}$$

the so-called Lane–Emden equation is obtained:

$$\frac{1}{\xi^2}\frac{d}{d\xi}\left(\xi^2\frac{d\phi}{d\xi}\right) = -\phi^n \tag{5.102}$$

ϕ is the Lane–Emden function. Analytical solutions of this differential equations only exist for $n = 0, 1$, and 5. For other values of n it is necessary to continue with numerical techniques, but with the advantage that all the above-mentioned types of star are studied simultaneously.

When $\phi = 0$, the surface of the star corresponds to the value of the radial coordinate termed ξ_1. (It can be shown that, in all cases of practical interest, ξ_1 exists if $n < 5$.)

As boundary conditions we have $\phi(\xi = 0) = 1$ and $d\phi/d\xi(\xi = 0) = 0$. Once the Lane–Emden equation is solved, $\phi(\xi)$ is determined. From (5.97), (5.100), and (5.101) we obtain $\rho(r)$ and thus $p(r)$. The solution is also given as a function of the central density λ. Within a given type of star, different values of $\rho_c (\equiv \lambda)$ generate a family of different models. It is preferable to use the stellar mass $\mathcal{M}$ rather than λ as the independent variable here, but the relation between $\mathcal{M}$ and λ can easily be obtained:

$$\mathcal{M} = \int_0^{a\xi_1} 4\pi r^2 \rho\, dr = 4\pi a^3 \lambda \int_0^{\xi_1} \phi^n \xi^2\, d\xi \tag{5.103}$$

The integral is easily solved now, because, taking (5.102) into account,

$$\mathcal{M} = -4\pi a^3 \lambda \int_0^{\xi_1} d\left(\xi^2\frac{d\phi}{d\xi}\right) = -4\pi a^3 \lambda \left[\xi^2\frac{d\phi}{d\xi}\right]_1 \tag{5.104}$$

where the subindex 1 denotes the surface. $[\xi^2(d\phi/d\xi)]_1$ may be tabulated: approximate values are -2.4 for $n = 1.5$ and -2 for $n = 3$. Note that a may also depend on λ. Therefore

$$\mathcal{M} = \frac{-1}{4\pi}\left(\frac{(n+1)K}{G}\right)^{3/2}\left[\xi^2\frac{d\phi}{d\xi}\right]_1 \lambda^{\frac{3-n}{2n}} \tag{5.105}$$

For $n = 3$, using a degenerate relativistic star as an example, we see that $\mathcal{M}$ does not depend on λ; it is, then, a constant. We now have a familiar result as this constant is the Chandrasekhar mass.

5.4   The metric of a star[1]

Curvature effects induced by the energy-momentum of a star can have a large influence in the late stages of its evolution, and in particular they can induce the formation of a black hole. Four quantities characterize a black hole: mass, angular momentum, electrical charge, and magnetic monopole charge. The simplest case, the Schwarzschild black hole, possesses only mass and it corresponds to a spherically symmetric star. Schwarzschild black holes will be considered here because this simple case clearly illustrates the most basic topological properties of these objects. Kerr black holes (with angular momentum), Kerr–Newman black holes (with electrical charge) and other more complete models will be excluded from this chapter. Quantum mechanical effects, such as Hawking radiation, will likewise not be considered.

The Schwarzschild metric will be discussed both outside and inside the star. The geodesics in the external Schwarzschild solution permit us to deal with some of the best-known relativistic astrophysical problems such as the deflection of light from a star by the curvature near the Sun, the advance of the perihelia of planets, gravitational lenses, and others. These are not true fluid problems, and we will concentrate mainly on the properties of the Schwarzschild surface and on the limit of the mass–radius ratio of a star, above which it will collapse gravitationally and become a black hole.

Black holes are not necessarily dead stars. Some were probably formed in the early Universe and others are probably located in the nuclei of galaxies. In what follows we will bear in mind the case of a star, but the whole analysis is valid for any object with spherical symmetry independent of its mass. The Schwarzschild metric is a solution of the Einstein field equations and is of interest in itself quite apart from its astrophysical application.

5.4.1   The metric around a star

Let us adopt the most simplified picture that is compatible with the actual conditions in the space around a star. If the star is static and spherically symmetric we expect a static and isotropic metric. Terms such as static, spherically symmetric, and isotropic all have well-known and precise meanings in a three-dimensional context but require to be reinterpreted in four-dimensions. To do this we must adopt a three-dimensional language. If we are looking for a useful expression for $d\tau^2 = -g_{\alpha\beta}\,dx^\alpha dx^\beta$, we accept that $g_{\alpha\beta}$ should depend on r but not on $\vec{r}$, and they should not depend on t. The differential products $dx^\alpha dx^\beta$, should be obtainable as functions of the three-dimensional invariants $\vec{r} \cdot d\vec{r}$, $d\vec{r} \cdot d\vec{r}$, and dt. As $d\tau^2$ is of second order we do not look for higher-order invariants. First, $\vec{r} \cdot d\vec{r}$ gives a second-order term if it is either multiplied by itself $(\vec{r} \cdot d\vec{r})^2$, or by dt, giving $\vec{r} \cdot d\vec{r}\,dt$. Both $(\vec{r} \cdot d\vec{r})^2$ and $(\vec{r} \cdot d\vec{r}\,dt)$ depend on $\vec{r}$, which is incompatible with the condition of isotropy. Therefore, $\vec{r} \cdot d\vec{r}$ may be excluded. To obtain a second-order term from dt, we can multiply it by $\vec{r} \cdot d\vec{r}$, which has been

---

1   Geometrized units ($c = 1$; $G = 1$; $k = 1$; $\epsilon_0 = 1$) will be used in this section.

rejected, or by itself, giving dt^2. Thus $dx^\alpha dx^\beta$ must be obtained from either $d\vec{r} \cdot d\vec{r}$, which is now ds_e^2, or dt^2. Then

$$d\tau^2 = B(r)dt^2 - C(r)ds_e^2 = B(r)dt^2 - C(r)(dr^2 + r^2(d\theta^2 + \sin^2\theta\,d\varphi^2)) \quad (5.106)$$

In general relativity we are allowed to choose a large variety of coordinates, and it is obvious that the simple expression (5.106) would not be conserved in any change of coordinates. We conclude that under the specified conditions, there is only one coordinate system for which the expression for $d\tau^2$ is given by (5.106). This form of $d\tau^2$ is called the 'isotropic' form, but there are other forms, which might be even more interesting for calculations and interpretations. One of them is the 'standard' form:

$$d\tau^2 = B(r)dt^2 - A(r)dr^2 - r^2(d\theta^2 + \sin^2\theta\,d\varphi^2) \quad (5.107)$$

where the radial coordinate r is now different. The change implies the substitution

$$C(r)dr^2 \rightarrow A(r)dr^2 \quad (5.108)$$

$$C(r)r^2 \rightarrow r^2 \quad (5.109)$$

which provides two equations for obtaining $A(r)$ and the new r as functions of $C(r)$ and the old r. Our problem now is to find the functions $A(r)$ and $B(r)$ in (5.107) so that the metric is completely determined. Einstein's field equations in the form of (2.86) will be used. Some initial and intermediate formulae for this derivation are given:

$$g_{\alpha\beta} = \begin{bmatrix} -B & 0 & 0 & 0 \\ 0 & A & 0 & 0 \\ 0 & 0 & r^2 & 0 \\ 0 & 0 & 0 & r^2\sin^2\theta \end{bmatrix} \quad (5.110)$$

$$g^{\alpha\beta} = \begin{bmatrix} -1/B & 0 & 0 & 0 \\ 0 & 1/A & 0 & 0 \\ 0 & 0 & 1/r^2 & 0 \\ 0 & 0 & 0 & 1/(r^2\sin^2\theta) \end{bmatrix} \quad (5.111)$$

The affine connection is then easily calculated:

$$\Gamma^0_{0r} = \frac{1}{2B}\frac{dB}{dr} \quad (5.112)$$

$$\Gamma^r_{00} = \frac{1}{2A}\frac{dB}{dr} \quad (5.113)$$

$$\Gamma^r_{rr} = \frac{1}{2A}\frac{dA}{dr} \quad (5.114)$$

$$\Gamma^\theta_{\theta r} = \Gamma^\varphi_{\varphi r} = \frac{1}{r} \quad (5.115)$$

$$\Gamma^\varphi_{\varphi\theta} = \frac{1}{\tan\theta} \quad (5.116)$$

$$\Gamma^r_{\theta\theta} = -\frac{r}{A} \quad (5.117)$$

$$\Gamma^r_{\varphi\varphi} = -\frac{r\sin^2\theta}{A} \quad (5.118)$$

$$\Gamma^{\theta}_{\varphi\varphi} = -\sin\theta\cos\theta \tag{5.119}$$

other components being zero. The non-vanishing components of the Ricci tensor are

$$R_{00} = -\frac{1}{2A}\frac{d^2B}{dr^2} + \frac{1}{4A}\frac{dB}{dr}\left(\frac{1}{A}\frac{dA}{dr} + \frac{1}{B}\frac{dB}{dr}\right) - \frac{1}{r}\frac{1}{A}\frac{dB}{dr} \tag{5.120}$$

$$R_{rr} = \frac{1}{2B}\frac{d^2B}{dr^2} - \frac{1}{4B}\frac{dB}{dr}\left(\frac{1}{A}\frac{dA}{dr} + \frac{1}{B}\frac{dB}{dr}\right) - \frac{1}{r}\frac{1}{A}\frac{dA}{dr} \tag{5.121}$$

$$R_{\theta\theta} = -1 + \frac{r}{2A}\left(-\frac{1}{A}\frac{dA}{dr} + \frac{1}{B}\frac{dB}{dr}\right) + \frac{1}{A} \tag{5.122}$$

$$R_{\varphi\varphi} = \sin^2\theta R_{\theta\theta} \tag{5.123}$$

Therefore, the Ricci tensor is diagonal. The scalar curvature is then expressed by

$$R = R^{\lambda}_{\lambda} = g^{\lambda\nu}R_{\nu\lambda} = g^{\lambda\lambda}R_{\lambda\lambda}$$
$$= \frac{1}{AB}\frac{d^2B}{dr^2} - \frac{1}{2AB}\frac{dB}{dr}\left(\frac{1}{A}\frac{dA}{dr} + \frac{1}{B}\frac{dB}{dr} + \frac{2}{r}\frac{1}{AB}\frac{dB}{dr}\right)$$
$$- \frac{1}{rA}\left(\frac{1}{B} + \frac{1}{A}\right)\frac{dA}{dr} + \frac{2}{r^2}\left(\frac{1}{A} - 1\right) \tag{5.124}$$

As we are dealing with the outside of the star, we have $\epsilon = p = 0$ and thus the energy-momentum tensor is zero. From Einstein's field equation we obtain

$$R_{\mu\nu} = 0 \tag{5.125}$$

To integrate (5.125) taking (5.120)–(5.123) into account, let us adopt a flat space for $r \to \infty$ as the boundary condition, so that $A(\infty) = B(\infty) = 1$. To obtain a rapid solution, we note that

$$AR_{00} + BR_{rr} = 0 \tag{5.126}$$

which is equivalent to

$$\frac{1}{A}\frac{dA}{dr} + \frac{1}{B}\frac{dB}{dr} = 0 \tag{5.127}$$

Therefore,

$$A = \frac{1}{B} \tag{5.128}$$

Substituting (5.128) in equation (5.122) we have

$$\frac{d(rB)}{dr} = 1 \tag{5.129}$$

which on integration gives

$$B = 1 - \frac{2M}{r} \tag{5.130}$$

where M is an integration constant. Of course, if $r \to \infty$, $B \to 1$. The solution is now complete and we should be able to identify the integration constant M in terms of the physical parameters of the star, such as mass and radius. To do so we note that for sufficiently large values of r the gravitational field must be weak (not completely flat) and thus the Einstein description must coincide with the Newtonian classical description. The orbit of a body is given by the geodesic equation, which for a body with moderate velocity is

$$\frac{d^2 x^\mu}{d\tau^2} + \Gamma^\mu_{00} \left(\frac{dt}{d\tau}\right)^2 = 0 \tag{5.131}$$

For $\mu = r$,

$$\frac{d^2 r}{dt^2} + \frac{1}{2A}\frac{dB}{dr} = 0 \tag{5.132}$$

and using (5.128),

$$\frac{d^2 r}{dt^2} + \frac{d}{dr}\left(\left(\frac{B}{2}\right)^2\right) = 0 \tag{5.133}$$

This must be compared with the Newtonian equation

$$\frac{d^2 r}{dt^2} = -\frac{\partial}{\partial r}\phi \tag{5.134}$$

where ϕ is the Newtonian gravitational potential. Then, for very large radii,

$$B^2 = 4\phi + constant \tag{5.135}$$

or

$$1 + \frac{4M^2}{r^2} - \frac{4M}{r} = 4\phi + constant \tag{5.136}$$

For large radii $4M^2/r^2$ is negligible as compared to $4M/r$. The constant must be equal to 1 if ϕ tends to zero for $r \to \infty$. Then

$$\phi = -\frac{M}{r} \tag{5.137}$$

This is the precise expression of the Newtonian potential, but with M now being the mass of the star. Therefore, the integration constant M in equation (5.130) can be identified with the mass of the star. This was the reason for denoting this constant initially by M.

Finally, we have the standard form of the Schwarzschild metric for the vacuum region outside the star:

$$d\tau^2 = \left(1 - \frac{2M}{r}\right)dt^2 - \left(1 - \frac{2M}{r}\right)^{-1}dr^2 - r^2(d\theta^2 - \sin^2\theta \, d\varphi^2) \tag{5.138}$$

It is interesting to note that the external Schwarzschild metric still holds even when the condition of static equilibrium does not. Only the condition of spherical symmetry is required, as first shown by Lemaître.

5.4.2 Properties of the external Schwarzschild metric. Black holes

The most noticeable property is the apparent singularity for $r = 2M$, where $g_{00} = 0$ and g_{rr} becomes ∞. The value of $2M$ is the Schwarzschild radius; it defines the Schwarzschild surface, which is sometimes called the Schwarzschild singularity. This surface divides the external space into two regions with very different topological characteristics.

The Schwarzschild surface at $r = 2M$ does not exist for most known objects in the Universe, because it usually lies inside them, where the external Schwarzschild solution does not hold. It is easily calculated that it is at about 3 km from the centre for a normal star such as the Sun, about 10 mm from the centre for a typical planet such as the Earth, and about 0.01 pc from the centre for a galaxy such as the Milky Way, much less in each case than the radius of the surface of these objects. But it may be located outside in highly compact objects, notably black holes.

It has been pointed out that the Schwarzschild surface is not a real singularity of the metric but, rather, it is a coordinate singularity. It can disappear when other coordinates are chosen and a freely falling observer can reach $r = 2M$ and be able to find his local Minkowskian system there. We first note that the coordinates used to establish the standard form are neither our time (i.e. the proper time of an observer at rest with respect to the star) nor our length (a fixed Δr is not the same distance when measured for different r). To find the relation between t, the coordinate time, and τ, the proper time of an observer at rest, with r, θ, and φ constant, we obtain

$$d\tau = \sqrt{1 - 2M/r}\ dt \qquad (5.139)$$

These two times coincide only for very large radii. That is, t is the proper time of a static observer far away from the star. This is compatible with the introduction of t. For $r = 2M$, dt becomes infinite. Similarly with θ, φ, and t constants, so that $d\tau^2$ corresponds to the radial variation measured by the observer at rest, that is, $d\rho$:

$$d\rho = \frac{dr}{\sqrt{1 - 2M/r}} \qquad (5.140)$$

At an infinite distance the coordinate radius and the distance radius, ρ, coincide, but when we get closer to the star dr is shorter than $d\rho$, becoming zero at the Schwarzschild radius. Are r and t appropriate for studing the Schwarzschild surface?

Before looking for another coordinate system, let us make an important commentary on equation (5.139). It also illustrates a well-known phenomenon in general relativity: when a clock is working in a gravitational field it runs slower. Suppose that an observer at an infinite distance from the star is observing the clock of another observer at rest in the gravitational field at some coordinate r. The time of this second observer would be τ, and the time for the distant observer in flat space would be t. Equation (5.139) is then applicable and states that $d\tau < dt$, that is, the observed time interval from outside is larger than the proper time. The clock inside runs more slowly, and stops completely at $r = 2M$. If the clock is replaced by an emitting atom with a proper frequency ν_0, there will be a redshift in the light emitted in the gravitational field given by

$$z = \frac{\lambda - \lambda_0}{\lambda_0} = \frac{\nu_0 - \nu}{\nu} = \frac{\nu_0}{\nu} - 1 = \frac{1}{\sqrt{1 - \frac{2M}{r}}} - 1 \qquad (5.141)$$

For large radii $(1 - 2M/r)^{1/2} \approx 1 - M/r$ and we obtain $z \approx M/r$. This gravitational redshift was one of the classical Einstein tests of general relativity which yielded good agreement between theory and observations. For a white dwarf $z \approx 2 \times 10^{-4}$, which produces $\Delta\lambda \approx 1A$ for a yellow light, and this is clearly detectable. For smaller r but still $r > 2M$, z can be very high and for $r = 2M$, $z = \infty$. Photons from the Schwarzschild surface are not observed because they are infinitely redshifted. Note also that the redshift produced by curvature effects has the general expression $z = (-g_{00})^{-1/2} - 1$.

Lemaître first proposed a coordinate system in which the apparent singularity at $r = 2M$ disappears, but the system which manifests this property and is most commonly used was discovered independently by Kruskal and Szekeres, and its use is very helpful in understanding the topology around the Schwarzschild surface. The elementary proper time of the Schwarzschild metric in the Kruskal–Szekeres coordinates is

$$d\tau^2 = \frac{32M^3}{r} e^{-r/2M}(dw^2 - dz^2) - r^2(d\theta^2 + \sin^2\theta \, d\varphi^2) \tag{5.142}$$

where the new coordinates are w (time-like) and z, related to the previous coordinates by

$$z^2 - w^2 = \frac{1}{2M}(r - 2M)e^{r/2M} \tag{5.143}$$

$$\frac{w}{z} = \tanh\frac{t}{4M} \tag{5.144}$$

Although r is present in (5.142), it must be understood as a function of z and w, $r(z, w)$ defined by (5.143).

The Kruskal–Szekeres system in fact permits a mathematical description for negative values of the radial coordinate z, but as these may not represent a region of physical interest, we discuss only regions with $z \geq 0$ here.

From (5.143) we see that the Schwarzschild radius $r = 2M$ corresponds to

$$z = \pm w \tag{5.145}$$

Both straight lines are plotted in Figure 5.2. We see in (5.144) that $z = w$ corresponds to $t = \infty$, whilst $z = -w$ corresponds to $t = -\infty$. The region between these two straight lines, to the right in the diagram, with $z > |w|$ and thus with $r > 2M$, is the region outside the Schwarzschild radius. In this region we consider an observer at rest with $r =$ constant. The curve $r =$ constant is also plotted in the diagram, its equation being given by (5.143). In addition, the curve $r = 0$ is plotted. Its equation

$$z^2 - w^2 = -1 \tag{5.146}$$

corresponds to the hyperbola shown in the diagram. Above and below there are two regions, the upper one corresponding to a black hole, the lower one to a white hole. The point $r = 0$ is a true metric singularity as seen in the isotropic and standard forms and also from equation (5.142) using the Kruskal–Szekeres coordinates. The names of black and white holes can be justified as follows.

Let us follow a light ray ($d\tau = 0$) which is now emitted ($t = 0$, hence $w = 0$) (at the $r =$ constant curve) radially inwards ($d\theta = d\varphi = 0$), shown as ray 1. From (5.142) we see that light rays are straight lines, either with slope 45°, or 135°. Rays with slope 135° are emitted inwards, as is our ray 1. We see that ray 1 traverses the Schwarzschild

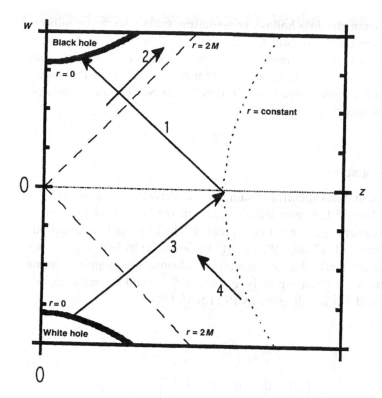

Figure 5.2 The black hole neighbourhood using the Kruskal–Szekeres coordinates.

surface at $t = \infty$, as viewed by an observer at the infinite (but in a finite proper time), and falls to the singularity at $r = 0$. However, a ray emitted at $r \leq 2M$ outwards (ray 2) can never reach us, that is, it can never reach the curve $r =$ constant. Therefore, we can send information through the Schwarzschild surface, but no information can come to us from inside. Light rays can traverse it in the inward direction. The curvature inside is so extreme that light cannot escape.

Exactly the opposite effect takes place at a Schwarzschild surface around a white hole. Any light ray emitted inside will eventually reach us (ray 3). However, no ray emitted by an external observer (ray 4) can reach the surface. White holes may be of less practical interest than black holes because ray 3 takes an infinite time to reach us from the Schwarzschild surface (measured by the far observer at rest), and thus if any white holes have been created after the Big-Bang, no signal is to be expected from them.

Black holes have almost certainly been created throughout the history of the Universe and, indeed, they are most probably the final stage of some massive stars. No signal from the Schwarzschild surface can reach us and any such signal would in any case be infinitely redshifted, so that black holes can be detected only indirectly, either through gravitational perturbation in the neighbourhood or through the energy released by accreting matter.

There is also spherical symmetry in a hollow, non-rotating sphere, so the conditions under which the Schwarzschild solution was found still prevail. In this case, however, there is no singularity at $r = 0$, and the mass of the hollow sphere must be zero; but when we consider $M = 0$ in the vacuum Schwarzschild metric we obtain a flat space. Thus in the same way that a hollow sphere is gravitation-free in Newtonian mechanics, it is a flat space in general relativity.

5.4.3   The metric inside a star

If the star is in equilibrium and has spherical symmetry, the derivation in Section 5.4.1 is still obtainable and (5.120)–(5.123) give Ricci's tensor in the interior of the star as well. But $T_{\mu\nu}$ is now non-vanishing and must be evaluated. The Einstein field equations will be used in the form of (2.86). We must therefore calculate the tensor $T_{\mu\nu} - 1/2g_{\mu\nu}T^{\lambda}{}_{\lambda}$. The calculation is shown systematically below, although only some initial and intermediate steps are given explicitly. $g_{\alpha\beta}$ and $g^{\alpha\beta}$ are still the same tensors as in (5.110) and (5.111) and $\Gamma^{\mu}_{\nu\lambda}$ is still given by (5.112)–(5.119).

Now,

$$U_0 = -\sqrt{B} \tag{5.147}$$

$$U_i = 0 \tag{5.148}$$

$$T_{\mu\nu} = g_{\mu\nu}p + (p + \epsilon)U_{\mu}U_{\nu} = \begin{bmatrix} \epsilon B & 0 & 0 & 0 \\ 0 & pA & 0 & 0 \\ 0 & 0 & pr^2 & 0 \\ 0 & 0 & 0 & pr^2 \sin^2\theta \end{bmatrix} \tag{5.149}$$

$$T^{\lambda}{}_{\lambda} = pg^{\lambda}{}_{\lambda} + (p + \epsilon)U^{\lambda}U_{\lambda} = 4p + (p + \epsilon)g^{\lambda\alpha}U_{\alpha}U_{\lambda}$$
$$= 4p + (p + \epsilon)\left(-\frac{1}{B}\right)(-\sqrt{B})(-\sqrt{B}) = 3p - \epsilon \tag{5.150}$$

where A and B are functions of r which must be determined. Then

$$R_{00} = \frac{1}{2}B(\epsilon + 3p) \tag{5.151}$$

$$R_{rr} = \frac{1}{2}A(\epsilon - p) \tag{5.152}$$

$$R_{\theta\theta} = \frac{1}{2}r^2(\epsilon - p) \tag{5.153}$$

The equation for $R_{\varphi\varphi}$ is not written because $R_{\varphi\varphi} = R_{\theta\theta}\sin^2\theta$ and the equation is not independent. Combining (5.151)–(5.153) and (5.120)–(5.123) we obtain

$$R_{00} = -\frac{1}{2A}\frac{d^2B}{dr^2} + \frac{1}{4A}\frac{dB}{dr}\left(\frac{1}{A}\frac{dA}{dr} + \frac{1}{B}\frac{dB}{dr}\right) - \frac{1}{Ar}\frac{dB}{dr} = -4\pi(\epsilon + 3p)B \tag{5.154}$$

$$R_{rr} = \frac{1}{2B}\frac{d^2B}{dr^2} - \frac{1}{4B}\frac{dB}{dr}\left(\frac{1}{A}\frac{dA}{dr} + \frac{1}{B}\frac{dB}{dr}\right) - \frac{1}{Ar}\frac{dA}{dr} = -4\pi(\epsilon - p)A \tag{5.155}$$

$$R_{\theta\theta} = -1 + \frac{r}{2A}\left(-\frac{1}{A}\frac{dA}{dr} + \frac{1}{B}\frac{dB}{dr}\right) + \frac{1}{A} = -4\pi(\epsilon - p)r^2 \qquad (5.156)$$

These are the three equations to be solved for the unknown radial functions A, B, p, ϵ. An equation of state $p(\epsilon)$ is thus also needed. Some help with the integration can be given. If we form the combination $R_{00}/2B + R_{rr}/2A + R_{\theta\theta}/r^2$ we obtain

$$-\frac{1}{rA^2}\frac{dA}{dr} - \frac{1}{r^2} + \frac{1}{Ar^2} = -8\pi\epsilon \qquad (5.157)$$

or

$$\frac{d}{dr}\left(\frac{r}{A}\right) = 1 - 8\pi r^2 \epsilon \qquad (5.158)$$

Just as we did for Newtonian stars, let us introduce the function $m(r)$:

$$m = \int_0^r 4\pi r^2 \epsilon \, dr \qquad (5.159)$$

which is slightly different here because $m(r)$ is now the total amount of mass–energy contained in a sphere of radius r. Then (5.158) can be integrated between 0 and r:

$$\frac{r}{A} = r - 2m \qquad (5.160)$$

obtained using the condition $A_0 = A(r = 0) \neq 0$. Then

$$A = \frac{1}{1 - \frac{2m}{r}} \qquad (5.161)$$

If we take m/r^3 as $(3/4\pi)$ times the mean energy density in a sphere of radius r, which we assume to be finite for $r \to 0$, then $A_0 = 1$. This conclusion must be accepted with caution, because some solutions for neutron stars give $\epsilon(r = 0) = \infty$. At the surface of the star, $m(r = R) = M$, the total mass–energy of the star. Then $A_s \equiv A(r = R)$ does coincide with the value of $A(r = R)$ in the external Schwarzschild solution. In this case we have

$$A_0 = 1 \qquad (5.162)$$

$$A_s = \frac{1}{1 - \frac{2M}{R}} \qquad (5.163)$$

Another simple equation that can be obtained from (5.154)–(5.156) can be derived directly from the fundamental equation of relativistic hydrostatics (2.103). In this case, only $\partial/\partial r$ is non-zero and assuming that $g_{00} = -B$ this equation gives

$$\frac{dp}{dr} + \frac{1}{2}(p + \epsilon)\frac{1}{B}\frac{dB}{dr} = 0 \qquad (5.164)$$

The equation for $R_{\theta\theta}$ (5.156) is also very simple and, together with (5.161), allows us to obtain B as a function of p and ϵ only:

$$\frac{1}{B}\frac{dB}{dr} = \frac{2A}{r^2}(m + 4\pi p r^3) \qquad (5.165)$$

which can be integrated between r and R:

$$B = B_s e^{-\int_r^R \frac{2A}{r^2}(m + 4\pi p r^3)dr} \qquad (5.166)$$

where $B_S = B(r = R)$ is calculated via

$$B_S = A_S^{-1} = 1 - \frac{2M}{R} \tag{5.167}$$

because the external and the internal metrics must coincide at the surface. Applying (5.165) at the surface where $m = M$ and $p = 0$, it is clear that

$$\left(\frac{dB}{dr}\right)_S = \frac{2M}{R^2} \tag{5.168}$$

Another interesting combination that can be obtained from (5.154)–(5.156) is $3R_{rr}B + R_{tt}A$, which yields B as a function of ϵ, but not of p:

$$\frac{d^2B}{dr^2} - \frac{1}{2}\frac{dB}{dr}\left(\frac{1}{A}\frac{dA}{dr} + \frac{1}{B}\frac{dB}{dr}\right) - \frac{3B\,dA}{rA\,dr} - \frac{1}{r}\frac{dB}{dr} = -16\pi AB\epsilon \tag{5.169}$$

which can be rewritten as

$$\frac{d}{dr}\left(\frac{1}{2r}\frac{1}{\sqrt{AB}}\frac{dB}{dr}\right) = \sqrt{AB}\frac{d}{dr}\left(\frac{m}{r^3}\right) \tag{5.170}$$

and which will be used later. Since the formulae (5.169) and (5.170) are equivalent we can see that $\sqrt{AB}$ on both sides of (5.170) must have the same sign, so we may adopt $\sqrt{AB} > 0$.

In general, numerical methods are required to solve these equations. One of the most important cases is that of a relativistic neutron star, where the curvature is very high and where an equation of state as simple as $\epsilon = 3p$ can be adopted.

5.4.4   Gravitational mass limit

Some conclusions can be obtained from the preceding equations without knowledge of the equation of state, and these are therefore general for all types of star. It is difficult to think of a Schwarzschild surface permanently inside a star. Let us consider a particle of the fluid just outside the surface. Due to its own thermal motion, or to any other cause, it could traverse this surface, but then it could not return. The external part of the star would be swallowed by the internal part in a finite time. Such a star would be unstable, so it may be assumed that a Schwarzschild surface in the internal Schwarzschild metric cannot exist in practice. Almost certainly, such surfaces do exist around the centres of galaxies and thus inside galaxies. But these are characterized by accretion and transient processes. Therefore, let us exclude $B = 0$, $A = \infty$ at some radius less than R.

A surface with $A = 0$ would also have exotic properties, and this possibility, too, may be excluded in reasonable scenarios. It will be further assumed that $d\epsilon/dr \leq 0$ because, otherwise, the star would be unstable. This does not mean a loss of practical generality because no actual star is suspected to have such an anomalous energy density distribution.

Let us consider equation (5.170) again. On the right-hand side m/r^3 is proportional to the mean energy density of a sphere of radius r. If $d\epsilon/dr \leq 0$, then also $d/dr(m/r^3) \leq 0$. This is easy to demonstrate:

$$\frac{d}{dr}\left(\frac{m}{r^3}\right) = \frac{1}{r^3}\frac{dm}{dr} - \frac{3m}{r^4} = \frac{4\pi\epsilon}{r} - \frac{3m/r^3}{r} = \frac{4\pi}{r}(\epsilon - \langle\epsilon\rangle)$$ (5.171)

where $\langle\epsilon\rangle$ is the mean density of a sphere of radius r. If $d\epsilon/dr < 0$, any value of ϵ for all radii less than r is higher than $\epsilon(r)$. Thus its mean value is also higher $\langle\epsilon\rangle \gg \epsilon$, and therefore $d/dr(m/r^3) \le 0$. We stated above that $\sqrt{AB}$ can be set as positive. Then the right-hand side of (5.170) is negative and thus

$$\frac{d}{dr}\left(\frac{1}{2r}\frac{1}{\sqrt{AB}}\frac{dB}{dr}\right) \le 0$$ (5.172)

Let us obtain a first integration of this inequality from r to R, taking (5.167) and (5.168) into account. Then

$$\frac{dB}{dr} \ge r\sqrt{AB}\frac{2M}{R^3}$$ (5.173)

In order to integrate once more it is convenient to rewrite

$$\frac{d}{dr}\sqrt{B} \ge r\sqrt{A}\frac{M}{R^3}$$ (5.174)

where $\sqrt{B}$ and $\sqrt{A}$ can both be taken as positive.

Let us integrate between the centre $r = 0$ and the surface $r = R$:

$$\sqrt{B_S} - \sqrt{B_0} \ge \frac{M}{R^3}\int_0^R \frac{r\,dr}{\sqrt{1 - \frac{2m}{r}}}$$ (5.175)

Therefore, there is an upper limit for B_0 given by

$$\sqrt{B_0} \le \sqrt{B_S} - \frac{M}{R^3}\int_0^R \frac{r\,dr}{\sqrt{1 - \frac{2m}{r}}}$$ (5.176)

The maximum value of $\sqrt{B_0}$ is obtained when both sides of (5.176) are equal, that is, for $m/r^3 = $ constant as deduced from (5.170). In this case, all the signs $\ge$ or $\le$ in the preceding discussion must be replaced by the sign $=$. Let us call $m_0(r)$ the function $m(r)$ for such a homogeneous energy density, so that $m_0 = (4/3)\pi\epsilon r^3$. Then:

$$\sqrt{B_0} \le \sqrt{B_S} - \frac{M}{R^3}\int_0^R \frac{r\,dr}{\sqrt{1 - \frac{2m_0}{r}}}$$ (5.177)

The integral is easily solved:

$$\sqrt{B_0} \le \sqrt{B_S} + \frac{1}{2}(\sqrt{B_S} - 1)$$ (5.178)

But $\sqrt{B_0}$ cannot be negative, so

$$\sqrt{B_S} + \frac{1}{2}(\sqrt{B_S} - 1) \ge 0$$ (5.179)

Therefore

$$B_S \ge \frac{1}{9}$$ (5.180)

and

$$\frac{M}{R} \le \frac{4}{9} \tag{5.181}$$

If the star (or any other body) is very massive it cannot be very small. If its radius becomes less than $\frac{9}{4}M$ the star becomes gravitationally unstable and collapses. If a massive star consumes its fuel and contracts further and further, its surface will never become smaller than its Schwarzschild surface if it passes through a series of equilibrium states, because before contracting to $R = 2M$ at $R = \frac{9}{4}M$ (or before) it will fall into the black hole state. In the argument used above, the radii are small, so we are probably speaking about some kind of neutron star, but our argument is general, and any body, whatever its equation of state, has this gravitational limit to the mass/radius ratio.

5.5   Stellar evolution

In preceding sections, main-sequence stars, white dwarfs, and black holes were considered in some detail. A descriptive view of stellar evolution connecting these basic types of star will now be outlined. The birth of a star is the isolation of a single fragment of the gravitational collapse of a gas cloud. How this collapse and fragmentation of the original cloud proceeds is a matter to be analysed in Chapter 7, where the interstellar medium will be studied. Adiabatic contraction goes on in the new-born star accompanied by a steady temperature rise.

If the mass of a star is lower than about $0.25 - 0.08$ $M_\odot$, contraction is prevented by the Fermi pressure of the degenerate electrons. Brown dwarfs have long lifetimes, emitting low quantities of radiative energy until they become black dwarfs. If the star has a mass in the range $0.25 - 70$ $M_\odot$, it becomes a main-sequence star. The lower limit was justified in Section 5.2.2. The smallest main-sequence star is also the smallest white dwarf and the largest brown dwarf. The upper limit can be justified as follows.

In Chapter 3 (Section 3.6, equation (3.172)) we found that under certain conditions a radiative flux can provide an outward force that is even higher than the inward gravitational force. In this case the outer layer is unstable and will be ejected. The radiative outward force was $(4\pi/c)H\kappa$, where H was the Eddington energy flux and κ the Rosseland opacity. H can be written in terms of the luminosity and radius of the star. When this force equals gravity, we have

$$\frac{GM}{R^2} = \frac{\kappa}{c} \frac{L}{4\pi R^2} \tag{5.182}$$

L is called the Eddington luminosity. By adopting a mass–luminosity relation of the type $L \propto M^a$ we obtain

$$\left(\frac{M}{M_\odot}\right)^{a-1} = 4\pi \frac{GM_\odot}{L_\odot} \frac{c}{\kappa} \approx 10^4 \tag{5.183}$$

where we have adopted a value for the opacity of order unity (in c.g.s. units). For a constant opacity, $a \approx 3$, and the limiting mass of a main-sequence star becomes $\mathcal{M} \leq 100 \, M_\odot$. This limit may in practice be slightly lowered. These stars lose a large quantity of material. For instance, a 30 $M_\odot$ star may eject $10^{-6} M_\odot$ year^{-1}, and this fact was not considered in our model of main-sequence stars. For very massive stars, the condition of equilibrium is no longer valid. Because of this large ejection of mass from the envelope, the hot inner layers become observable; this type of star is called a Wolf–Rayet star. A $\sim 60 \, M_\odot$ star is subject to oscillational instability. The most luminous stars observed have $L \sim 10^6 L_\odot$.

A main-sequence star would maintain a constant luminosity throughout its life if its radius were reduced by a small amount. In practice these stars increase their luminosity slightly and their radius increases slightly. This produces a short displacement towards the upper-right region in the HR-diagram (to higher luminosities and higher effective temperatures). A star such as the Sun has about 36% hydrogen and 62% helium in its core. In the outermost layers it preserves the original composition of about 73% hydrogen and 25% helium. Radiative flux dominates the energy transport to the surface up to 0.85 $R_\odot$, where the convective energy flux becomes more effective. However, a 9 $M_\odot$ zero-age, main-sequence star has a convective core up to 0.22 times the stellar radius, and is radiation dominated in the external layers. When hydrogen is exhausted in the core the star leaves the main sequence.

A complicated evolution then follows. For $\mathcal{M} < 0.5 \, M_\odot$, the onset of He combustion never arises and the star becomes a white dwarf after about 10^{12} years. If the mass lies in the range 0.5–10 $M_\odot$, fusion of helium to produce carbon and oxygen follows the main-sequence stage. The mass of the star is sufficiently large to produce a further collapse once hydrogen is consumed to reach the temperature needed to ignite helium fusion. For masses higher than about 2.2 $M_\odot$ such stars are able to ignite more massive nuclei (especially for $\mathcal{M} > 8 \, M_\odot$), and their cores at the end of the nuclear reaction phase contain a large concentration of iron. Higher elements cannot be produced by exothermic reactions.

During the burning of helium and more massive nuclei, the star becomes a giant, a red giant for moderate masses such as that of the Sun, or a supergiant for very large masses. During its red giant phase the Sun will become 50 times larger than its present size. A supergiant has a radius of the order of $10^3 R_\odot$. During this phase, and due to mechanisms not yet perfectly understood, mass loss is important. For $\mathcal{M} \leq 10 \, M_\odot$, a planetary nebula is formed following ejection rates of the order of $10^{-6} M_\odot$/year. In the centre of the planetary nebula a hot ($\sim 3 \times 10^4$ K effective temperature) star remains, becoming a white dwarf. Then a long cooling period follows, terminating in a black dwarf. If the mass loss is not efficient enough to end up with a mass lower than the Chandrasekhar mass, the star cannot become a white dwarf, but follows the evolutionary track of more massive stars, exploding as a supernova.

The explosion mechanism is not understood in all its details. At a temperature of about 5×10^9 K, photons are able to disintegrate nuclei and produce neutrons and protons in endothermic processes. This produces a further collapse and a further increase in temperature. At sufficiently high temperatures electrons combine with protons to produce neutrons. The core of the massive star becomes a neutron star. The outer layers collapse down, the gas becomes overheated, and the reaction rates of nuclear reactions abruptly increase for the still unburned outer layers, which are ejected at velocities of the order of 0.01 c. Magnetic fields are probably involved in the ejection mechanism. Energies of the order of 10^{51} erg or more are characteristic of supernova explosions. The photometric magnitude may decrease by more than 18 magnitudes. A large number of neutrinos, $\sim 10^{58}$, with an energy spectrum corresponding to a temperature of $\sim 10^{10}$ K are also emitted and escape. During neutronization of the core, some neutrons, captured by nuclei in the envelope, give rise to nuclei heavier than iron.

Neutron stars or pulsars are stars that are made up of degenerate neutrons. When the electron star has a mass higher than the Chandrasekhar limit, and the formation of a planetary nebula is not a sufficiently efficient mass loss mechanism, it contracts further to reach a density comparable to nuclear densities. Electrons merge with protons to form neutrons. Neutrons are also fermions and also become degenerate. The previous analysis of an electron star cannot be used to study neutron stars, since their space-time curvature is now very high and the internal structure becomes complex. Nevertheless, it provides some basic orders of magnitude. Equation (5.71) gave us the maximum concentration of electrons required to reach a relativistic system. Now n_e must be replaced by n_n, the neutron concentration, and m_e by m_n, the neutron mass. Therefore,

$$\frac{n_n}{n_e} = \left(\frac{m_n}{m_e}\right)^3 \approx 6 \times 10^9 \tag{5.184}$$

Therefore, the neutron density must be of the order of 10^{39}cm^{-3}. The characteristic distance between neutrons is about $10^{-13}\text{cm} = 1$ fermi. The densities are also related by this rate and thus the neutron star has a density of about 10^{15}g cm^{-3}. The constant C_R in (5.87) is now of the same order of magnitude, and therefore the limiting mass must now retain approximately the same value. A more detailed analysis gives a limiting value of 3 $M_\odot$, called the Landau–Oppenheimer–Volkoff limit. As the mass of neutron and electron stars is similar the characteristic radius of a neutron star is easily calculated to be about 20 km.

When the limiting value of the mass/radius ratio is surpassed, the star becomes a black hole.

6 The fluid of stars

6.1   The fluid of stars

In contrast with the fluid in a star, defined as the fluid of which stars are constituted, we will now deal with a fluid of stars, defined as a fluid whose microscopic particles are stars. Our first example will be a galaxy. It is true that a galaxy is not composed solely of stars. It also possesses gas and dust in different amounts, and gas influences the dynamics of the star system decisively. However, as an introduction to galactic dynamics we will first study the dynamics of the stellar system, then that of the inter-stellar gas, and will finally try to combine the two. In this chapter, those terms in the hydrodynamical equations that are not influenced by the presence of gas will be considered.

Galaxies are not the only examples of fluids of stars. Globular clusters and even open clusters can be systems with these characteristics. Clusters of galaxies can also be analysed using the same theoretical approach, with the stars being replaced by galaxies, that is, assuming a fluid of galaxies. The galactic scenario is assumed in general here, but other types of stellar system will be considered. The number of stars in our Galaxy is of the order of 10^{11}, a sufficiently large number to justify the use of hydrodynamics.

A great simplification in the hydrodynamical equations consists of azimuthal or axial symmetry, which is appropriate for the majority of galaxies. Unfortunately, irregular galaxies ($\sim 3\%$) and peculiar galaxies ($\sim 1\%$) would have to be excluded in a first analysis. In any case, the gas content of irregulars is an important fraction of the total mass ($\sim 30\%$) and a pure fluid of stars does not represent these galaxies properly. At first glance, spirals do not present axisymmetry. However, spiral arms are more of a luminous phenomenon than a non-symmetric mass distribution, and the condition of axisymmetry in spiral galaxies is better met than in any other type of galaxy. Also, many spirals have no well-defined spiral arms, but, rather, have spiral 'hair' (flocculent galaxies). This distinction is important as most non-barred and isolated or field spirals are flocculent. Indeed, the optical appearance of flocculent galaxies suggests a large degree of axisymmetry in the mass. Though barred spirals are not perfectly axisym-metric, and ellipticals and bulges of spirals are often triaxial, they have rather a large degree of axisymmetry. Therefore, the assumed condition of axisymmetry is reasonable for most ellipticals ($\sim 13\%$), lenticulars ($\sim 22\%$), and barred and normal spirals ($\sim 61\%$). A plane of symmetry, called the galactic plane, perpendicular to the axis of symmetry will also be assumed. These assumptions provide a first, rather realistic introduction.

To give some parameters that characterize the distribution of stars within a galaxy, let us remember some parameters defining the morphology of the Milky Way. There is a central bulge, that is, a spheroidal component for which the intensity I_B decreases with galactocentric radius r, following a law that is characteristic of ellipticals:

$$I_B = I_{B_0} e^{-(r/B)^{1/4}} \tag{6.1}$$

where I_{B_0} is $I_B(r = 0)$. The constant B is about 5 kpc in a radial direction within the galactic plane, and about 3 kpc in the vertical (normal to the galactic plane) direction.

There is also a flat disc ($\sim$ 300 pc thick) that is much bluer than the bulge, the luminosity I_D, a function of r, being presumed to be exponential (by analogy with observed external galaxies):

$$I_D = I_{D_0} e^{-r/D} \tag{6.2}$$

where $I_{D_0} = I_D(r = 0)$ and D is a scalelength of about 4 kpc. I_{D_0} is noticeably independent of the galaxy type (equivalent to a magnitude/arcsec^2 of 21.65 in blue). Deviations from the exponential law of (6.2) are important. It is very common to find higher values with respect to the filled exponential for intermediate values of r, sometimes constituting a true ring, coinciding with a gaseous ring. The flatness, colour, and population of the ring are not very different from those in the disc. In the vertical direction a gaussian law or a $(\cosh^{-1} z)^2$ function fits the observations, with z being the distance from the galactic plane. The disc is a component of all spirals and lenticular galaxies (the difference is that lenticulars are gas deplete).

The existence of a spherically symmetric dark halo with an 'isothermal' density distribution ($\rho = \rho_0(1 + (r/R_c)^2)^{-1}$) is usually assumed, where ρ_0 and R_c are constants. The existence and cosmological implications of a massive dark halo will be considered in Chapter 10.

At optical wavelengths the emission is mainly stellar, with a small and easily identifiable contribution from HII regions. As we are now interested in stellar distribution and dynamics, our attention is focussed on optical photometry.

6.2   Observations in our Galaxy

The observational determination of n, the number of stars per unit volume, and $\vec{v}_0$, the macroscopic velocity, is relatively easy in other galaxies, where we mainly observe the macrostate. In our own Galaxy, we observe peculiar velocities of stars in detail, as we are able to observe the microstate in the solar neighbourhood; but macroscopic quantities such as n and $\vec{v}_0$ are more difficult to obtain.

6.2.1 Density

In directions close to the galactic plane the density of stars can be obtained for a region of only a few kpc around the Sun. Absorption by dust prevents observations at larger distances. In other directions the absorption is much less and distant galaxies may be observed. Suppose that a star with luminosity L is observed with a flux q at the Earth. The relation between L, q, and the distance r to the star will be

$$q = \frac{L}{4\pi r^2} e^{-\int_0^r \rho\kappa \, dr} \tag{6.3}$$

where ρ is the density of the absorbing gas and κ is the opacity coefficient. Suppose now that radio observations permit a good estimate of $\rho(r)$ for every line-of-sight, so that the opacity is not a problem, and the relation (q, L, r) is well known for any observable star. Equation (6.3) was obtained in Chapter 3.

It must be pointed out that absorption is not produced by gas, but, rather, by dust. Observations indicate that the relative proportion of gas and dust varies little throughout the Galaxy. Then κ, scaled by the gas density, should be understood as an equivalent opacity. (See also Chapter 3.)

Let $\phi(L)$ be the distribution function for L, that is, $\phi(L)dL$ gives the fraction of stars with luminosity between L and $L + dL$. We then assume that $\phi(L)$ is normalized, so that $\int_0^\infty \phi(L)dL = 1$. This distribution function is observed in the solar neighbourhood, and depends in principle on the region considered within the Galaxy. However, ϕ is basically determined by the processes of star formation and the time elapsed since then. These processes may not be too different from region to region; so let us assume as first approximation that ϕ does not depend on the vector position from the Sun.

When we observe in a given direction with a unit solid angle, we may obtain the distribution function $A(q)$, defined so that $A(q)dq$ gives the number of observed stars with a flux in the range q and $q + dq$ in a unit solid angle. A volume element at distance r limited by the unit solid angle would be $r^2 dr$. Therefore,

$$A(q) = \int_0^\infty n(r)\phi(L)r^2 dr \tag{6.4}$$

Note that $A(q)$ is a distribution function that is not normalized. Now L may be written as a function of q and r, using (6.3) and therefore

$$A(q) = \int_0^\infty n(r)\phi(4\pi r^2 q e^{\int_0^r \rho\kappa \, dr})r^2 dr \tag{6.5}$$

Since $A(q)$ is determined by observations, and ϕ is a known function, equation (6.5) may provide $n(r)$ in any direction. Numerical procedures are required to solve this star count equation.

Suppose, now, that we are not dealing with stars in our Galaxy but with galaxies in the Universe. This calculation has little relation to the problem under study, but it uses the same equations and is an interesting extension. Suppose that $n(r)$ is now constant in the Universe and that no significant opacity affects our observations. Then (6.4) may be used, but let us now write r as a function of L and q, with the help of (6.3) (where $\kappa = 0$):

$$A(q) = n\frac{(4\pi q)^{-3/2}}{2}\int_0^\infty \phi(L)L^{1/2}dL \tag{6.6}$$

The integral yields a non-vanishing constant value, and the equation says that under the assumed hypotheses $A = kq^{-3/2}$ (k being a constant). We now ask about the total flux observed in a given direction, in a unit solid angle, and due to all galaxies. This will be

$$Q = \int_0^\infty qA(q)dq = \int_0^\infty kq^{-1/2}dq = \left[2kq^{1/2}\right]_0^\infty = \infty \tag{6.7}$$

This surprising result is Olber's paradox, which has important implications for cosmology. If extinction were not neglected, the paradox would persist, because the intergalactic absorbing medium would be heated and would eventually re-emit the absorbed energy. The simplest way to avoid Olber's paradox is to assume that the Universe is finite in space and/or time.

6.2.2 Peculiar velocities

It is possible to deduce the peculiar velocity of each star in the solar neighbourhood, as well as the peculiar velocity of the Sun itself. Following the nomenclature proposed in Chapter 1, let us denote by $\vec{v}$ the velocity of a given star as measured by an observer at rest who is placed for instance in the centre of the galaxy and by $\vec{v}_0$ the macroscopic mean velocity which may be a function of $\vec{r}$, the position vector, and the time t. Calculating the average of $\vec{v}$ in a given volume element, the mean velocity is obtained:

$$\langle\vec{v}\rangle = \vec{v}_0 \tag{6.8}$$

This is the same definition as the definition given in Chapter 1 (1.5), because for simplicity we now assume that all stars have the same mass. The peculiar velocity $\vec{V}$ of a star would be defined by

$$\vec{v} = \vec{v}_0 + \vec{V} \tag{6.9}$$

Therefore

$$\langle\vec{V}\rangle = 0 \tag{6.10}$$

as in (1.7). As a particular case (6.9) will be expressed for the Sun:

$$\vec{v}_\odot = \vec{v}_{0\odot} + \vec{V}_\odot \tag{6.11}$$

where $\vec{v}_\odot$ is the velocity of the Sun with respect to the observer at the centre of the Galaxy, $\vec{v}_{0\odot}$ is the mean velocity of the volume element to which the Sun belongs, the so-called LSR (local standard of rest), and $\vec{V}_\odot$ is the peculiar velocity of the Sun. One of the stars contained in this volume element associated with the solar neighbourhood would have a velocity $\vec{v}_*$ with respect to the Sun:

$$\vec{v}_* = \vec{v} - \vec{v}_\odot = \vec{V} - \vec{V}_\odot \tag{6.12}$$

Taking averages:

$$\langle \vec{v}_\star \rangle = \langle \vec{V} \rangle - \vec{V}_\odot = -\vec{V}_\odot \tag{6.13}$$

$\vec{v}_\star$ is an observable quantity from doppler shifts and proper motions. Therefore, $\langle \vec{v}_\star \rangle$ is obtainable from the observations, and therefore (6.13) enables us to determine the Sun's peculiar velocity $\vec{V}_\odot$. Then (6.12) enables us to obtain the peculiar velocity $\vec{V}$ of any star in the solar volume element.

We thus know that, with respect to the LSR, the Sun moves towards a point with galactic coordinates $l = 51°$, $b = 23°$ with a speed of the order of 15 km s^{-1}. This is the solar peculiar velocity of components: $V_{\odot r} = -9$ km s^{-1} (radial direction), $V_{\odot \varphi} = 11$ km s^{-1} (azimuthal direction), and $V_{\odot z} = 6$ km s^{-1} (vertical direction, that is, perpendicular to the plane of symmetry). This means that the Sun moves towards the galactic centre, has a rotation velocity about it that is greater than the surrounding stars, and has a component towards the galactic north pole.

The peculiar velocities of other nearby stars do not correspond at all to a Maxwellian distribution. Whilst the observed distribution in a given direction is more or less gaussian, the velocity half-width depends closely on the direction. The distribution function is clearly anisotropic. As a result, the pressure tensor $\mathcal{P} = mn\langle \vec{V}\vec{V} \rangle$ does not have the simple equilibrium form $p\delta$, but, rather, once diagonalized, it takes the form:

$$\mathcal{P} = \begin{pmatrix} \mathcal{P}_{rr} & 0 & 0 \\ 0 & \mathcal{P}_{\varphi\varphi} & 0 \\ 0 & 0 & \mathcal{P}_{zz} \end{pmatrix} \tag{6.14}$$

The coordinate system in which this tensor is diagonal is close to the one suggested by intuition, (r, φ, z) (radial, azimuthal, vertical), considering the symmetry of most galaxies. No large deviations with respect to this coordinate system have been found. For example, the deviation of the proper motion axis close to the radial direction is called the vertex deviation and is less than $\sim 10°$ (though it is much higher for O and B stars). The vertex deviation must be attributed mainly to the presence of asymmetries on a local scale peculiar to the presence of spiral arms. In a first analysis it may be ignored, and we may assume that the (r, φ, z) coordinate system is the proper system of the pressure tensor. Of the three components, $\langle V_r^2 \rangle^{1/2}$ (of the order of peculiar velocities in the radial direction) is the largest, in the range 25–30 km s^{-1}; $\langle V_\varphi^2 \rangle^{1/2} \approx 15-20$ km s^{-1}. The value $\langle V_z^2 \rangle^{1/2}$ is the smallest, but not very different from $\langle V_\varphi^2 \rangle^{1/2}$ once high velocity stars are eliminated from the sample. These high velocity stars have $\langle V_z^2 \rangle^{1/2}$ of the order of 80 km s^{-1}, they belong to the halo, and need not be considered here as they do not belong to the disc. Their spectral characteristics are very different since they are very old stars with very low metallicities.

The anisotropy of peculiar velocities, clearly observed in our own neighbourhood, is a common characteristic of all galaxies. A velocity distribution which consists of a gaussian for every direction, but with the half-width being different for different directions, is called a Schwarzschild distribution or Schwarzschild ellipsoid.

In this velocity distribution, the velocity half-width in a given direction may be obtained as a function of the standard deviations for the proper directions of the pressure tensor. The relation can be obtained from the condition that the pressure tensor is a true tensor:

$$\mathcal{P}_{ij} = a_{ik} a_{jm} \mathcal{P}'_{km} \tag{6.15}$$

where the a_{ik} are the components of the orthogonal matrix of the axis rotation, from the (r, θ, z) coordinate system to another one. $\mathcal{P}'_{km}$ is diagonal as in (6.14). Suppose we need to calculate $\langle V_1^2 \rangle$ in a given direction which is chosen as OX_1 in the new coordinate system. From (6.15) we obtain

$$\mathcal{P}_1 = mn \langle V_1^2 \rangle = a_{11}^2 \mathcal{P}'_{rr} + a_{12}^2 \mathcal{P}'_{\varphi\varphi} + a_{13}^2 \mathcal{P}'_{zz} \tag{6.16}$$

where the (a_{11}, a_{12}, a_{13}) are the components of the unitary vector in the direction considered, with respect to the coordinate system with directions [radial–azimuthal–vertical].

The fluid of stars may be considered as a perfect gas, with equation of state

$$p = nkT \tag{6.17}$$

where the hydrostatic pressure is

$$p = \frac{1}{3}(\mathcal{P}_{rr} + \mathcal{P}_{\varphi\varphi} + \mathcal{P}_{zz}) = \frac{1}{3} mn \langle V_r^2 + V_\varphi^2 + V_z^2 \rangle = \frac{1}{3} mn \langle V^2 \rangle \tag{6.18}$$

where m is the mass of a star. Then

$$\frac{3}{2} kT = \frac{1}{2} m \langle V^2 \rangle \tag{6.19}$$

T would then be the temperature of the star fluid. It is of the order of 10^{61} K and is therefore not comparable with the temperature of more conventional thermodynamic systems, in particular with the temperature of an individual star.

6.2.3   Mean velocities

To obtain $\vec{v}_\odot$, or the macroscopic velocity $\vec{v}_{0\odot}$ at the Sun's position by using (6.11), we observe the velocities of globular clusters which have a spherical distribution about the galactic centre or of nearby galaxies. We may then measure the rotation around the galactic centre, with $\theta \approx 230$ km s^{-1}, since expansion or vertical motions are small. (In the convential nomenclature II, θ, and Z are the components of $\vec{v}_0$ in the radial, azimuthal, and vertical directions.)

Radio observations provide much more complete information about the rotation curve $[\theta, r]$ of the gas, which would be identical to the rotation curve of the star system under reasonable dynamical assumptions. At 21 cm, for example, the interstellar extinction is very small. The problem is not (as at optical wavelengths) that the farthest regions are unobservable, but, rather, that we observe all regions simultaneously. An early technique surmounts this difficulty for galactocentric radii $r < r_\odot$. From the observed profile of the 21 cm line in a given direction, given the assumption that galactic motion is pure rotation, the highest redshift corresponds to point P in Figure 6.1, for which the line-of-sight is tangent to the circle of radius r. We then determine $\theta(r)$. By observing along different lines of sight in the galactic plane, we

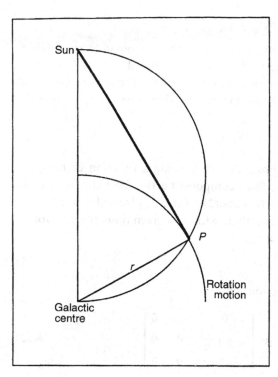

Figure 6.1 Estimation of the galactic rotation curve by radio observations.

obtain other points on the curve $\theta(r)$. The different points P form a semicircle, as is shown in Figure 6.1. At the very centre, $\theta(r = 0) = 0$, then θ increases until reaching a maximum of 250 km s^{-1} at about $r = 1$ kpc; it also has a minimum of 200 km s^{-1} for $r = 3$ kpc and it increases up to 250 km s^{-1} at about the solar galactocentric radius, beyond which $\theta(r)$ decreases slowly. For very large radii, the rotation curve should become Keplerian, $\theta \propto r^{-1/2}$, but these regions are difficult to observe and no evidence of a Keplerian decrease has been found, nor in fact any rapid decrease at all. A simple calculation allows us to find the total mass of the Galaxy from $\theta(r)$, because if the gravitational force were balanced by the centrifugal force, the mass $M(r)$ contained inside r would be $\theta^2 r/G$. For large r, $M(r)$ would provide the galactic mass. The absence of a decrease in the function $\theta(r)$ for large radii would indicate a very large mass of the Galaxy associated with the problem of dark matter in the Universe as a whole. The absence of a decrease of $\theta(r)$ at large radii is common in many other spirals, and is taken as one of the most persuasive pieces of evidence for the existence of dark matter. Other interpretations, however, cannot be ruled out. The simple calculation will be replaced in brief by a more precise one, as stellar dynamics is now our objective, but the existence of dark matter within galaxies is widely accepted.

6.2.4   The velocity gradient

By optical observations of the stars in our neighbourhood, it is possible to obtain a value for the macroscopic velocity gradient. This method, carried out by Oort, permitted the early determination of the direction in which the galactic centre is found, and introduced some functions which are widely used in galactic dynamics. Suppose that the velocity is pure rotation:

$$\vec{v}_0 = \vec{\omega} \times \vec{r} \tag{6.20}$$

where $\vec{\omega}$ is of the type $(0, 0, -\omega(r))$; the possibility of ω being a function of r implies differential rotation. The minus sign in the third component arises from the fact that in Figure 6.2 OZ is directed outwards from the paper, whilst $\vec{\omega}$ is directed inwards. The coordinates x and y are chosen in the galactic plane with, at a given time, the position of the solar neighbourhood being $(x = 0; y = r_\odot)$. Then

$$\vec{v}_0 = \vec{\omega} \times \vec{r} = [y\omega, -x\omega, 0] \tag{6.21}$$

without a component in the vertical direction. Now

$$\nabla \vec{v}_0 = \begin{bmatrix} y\frac{\partial\omega}{\partial x} & -\omega - x\frac{\partial\omega}{\partial x} & 0 \\ \omega + y\frac{\partial\omega}{\partial y} & -x\frac{\partial\omega}{\partial y} & 0 \\ y\frac{\partial\omega}{\partial z} & x\frac{\partial\omega}{\partial z} & 0 \end{bmatrix} \approx \begin{bmatrix} 0 & -\omega & 0 \\ \omega + y\frac{\partial\omega}{\partial y} & 0 & 0 \\ 0 & 0 & 0 \end{bmatrix} \tag{6.22}$$

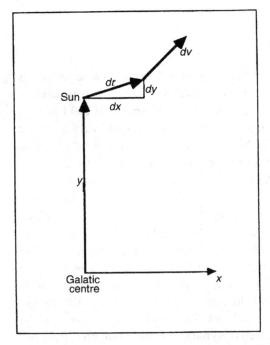

Figure 6.2   Differential rotation in the solar neighbourhood.

because $x = 0$, and $\partial\omega/\partial z$ is negligible. Also, $\partial\omega/\partial x$ has been ignored due to the large distance between the Sun and the galactic centre.

With respect to the centre of the volume element (i.e., with respect to the Sun, once its peculiar velocity has been corrected), there will be a non-vanishing small macroscopic velocity, calculated by

$$d\vec{v}_0 = d\vec{r} \cdot \nabla\vec{v}_0 = (dx, dy, dz) \begin{bmatrix} 0 & -\omega & 0 \\ \omega + y\frac{\partial\omega}{\partial y} & 0 & 0 \\ 0 & 0 & 0 \end{bmatrix}$$

$$= \left(dy(\omega + y\frac{d\omega}{dy}), -\omega dx, 0 \right)$$

(6.23)

Or by

$$\theta = \omega y$$

(6.24)

$$\frac{d\theta}{dy} = \omega + y\frac{d\omega}{dy}$$

(6.25)

In (6.23),

$$d\vec{v}_0 = \left(\frac{d\theta}{dy}dy, -\frac{\theta}{y}dx, 0 \right)$$

(6.26)

Now, $d\vec{r}$ is

$$dx = dr\cos(l - 90) = dr\sin l$$

(6.27)

$$dy = dr\sin(l - 90) = -dr\cos l$$

(6.28)

where dr is the distance of a star in the galactic plane, with a galactic longitude l, from the Sun. Therefore,

$$d\vec{v}_0 = \left(-\frac{d\theta}{dr}\cos l, -\frac{\theta}{r}\sin l, 0 \right)dr$$

(6.29)

where y has been replaced by r.

The component of $d\vec{v}_0$ along the line of sight dv_R is observable by means of the doppler effect:

$$dv_R = \frac{d\vec{v}_0 \cdot d\vec{r}}{dr} = dr\left(-\frac{d\theta}{dr} + \frac{\theta}{r} \right)\sin l\cos l$$

$$= \frac{1}{2}dr\left(-\frac{d\theta}{dr} + \frac{\theta}{r} \right)\sin 2l = Adr\sin 2l$$

(6.30)

where A is called A-Oort's constant

$$A = \frac{1}{2}\left(-\frac{d\theta}{dr} + \frac{\theta}{r} \right)$$

(6.31)

A is a function of r. Oort's A constant is the value of this function at the solar galacto-centric radius. A-Oort's constant can be determined by fitting the curve $\left[\frac{dv_R}{dr}, l\right]$. Also this curve must present two maxima for $l = 45°$ and $l = 225°$, which would permit the determination of the centre–Sun direction if it were unknown.

Additionally, for some of the nearest stars, the 'proper' motions (i.e. those in a direction perpendicular to the line of sight in the galactic plane) can be observed. This tangential velocity is

$$
dv_T = \sqrt{dv_0^2 - dv_R^2} = \pm dr \left(\frac{d\theta}{dr} \cos^2 l + \frac{\theta}{r} \sin^2 l \right)
$$

$$
= \pm dr \left(\left(\frac{d\theta}{dr} - \frac{\theta}{r} \right) \cos^2 l + \frac{\theta}{r} \right)
\tag{6.32}
$$

Taking into account

$$
\cos^2 l = \frac{1}{2}(1 + \cos 2l)
\tag{6.33}
$$

$$
dv_T = \mp (B + A \cos 2l) dr
\tag{6.34}
$$

where B is called B-Oort's constant

$$
B = -\frac{1}{2} \left(\frac{d\theta}{dr} + \frac{\theta}{r} \right)
\tag{6.35}
$$

B is a function of r. Oort's B constant is the value of this function at the solar galactocentric radius. Equation (6.34) can also provide information about the direction of the galactic centre. (This is in fact well determined from radio observations.) Both constants can be determined after fitting the curve $[dv_T/dr, l]$. It should also be noted that proper motion observations give dv_T/dr directly, in which case a knowledge of dr is not required. The determination of the constants A and B is important as

$$
A - B = \frac{\theta}{r}
\tag{6.36}
$$

$$
A + B = -\frac{d\theta}{dr}
\tag{6.37}
$$

Equation (6.37) provides the differential rotation, or velocity gradient, in our neighbourhood. Equation (6.36) gives θ if r is known, or r if θ is known. In the solar neighbourhood $A \approx 15$ km s^{-1} kpc^{-1}, $B \approx -10$ km s^{-1} kpc^{-1}, $r_\odot \approx 10$ kpc (or slightly less). Constants A and B can be calculated at any point in the Galaxy, so that (6.30) and (6.34) in fact define Oort's functions, which appear frequently in galactic dynamic expressions.

6.3   Relaxation time

This time is defined as the time in which the trajectory of a star becomes substantially modified due to encounters with other stars. As the modification induced by one star may compensate for that induced by another star, our criterion for finding the relaxation time will be that the sum of the squared increments of kinetic energy due to different encounters becomes of the order of the initial squared kinetic energy over the relaxation time τ.

If we consider the motion of a given star, the trajectory deviates in the individual encounter by an angle ψ, given by

$$\tan\frac{\psi}{2} = \frac{2mG}{V^2 D} \tag{6.38}$$

where m is the stellar mass, V is the speed of our star, being of the order of the characteristic peculiar velocity, and D is the impact parameter (i.e. the distance between a star and the straight line defining the initial ($t = -\infty$) unmodified trajectory of our star). The speed of our star is modified from V to V', where

$$V' = V \cos \psi \tag{6.39}$$

Therefore, in one collision the increment of kinetic energy ΔE is found from

$$\Delta E = \frac{1}{2}\mathcal{M}(V'^2 - V^2) = -\frac{1}{2}\mathcal{M}V^2 \sin^2 \psi = -E \sin^2 \psi \tag{6.40}$$

so that

$$(\Delta E)^2 = E^2 \sin^4 \psi \tag{6.41}$$

In a given t, our star will interact with $n2\pi DVt\,dD$ stars, with an impact parameter between D and $D + dD$. Writing D as a function of ψ, using (6.38), the total $(\Delta E)^2$ will be of the order of

$$\Sigma(\Delta E)^2 = -\frac{4\pi E^2 n\mathcal{M}^2 G^2 t}{V^3} \int_0^\pi \frac{\sin^4 \psi}{\tan^2\frac{\psi}{2}}\left(1 + \tan^2\frac{\psi}{2}\right)d\psi \tag{6.42}$$

Values higher than $\psi = \pi$ are obviously very infrequent.

We assume that E and V are not greatly modified after a large number of collisions. The integral is -2π. Therefore, when $\Sigma(\Delta E)^2 = E^2$, then $t = \tau$

$$\tau = \frac{V^3}{4\pi\mathcal{M}^2 G^2 n} \tag{6.43}$$

In the solar neighbourhood, V is of the order of 25 km s^{-1} and τ of the order of 10^{14} years is obtained. As this is much greater than the galactic lifetime, we infer that stellar–stellar collisions are unimportant in the solar neighbourhood. This property is found over the whole Galaxy except perhaps, in the galactic centre. In a globular cluster, $n \approx 100$ pc^{-3}, and V is of the order of 10 km s^{-1}, so we obtain $\tau \approx 0.5 \times 10^{10}$ years: shorter than the globular cluster lifetime. Therefore, in a globular cluster, collisions cannot be neglected and they are even frequent enough for the stars to attain a Maxwell distribution. In an open cluster $n \approx 1$ pc^{-3}, peculiar velocities are small, $V \sim 0.5$ km s^{-1}, so that $\tau \approx 10^8$ years, but in this case the lifetime of the clusters themselves is also very short.

Thus interactions among stars may be neglected in considering the internal dynamics of galaxies. It is not surprising that the stellar fluid is not in thermodynamic equilibrium, and that it is characterized by an anisotropic distribution of velocities. Another important consequence of the virtual absence of collisions is that several stellar populations can share a common place in space, without loss of their identity. A galaxy actually possesses different populations, which originate at different stages of galactic evolution and have different geometrical parameters in their spatial distributions. They are perfectly distinguishable spectroscopically, but they nevertheless cohabit with, produce, and are subject to a common gravitational potential. The bulge, disc, and halo are coexisting stellar systems.

6.4   Stellar hydrodynamics

In cylindrical coordinates, the hydrodynamical equations (1.77)–(1.79) become:

(a) Continuity

$$\frac{\partial \rho}{\partial t} + \frac{\partial}{\partial r}(\rho \Pi_0) + \frac{\rho \Pi_0}{r} + \frac{1}{r}\frac{\partial(\rho \Theta_0)}{\partial \varphi} + \frac{\partial(\rho Z_0)}{\partial z} = 0 \tag{6.44}$$

(b) Radial motion

$$\rho \frac{\partial \Pi_0}{\partial t} + \rho \Pi_0 \frac{\partial \Pi_0}{\partial r} + \rho \frac{\theta_0}{r}\frac{\partial \Pi_0}{\partial \varphi} + \rho Z_0 \frac{\partial \Pi_0}{\partial z} - \frac{\rho \theta_0^2}{r}$$

$$+ \frac{\partial P_{rr}}{\partial r} + \frac{1}{r}\frac{\partial P_{r\varphi}}{\partial \varphi} + \frac{\partial P_{rz}}{\partial z} + \frac{P_{rr}}{r} - \frac{P_{\varphi\varphi}}{r} + \rho \frac{\partial \mathcal{F}}{\partial r} = 0 \tag{6.45}$$

(c) Rotational motion

$$\rho \frac{\partial \theta_0}{\partial t} + \rho \Pi_0 \frac{\partial \theta_0}{\partial r} + \rho \frac{\theta_0}{r}\frac{\partial \theta_0}{\partial \varphi} + \rho Z_0 \frac{\partial \theta_0}{\partial z} + \rho \frac{\Pi_0 \theta_0}{r}$$

$$+ \frac{\partial P_{r\varphi}}{\partial r} + \frac{1}{r}\frac{\partial P_{\varphi\varphi}}{\partial \varphi} + \frac{\partial P_{\varphi z}}{\partial z} + 2\frac{P_{r\varphi}}{r} + \frac{\rho}{r}\frac{\partial \mathcal{F}}{\partial \varphi} = 0 \tag{6.46}$$

(c) Vertical motion

$$\rho \frac{\partial Z_0}{\partial t} + \rho \Pi_0 \frac{\partial Z_0}{\partial r} + \rho \frac{\theta_0}{r}\frac{\partial Z_0}{\partial \varphi} + \rho Z_0 \frac{\partial Z_0}{\partial z}$$

$$+ \frac{\partial P_{rz}}{\partial r} + \frac{1}{r}\frac{\partial P_{\varphi z}}{\partial \varphi} + \frac{\partial P_{zz}}{\partial z} + \frac{P_{rz}}{r} + \rho \frac{\partial \mathcal{F}}{\partial z} = 0 \tag{6.47}$$

(e) Energy balance

$$\frac{3}{2}\rho k \frac{\partial T}{\partial t} + \frac{3}{2}\rho k \Pi_0 \frac{\partial T}{\partial z} + \frac{3}{2}\rho k \theta_0 \frac{1}{r}\frac{\partial T}{\partial \varphi} + \frac{3}{2}\rho k Z_0 \frac{\partial T}{\partial z}$$

$$+ P_{rr}\frac{\partial \Pi_0}{\partial r} + P_{r\varphi}\left(\frac{1}{r}\frac{\partial \Pi_0}{\partial \varphi} - \frac{\theta_0}{r} + \frac{\partial \theta_0}{\partial r}\right) + P_{rz}\left(\frac{\partial \Pi_0}{\partial z} + \frac{\partial Z_0}{\partial r}\right)$$

$$+ P_{\varphi\varphi}\left(\frac{1}{r}\frac{\partial \theta_0}{\partial \varphi} + \frac{\Pi_0}{r}\right) + P_{\varphi z}\left(\frac{\partial \theta_0}{\partial z} + \frac{1}{r}\frac{\partial Z_0}{\partial \varphi}\right) + P_{zz}\frac{\partial Z_0}{\partial z}$$

$$+ \frac{\partial q_r}{\partial r} + \frac{1}{r}\frac{\partial q_\varphi}{\partial \varphi} + \frac{\partial q_z}{\partial z} + \frac{q_r}{r} = 0 \tag{6.48}$$

These equations have already been formulated using some simplifying hypotheses. Firstly, stellar viscosity has been neglected. However, even if collisions are unimportant, stellar viscosity might be sufficiently important to affect the dynamics of actual galaxies. The viscosity coefficient is independent of the density (see equation (1.141)). A naïve calculation would indicate that as the mean-free-path is very large, viscosity is so important that it could deform a galaxy totally in a short time. A more careful analysis, taking into account the fact that stars move in bound orbits and are unable to interchange momentum over large distances greater than typical galactic diameters, indicates that stellar viscosity should become important in characteristic times in the range $10^9 - 10^{10}$ years. Nevertheless, this has been excluded above. The effects of the births and deaths of stars, which obviously modify all the above equations, have also been ignored, in accordance with the guidelines of this chapter.

Let us further assume the following simplifying hypotheses: P becomes diagonal with these coordinates; $\vec{q}$ is zero, which is easily deduced using the Schwarzschild distribution; the motion consists of pure rotation $\Pi_0 = Z_0 = 0$; perfect axisymmetry exists, $\partial/\partial\varphi = 0$; the galaxy is in a steady-state condition. Then only two non–trivial equations are obtained:

$$-\frac{\rho\theta^2}{r} + \rho\frac{\partial\mathcal{F}}{\partial r} + \frac{\partial\mathcal{P}_{rr}}{\partial r} + \frac{\mathcal{P}_{rr}}{r} - \frac{\mathcal{P}_{\varphi\varphi}}{r} = 0 \tag{6.49}$$

and

$$\frac{\partial P_{zz}}{\partial z} + \rho\frac{\partial\mathcal{F}}{\partial z} = 0 \tag{6.50}$$

Let us write the pressure tensor as

$$P = \frac{1}{2}\mathcal{M}n\begin{bmatrix} \sigma_\pi^2 & 0 & 0 \\ 0 & \sigma_\varphi^2 & 0 \\ 0 & 0 & \sigma_z^2 \end{bmatrix} \tag{6.51}$$

where

$$\sigma_\pi^2 = 2\langle V_r^2\rangle \tag{6.52}$$

and similarly for each component.

6.4.1 The equation of vertical motion

Equation (6.50) is equivalent to

$$\frac{1}{2}\sigma_z^2\frac{\partial n}{\partial z} + n\sigma_z\frac{\partial\sigma_z}{\partial z} + n\frac{\partial\mathcal{F}}{\partial z} = 0 \tag{6.53}$$

Let us suppose that $\partial\sigma_z/\partial z = 0$. For instance, if we are dealing with a thin disc, in the absence of collisions stars will span the complete height range, rendering vertical thermal motions more or less homogeneously distributed. Even with the help of this approximation, we have two unknowns, n and $\mathcal{F}$. This problem is typical in galactic dynamics. To obtain a closed system, it would be necessary to add Poisson's equation, relating z and $\mathcal{F}$, to our system. This is dangerous however, as the gravitational field is not produced by stars alone. Gas, and (perhaps more significantly) dark matter, may also contribute. Then

$$n(r,z) = n(r,0)e^{-\frac{2}{\sigma_z^2}\mathcal{F}(r,z)} \tag{6.54}$$

if we assume that $\mathcal{F}(r, z = 0) = 0$. However, $\mathcal{F}(r, z)$ remains unknown. As the disc is very thin, we may write for a given r

$$\mathcal{F}(z) = \mathcal{F}(0) + \left(\frac{\partial \mathcal{F}}{\partial z}\right)_0 z + \frac{1}{2}\left(\frac{\partial^2 \mathcal{F}}{\partial z^2}\right)_0 z^2 + \dots \tag{6.55}$$

$\mathcal{F}(0)$ was set equal to zero, but $(\partial \mathcal{F}/\partial z)_0$ is also zero because of the symmetry about the plane of the system. Then, using (6.54)

$$n(r, z) = n(r, 0)e^{-\frac{1}{\sigma_z^2}\left(\frac{\partial^2 \mathcal{F}}{\partial z^2}\right)_0 \frac{z^2}{2}} \tag{6.56}$$

As $(\partial^2 \mathcal{F}/\partial z^2)_0$ is a constant, (6.56) is a gaussian, as was suggested in Section 6.1.1. This gaussian has a height to half density of about 300 pc.

6.4.2   Radial motion

Now (6.49) may be written as

$$-\frac{\theta_0^2}{r} + \frac{\partial \mathcal{F}}{\partial r} + \sigma_\pi \frac{\partial \sigma_\pi}{\partial r} + \frac{1}{2}\frac{1}{n}\frac{\partial n}{\partial r}\sigma_\pi^2 + \frac{\sigma_\pi^2}{2r} - \frac{\sigma_\theta^2}{2r} = 0 \tag{6.57}$$

We first note that gravitational forces are not perfectly balanced against the centrifugal force: there are other effects, such as the pressure gradient force (3rd and 4th terms) and the influence of the anisotropic peculiar velocity distribution, because $\sigma_\pi \neq \sigma_\theta$. Therefore, estimates of the galactic mass, taking only the first two terms in (6.57) into account, must be reconsidered. However, the other terms are probably less important. It is clear that $\frac{\sigma_\pi^2}{2r}$ and $\frac{\sigma_\theta^2}{2r}$ are much less than $\frac{\theta_0^2}{r}$, because σ_π and σ_θ are much less than θ, at least in the Milky Way disc. The pressure gradient force may also be neglected: $\sigma_\pi \frac{\partial \sigma_\pi}{\partial r}$ is of the order of $\frac{\sigma_\pi^2}{r}$. To estimate the fourth term, note that $L \propto \mathcal{M}^\alpha$ (the mass–luminosity relation) implies that

$$\frac{1}{I}\frac{dI}{dr} = -\frac{1}{D} = \alpha\frac{1}{n}\frac{dn}{dr} \tag{6.58}$$

D was defined in (6.2) and is about 4 kpc. The constant α was defined as having a value of about 4. Therefore, $\frac{1}{n}\frac{dn}{dr}$ is $-1/(16\text{ kpc})$ and the fourth term is as small as the other neglected terms. We can therefore conclude that

$$\frac{\partial \mathcal{F}}{\partial r} = \frac{\theta_0^2}{r} \tag{6.59}$$

is a reasonable approximation. For a sphere,

$$GM(r) = \theta_0^2 r \tag{6.60}$$

with $M(r)$ being the mass in the volume with galactocentric radius less than r. For a flat disc, other expressions for the potential must be used. For sufficiently large galactocentric distances a central point mass potential should be a good aproximation, so that $GM = \theta_0^2 r = $ constant. As stated above, $\theta_0^2 r$ does not seem to converge for $r \to \infty$, either in our Galaxy or in many others, at least for those values of r with detectable star or gas densities. The dynamical mass of a galaxy is apparently much greater than the mass deduced from the observed light, and this is normally taken as one of the clearest indications of the existence of dark matter.

7 Astrophysical plasma fluids

7.1 Introduction

Most cosmic fluids are plasmas. As was shown in Chapter 4, the internal motions of the fluid produce magnetic fields, which in turn affect this motion. Magnetic forces are, then, fundamental to the birth, structure, and evolution of most cosmic systems. However, considerable effort was made in the past to explain everything in astrophysics by taking only the gravitational force into account. This approach often yielded inadequate results, and the importance of magnetism as a basic cosmic interaction has now begun to be fully appreciated. In attemping to explain any astrophysical system, systematically ignoring magnetic fields is simply naïve.

In this chapter particular attention will be paid to two topics: the Sun and the interstellar medium. A basic introduction to the physics of active galactic nuclei is also included. Observations clearly show that we are dealing with complicated systems. Because of the proximity of the Sun, we are able to appreciate how complex a star can be. There is a large variety of transient phenomena at all observable outer layers in the Sun which are now interpreted as structures in which magnetism plays a leading role. Also, the morphology of galaxies and the complex structures of interstellar clouds may reveal the presence of magnetism. Steady-state phenomena, both in the Sun and in the galactic gas, will also be treated here, taking magnetism into account.

7.2 The Sun

In order to determine the general properties of the solar atmosphere plasma, we must determine the characteristic quantities defined in Chapter 4. Their values vary considerably with depth, so that the different layers may behave very differently. The solar atmosphere is usually divided into three parts: photosphere, chromosphere, and corona.

The photosphere is the layer that is usually observed in the visible part of the spectrum. The temperature decreases with height, the effective temperature being just under 6000 K. The Landau length is then about 3×10^{-7} cm. This is less than the interparticle distance of about $n_e^{-1/3} = 10^{-5}$ cm for a typical density of 10^{15} cm^{-3}, which means that recombination and ionization are relatively unimportant through most of the photosphere. Nevertheless, n_e varies considerably with height, so that recombination becomes less effective in the upper photosphere. From the formulae given in Chapters 1 and 4, we obtain a mean-free-path of the order of 2×10^{-4} cm, a collision frequency of $10^{11} \mathrm{s}^{-1}$ and a scalar conductivity of $2 \times 10^{12} \mathrm{s}^{-1}$, though these quantities again vary rapidly with height. The magnetic field strength is about 1 G in quiet regions, but it is enhanced to about 3000 G in special zones called sunspots. In these sunspots the gyrofrequency becomes of the order of $5 \times 10^{10} \mathrm{s}^{-1}$. This is less than the collision frequency, indicating that the photospheric plasma is, in general, isotropic.

To calculate the magnetic Reynolds number, different convective scales and velocities must be considered. Observations reveal a granulation, a mesogranulation, and a supergranulation, comprising convective cells of different sizes of the order of 2×10^8, 5×10^8, and 3×10^9 cm. The granulation manifests associated overshooting, probably due to the latent heat liberated in the recombination of atomic hydrogen. Mesogranulation overshooting is due to He^+ recombination and that of supergranulation to He^{++} recombination. The velocities involved are not very different from one scale to another, with $4 \times 10^4 \mathrm{cm}\ \mathrm{s}^{-1}$ being a characteristic value. A magnetic Reynolds number of about 2×10^5 is then obtained at the level of granulation. Therefore, frozen-in conditions may be assumed throughout the photosphere.

Due to mechanisms not completely understood at present (probably, damping of sound waves but possibly of hydromagnetic waves generated in the photosphere), the temperature increases with altitude above the photosphere, becoming of the order of 10^4 K in the chromosphere. Above a sudden transition region, the temperature increases further to 10^6 K in the corona. If, as is the convention, the zero altitude level is taken at $\tau = 1$ for 500 nm, the photosphere–chromosphere transition takes place at about 400 km, the chromosphere–corona at 3000 km, and there is no clear boundary to the corona, which extends in the form of the solar wind to planetary distances. This occurs principally in the coronal 'holes': regions with magnetic field lines open to space.

The photospheric supergranulation corresponds to a chromospheric network (cells of about 3×10^9 cm) with typical values of magnetic strength of 25 G. Other zones in the chromosphere, with enhanced magnetic field, called plages, have fields of approximately 200 G. The network gyrofrequency is in general $4 \times 10^8 \mathrm{s}^{-1}$ and in plages it is $4 \times 10^9 \mathrm{s}^{-1}$. Clearly, this means that the chromospheric plasma is anisotropic as n_e in the chromosphere is no greater than about $10^{11} \mathrm{cm}^{-3}$.

The anisotropy becomes still higher in the corona. There, the magnetic field structure is very complicated, the field strength reaching high values of approximately 10 G in the large, looped structures called prominences. A typical value of the electron number density in the corona is $10^8 \mathrm{cm}^{-3}$, which corresponds to a collision frequency of only $5 \mathrm{s}^{-1}$. The magnetic Reynolds number also becomes very high. This may be checked using the following characteristic values: $\sigma_0 \approx 5 \times 10^{15} \mathrm{s}^{-1}$, $v = 6 \times 10^7 \mathrm{cm}\ \mathrm{s}^{-1}$, $L = 5 \times 10^8$ cm (prominence thickness). These are only guideline figures because prominences are not very representative of coronal conditions in general.

Summarizing, the photospheric plasma has frozen-in magnetism and is isotropic. The plasma in the chromosphere and the corona have frozen-in magnetism but are anisotropic.

7.2.1 The mean field

Dynamo and turbulent diffusion

Because of convective motions, magnetic field lines become tangled, giving rise to many exotic solar structures. So a mean 'seed' magnetic field amplified in this way must be present in all visible parts of the Sun.

Convective and turbulent motions may require major modifications to the MHD equations proposed in Chapter 4. Turbulence is a potential source of amplified and organized magnetic field patterns on a large scale, as was first considered by Larmor in 1919. This effect is called the dynamo effect. On the other hand, turbulence also provides an additional mechanism for magnetic field loss. This effect is called turbulent diffusion. Both effects are subject to rather sophisticated theories which are outside the scope of this book. Spatial and time variations of the mean solar magnetic field also give rise to complicated phenomenology. Our aim now is to justify how irregular motions can lead to regular fields, rather than to explain their morphology in detail. A basic theory of dynamo and turbulent diffusion is introduced here from the macroscopic point of view, and will later be extended to galaxies. Later, in Section 7.3.1, some microscopic processes explaining the dynamo action in galaxies will be analysed. Though they are galactic processes, Section 7.3.1 in fact completes the present section.

Let us assume that the velocity field is known, at least in respect of its statistical properties. This motion modifies the magnetic field, but the field is assumed to be incapable of modifying the motion. Consider the induction equation (4.75). At the granulation and supergranulation levels, the magnetic field was shown to provide isotropic transport, so we may ignore the term containing x_e. The equation is rewritten as

$$\frac{\partial \vec{B}}{\partial t} = \nabla \times (\vec{v}_0 \times \vec{B}) - \frac{c^2}{4\pi\sigma_0} \nabla \times (\nabla \times \vec{B}) \tag{7.1}$$

The decomposition considered in Chapter 1, where turbulence was considered, is also taken into account here:

$$\vec{v}_0 = \vec{v}_1 + \vec{v}' \tag{7.2}$$

where $[\vec{v}'] = 0$, and $[\vec{v}_0] = \vec{v}_1$. The meaning of the symbol $[\ ]$ for the mean values is the same as was introduced in Section 1.5.2. For the magnetic field we have

$$\vec{B} = \vec{B}_1 + \vec{B}' \tag{7.3}$$

Again, the mean value of the fluctuating term is zero, $[\vec{B}'] = 0$, and $\vec{B}_1$ is defined as $[\vec{B}] = \vec{B}_1$. The induction equation becomes

$$\begin{aligned}
\frac{\partial \vec{B}_1}{\partial t} + \frac{\partial \vec{B}'}{\partial t} = {} & \nabla \times (\vec{v}_1 \times \vec{B}_1) + \nabla \times (\vec{v}_1 \times \vec{B}') \\
& + \nabla \times (\vec{v}' \times \vec{B}_1) + \nabla \times (\vec{v}' \times \vec{B}') \\
& - \frac{c^2}{4\pi\sigma_0} \nabla \times (\nabla \times \vec{B}_1) \\
& - \frac{c^2}{4\pi\sigma_0} \nabla \times (\nabla \times \vec{B}')
\end{aligned} \tag{7.4}$$

Taking mean values:

$$\frac{\partial \vec{B}_1}{\partial t} = \nabla \times (\vec{v}_1 \times \vec{B}_1) + c\nabla \times \vec{\mathcal{E}} - \frac{c^2}{4\pi\sigma_0}\nabla \times (\nabla \times \vec{B}_1) \tag{7.5}$$

where

$$\vec{\mathcal{E}} = \left[\frac{\vec{v}'}{c} \times \vec{B}'\right] \tag{7.6}$$

is an electric field induced by the turbulence, which is in practice the dominant source of magnetic fields. The presence of the turbulent electric field makes this equation different from the initial equation (7.1). Dynamo theories are developed in order to obtain an expression for $\vec{\mathcal{E}}$ which permits the integration of equation (7.5). As the statistical properties of the fluctuating velocity $\vec{v}'$ are known, those of $\vec{B}'$ may be derived.

By subtracting (7.5) from (7.4), the induction equation for the fluctuating magnetic field component is obtained:

$$\frac{\partial \vec{B}'}{\partial t} = \nabla \times \left(\vec{v}_1 \times \vec{B}' + \vec{v}' \times \vec{B}_1 + \vec{v}' \times \vec{B}' - c\vec{\mathcal{E}} - \frac{c^2}{4\pi\sigma_0}\nabla \times \vec{B}'\right) \tag{7.7}$$

From this formula, $\vec{B}'$ may be obtained. In practice, this is only possible if one of a set of simplifying hypotheses is adopted.

Let us look first at a brief argument that gives an approximate account of the turbulent electric field.

If the turbulence is steady, isotropic, and homogeneous, $\vec{\mathcal{E}}$, as a macroscopic quantity, should be a function only of $\vec{v}_1$, $\vec{B}_1$ and their spatial derivatives on a large scale. Let us assume that the term $\nabla \times (\vec{v}' \times \vec{B}_1)$ on the right-hand side of (7.7) is dominant. Though this approximation is probably not really adequate for the Sun, it implies that $\vec{B}'$ is a function of $\vec{B}_1$ but not of $\vec{v}_1$, and therefore from (7.6) $\vec{\mathcal{E}}$ would be a function only of the macroscopic quantities $\vec{B}_1$ and their spatial derivatives. This conclusion is reinforced when we consider that convection in the Sun arises as a consequence of the vertical temperature gradient, rather than of the mean velocity, which is very small on the large scale. The simplest course is to assume that $\vec{B}'$ depends only on $\vec{B}_1$ and on its first derivatives, and to look for a linear relation of the type

$$c\vec{\mathcal{E}} = \alpha\vec{B}_1 - \beta\nabla \times \vec{B}_1 \tag{7.8}$$

This equation must be a true tensor relation (pseudotensors may also be allowed). The pseudovector $\nabla \times \vec{B}_1$ is the only one that can be formed from the first derivatives of $\vec{B}_1$. Here α and β are scalars which depend only on the macroscopic properties of the fluctuating velocity field. Inserting (7.8) into (7.5) we obtain

$$\frac{\partial \vec{B}_1}{\partial t} = \nabla \times (\vec{v}_1 \times \vec{B}_1) + \alpha\nabla \times \vec{B}_1 - \left(\frac{c^2}{4\pi\sigma_0} + \beta\right)\nabla \times \nabla \times \vec{B}_1 \tag{7.9}$$

The second term on the right-hand side represents the dynamo action in this approximation. We also see that the diffusion coefficient of the magnetic field has an additive coefficient β, which is therefore called turbulent diffusivity. It is easy to show that the new terms are very important. From (7.8) and (7.6), α is of the order of $v'B'/B_1$. Values of α in the range $1-10^4$ cm s^{-1} have been proposed by different authors. In order to estimate the relative importance of the first term with respect to the second term in (7.9), we see that the 'frozen-in' term $(\nabla \times (\vec{v}_1 \times \vec{B}_1))$ is v_1/α multiplied by the dynamo term, that is, both terms are roughly of the same order of magnitude. The dynamo term is, then, by no means negligible.

The turbulence greatly increases the ohmic diffusivity since

$$\frac{\beta}{c^2/4\pi\sigma_0} \approx \frac{\frac{v'B'L}{B_1}}{\frac{c^2}{4\pi\sigma_0}} \approx 10^7 \tag{7.10}$$

The turbulent diffusivity may be seven orders of magnitude higher. All these figures indicate that a magnetized turbulent medium exhibits quite different properties from purely laminar flows. The effects of dynamo and turbulent diffusivity are important not only in the Sun but also in any turbulent plasma, and it should be borne mind that most cosmic fluids are turbulent.

Basic morphology

Cowling, Parker, Steenbeck, Krause, Ruzmaikin, and other authors have made important contributions to dynamo theory, which is now able to explain the most basic features of the main solar field. This field is of the order of 1 G, but much higher values are reached in some zones. A toroidal and a poloidal field may be observed in the Sun. In the toroidal component the lines of force are circles around the solar axis. A poloidal field has the lines of force in meridional planes. In a region of latitudes in the range 10°–20°, the poloidal field periodically reaches high values. The period is that of the well-known solar cycle of 11 years. As a given solar cycle evolves, the maximum magnetic field migrates to the equator, 20° being the approximate latitude of maximum strength at the beginning of the cycle and 10° at the end. The magnetism in both hemispheres is antisymmetric with opposite polarities. In the next cycle polarities are interchanged, and therefore the true solar cycle lasts for 22 years. The dynamo theory is at present perfectly able to explain such an oscillating solution, though we refer the interested reader to more specialized books. The periodic behaviour of the solar magnetic field may be followed directly by magnetograms, but other indicators may be used. When the mean field is large, the development of small magnetized structures can be observed relatively easily. The number of sunspots is probably the best example. Historical observations of sunspot numbers may be used to obtain knowledge of past solar magnetism. Prominences and polar faculae also follow the 11-year solar cycle.

7.2.2 Transient phenomena

Photosphere

Sunspots have been observed on the solar disc for 2000 years as dark structures that become more frequent during solar maximum. Their size is about that of mesogranulation cells. They manifest internal substructure on the granulation scale and are often present in groups of similar size to that of supergranulation cells. They have large magnetic field strengths of the order of 3000 G, and they usually have a companion spot with opposite polarity of the magnetic field. They are rather stable, having characteristic lifetimes from days to weeks.

Sunspots have been interpreted as concentrations of magnetic flux tubes. The vertical convective motions below the photosphere tangle the magnetic field lines, giving rise to small-scale structures of high field strength, in which the magnetic Reynolds number is no longer higher than unity. Dissipation then prevents the increase of magnetic pressure and the number of field lines is destroyed, which produces the macroscopic appearance of reconnection of field lines. In the region of contact of two adjoining convection cells the magnetic field strength may be several times the strength at the centres of each cell. A large concentration of vertical field lines is formed along the common boundaries of the cells (which are observed to have a quasi-polygonal structure). All numerical simulations confirm this point. This large concentration of, normally vertical, field lines constitutes a flux tube. Horizontal flux tubes may also be formed and are subject to magnetic buoyancy, as mentioned in Section 4.4.1, but these are unstable.

Sunspots are dark because they are cooler. This is also due to a magnetic effect explained by Biermann. There is a blocking of convection by intense vertical magnetic fields. This is due to the fact that the horizontal motions which must be present in convective cells are perpendicular to the field, and hence require much work. Convection is halted when the magnetic field reaches the equipartition value B_e, corresponding to a magnetic energy equal to the macroscopic kinetic energy

$$\frac{B_e^2}{8\pi} = \frac{1}{2}\rho v^2 \tag{7.11}$$

or when it reaches a value B_p equivalent to the microscopic kinetic energy, in orders of magnitude

$$\frac{B_p^2}{8\pi} = nkT \tag{7.12}$$

Typical values are $B_e \approx 1$ G and $B_p \approx 100$ G in the photosphere. Clearly, the magnetic fields are higher in sunspots and magnetic fields therefore impede convection there. The vertical energy flux is in turn impeded and the sunspot becomes darker than its surroundings. The temperature is about 4000 K and the sunspot presents a radiative flux and spectrum similar to that of a K or an M star.

As another consequence, and due to the similarity with a small K star, the $\tau = 1$ level is much deeper than the $\tau = 1$ level of the surroundings. The difference in altitude may be as much as 800 km. This effect is known as the Wilson effect.

To see how slender a flux tube is, let us calculate its diameter d using a semi-quantitative argument. At a point of convergence in the convective pattern, the dissipation time is of the order of $(4\pi\sigma_0 d^2)/c^2$. Let us suppose that we perform the calculation in a small zone of the convection cell, where no other smaller-scale turbulence introduces enhanced diffusivity of the magnetic field. If the flux tube is stable, this time must be equal to the time required to provide matter and magnetic field by convection: L/v, L being the length scale of the cell. Then

$$\frac{d^2}{L^2} = \frac{c^2}{4\pi\sigma_0 vL} = \frac{1}{\mathcal{R}_m} \approx 10^{-6} \tag{7.13}$$

Therefore $d \approx 10^{-3} L \approx 10$ km, which is indeed very thin, compared with the diameter of a typical convection cell, which is of the order of a few hundred km.

The order of magnitude of B' could, in principle, be calculated, taking account of the fact that the total magnetic flux emerging in a cell must be more or less concentrated in the flux tube:

$$B_1 L^2 = B' d^2 \tag{7.14}$$

where B' is the field strength in the sunspot and B_1 is the mean field in the surroundings, that is, about 1 G. A very high value of B' is obtained, of about 10^6 G, which is not confirmed by observations. Values higher than 4000 G are very unusual. This crude argument may fail in part because turbulence on a smaller scale cannot be ruled out, and in part because it is a kinematic argument, that is, the magnetic field is transported by convective motion but has no influence on it. In practice, when $B = B_e$ or $B = B_p$ we know that convection ceases.

Although we have assumed that the magnetic diffusivity is not influenced by turbulence in a small corner of a convective cell, this approximation is no longer valid for the whole cell. If the decay of the cell were due to the turbulent diffusion of the magnetic field, at a rate 10^7 times that due to the finite conductivity (7.10), a sunspot would decay in L^2/β seconds, that is, about eight hours. This estimate gives the correct order of magnitude for other photospheric structures known to have a flux tube geometry. Examples of these smaller-diameter structures are pores, magnetic knots, the filigree, and points in faculae. These have lifetimes in the range one hour to one day. For the details of the morphology of these small structures, we refer the reader to descriptive texts. Here, it is emphasized that they probably belong to a large family of flux tube structures. But the value of eight hours is too short for sunspots.

If the condition $\nabla \cdot \vec{B} = 0$ is satisfied, the upward polarity of a sunspot may be acompanied by another sunspot with downward polarity. This is the main reason for the bipolarity of spot pairs. Some unipolar sunspots must then be accompanied by a number of much smaller structures with inverse polarity. This is probably the nature of the pores mentioned above. In fact, pores are present in the neighbourhood of unipolar spots, have opposite polarity, and the sum of the magnetic fluxes of all satellite pores should equal the sunspot magnetic flux.

Chromosphere and corona

If magnetic fields are important in explaining transient photospheric phenomena, their influence becomes even greater in the chromosphere and the corona. Chromospheric dark mottles are local enhancements of the magnetic field, which outline a network that is clearly related to supergranulation structures. Plages are extended over the network and correspond to active regions, that is, regions of higher mean magnetic flux. But it is in the corona where magnetism is dominant and is responsible for many complicated, energetic, and beautiful structures. Matter, heat, and electrical currents are forced to follow field lines, causing the anisotropy of the coronal plasma. For instance, images of the corona at the minimum of the 11-year cycle show polar plumes that suggest the field lines of a bar magnet. The corona maximum is more complicated showing 'helmets', 'streamers', and other features which may be present above any part of the Sun, and which show the direction of the magnetic field directly.

   The field is so dominant that structures in a quasi-steady state may be adequately explained as force-free magnetic field systems, in which only magnetic forces are important, and for which

$$\vec{j} \times \vec{B} = 0 \tag{7.15}$$

is obeyed.

   This equation is not simple. Using (4.73) we have

$$\nabla \times \vec{B} = K\vec{B} \tag{7.16}$$

where K may be a function of the spatial coordinates. Taking the operator $\nabla \cdot$ on both sides:

$$\vec{B} \cdot \nabla K = 0 \tag{7.17}$$

∇K is perpendicular to $\vec{B}$, hence K remains constant along a field line. Its value should, in principle, be determined at its 'foot' in the chromosphere. The most striking feature of (7.16) is that it suggests helicity. Note, for instance, that $\vec{B} = (0, B\sin kx, B\cos kx)$, where B and k are constants, satisfies (7.16). When helicity develops in a system of field lines, these become tangled and the system may end as an eruptive event.

   Other noticeably stationary structures are quiescent prominences. These are seen projected on the solar disc as filaments of chromospheric material, with the whole structure lying in the corona. They are connected to the chromosphere at both ends, forming an arch. The arch is clearly visible when prominences are observed at the solar limb. Prominences are very large. The part of the structure lying in the corona is typically $200\,000 \times 50\,000 \times 6000$ km. They have typical masses of 3×10^{15} g and can have lifetimes of several months. During this time magnetic fields and gravity are in equilibrium. Probably the foot points correspond to bipolar chromospheric structures. The matter in a prominence may be considered chromospheric as its temperature is only some 10^4 K (a typical chromospheric temperature), which is very low compared to the 10^6 K typical in the corona. At the prominence boundaries the pressure in the prominence and the corona must be equal and thus a sudden decrease in density must accompany a sudden increase in temperature. So the density in the prominence material is much larger, with typical chromospheric values. The magnetic strength is about 10 G in the lower parts.

   Prominence filaments often become eruptively unstable, producing flares. Flares consist of ejections of matter and magnetic fields into space, enhancing the solar wind; and large amounts of radiative energy are also released, mainly at short wavelengths. The γ-ray emission shows the electron–positron annihilation line at 0.511 MeV.

7.3   The galactic gas

The galactic gas also provides an interesting cosmic system in which the magnetic field plays an important role in configuring both the large-scale morphology and the small-scale effects. Some star formation regions present a filamentary structure typical of systems dominated by magnetic fields, and the last stages of star formation constitute one of the classical (though not yet fully understood) topics of cosmic magnetism. The nuclei of galaxies contain large amounts of magnetic energy, which must be taken into account in order to explain their activity. This is also the case for the large lobes of radiogalaxies, in which the magnetically generated synchrotron emission is in fact the only means of observing them. Thus one realizes the importance of magnetism in the birth and evolution of galaxies and possibly in deciding many characteristic morphological features on a large scale. For instance, rings, flaring, corrugations, warps, spiral arms, and many other effects may be affected by magnetic fields. The discrete distribution of the galactic gas, forming clouds of either molecular or atomic hydrogen flowing in a less dense and hotter intercloud medium, is, at least in part the result of magnetic instability.

As in the case of the Sun, we can often observe directly how beautiful and complicated magnetic instabilities may be. The similarity to the Sun is apparent in some aspects, but not in others.

The galactic magnetic field is also frozen-in. To calculate the magnetic Reynolds number let us consider some characteristic numbers in formulae (4.78) and (4.79). Let us concentrate on spiral galaxies. For a number density in the solar neighbourhood of the order of 1 cm^{-3}, even if the ionization is only about 1%, a scalar conductivity of about 4×10^{15}s^{-1} is obtained. The frozen-in condition mainly arises from the vast dimensions involved. For a characteristic length L of about 1 kpc for the ionized component (the typical height above the symmetry plane is about 400 pc, larger than that for the neutral component) and a rotation velocity of about 200 km s^{-1}, the magnetic Reynolds number can be as high as 10^{22}. The magnetic diffusion time is 10^{31} years.

Note that the solar plasma is isotropic in the lower layers and moderately anisotropic in the higher ones. In contrast, the galactic gas is highly anisotropic. The magnetic field strength is 1–10 μ G, hence equation (4.3) gives a cyclotron frequency of 175 s^{-1}. Using (1.112) or (1.139) for a temperature of about 100 K, the collision frequency can be estimated to be of the order of 7×10^{-10}s^{-1}. An electron collides with a hydrogen atom once every 50 years. The anisotropy parameter x_e defined in (4.6) takes values as high as 2.5×10^{11}. These extreme values permit some important simplifications both in the equations and in qualitative descriptions. Because of the high anisotropy, the effective magnetic Reynolds number becomes much lower, $\mathcal{R}_m/(1 + x_e) \approx \mathcal{R}_m/x_e \approx 10^{11}$, which is in any case a high value, with the frozen-in condition being well satisfied. Molecular clouds have lower ionization coefficients, with major spatial variations, but the condition of frozen-in magnetic field lines is in any case valid.

As in the case of the Sun, calculations of the order of magnitude of the characteristic time for magnetic diffusion in galactic gas may not be adequate, especially when the equations are applied to the whole galactic system, as the turbulence Reynolds number is very high. By using (1.155) with (1.141) (which gives $\eta = 3 \times 10^{-3}$ g cm^{-1} s^{-1}), we obtain a turbulence Reynolds number of $\mathcal{R} = 2 \times 10^7$. The turbulence makes the motion of a plasma very complicated. On the one hand it makes the action of turbulent dynamos possible. And on the other hand it enhances magnetic diffusivity by orders of magnitude with respect to the normal magnetic diffusivity, $c^2/4\pi\sigma_0$. Both the galactic dynamo and the diffusivity will be considered briefly, and we will see that both effects are by no means negligible.

The presence of so much turbulence also modifies the interpretation and order of magnitude of other dynamic quantities, such as the pressure and the viscosity coefficient. It is well known from observation that the interstellar matter is distributed in clouds. Typical dimensions of HI clouds are 5 pc. The number density inside these clouds is about 30 cm^{-3} and the temperature is ~ 100 K. Another component of the interstellar medium (ISM), the warm neutral gas, may be considered along with an intercloud continuum medium through which HI clouds move. In this intercloud the density is two orders of magnitude less. Since clouds and intercloud regions are in dynamic equilibrium, having approximately the same pressure, the temperature in the intercloud regions is two orders of magnitude higher. This intercloud medium probably has a certain degree of clumpiness. Moreover, other ISM components can be identified, the warm ionized gas, with $T \sim 10^4$ K and the hot ionized gas with $T \sim 10^6$ K. It is obtained that $nT \approx 10^3$K cm^{-3}, indicating a noticeable dynamic equilibrium between all ISM components, though magnetic fields could also play a role. There exist H_2 clouds (or molecular clouds), also in dynamic equilibrium, with temperatures of about 10 K. As contrast, HII regions are not in equilibrium at all.

The cloud-to-cloud distance is about ~ 10 pc, and their relative velocities are about 5 km s^{-1}. Pressure means momentum transport, and the random cloud motions imply an additional mechanism of momentum transport. The mean-free-path of clouds is easily calculated: $l = (N\pi D^2)^{-1} \approx 100$ pc. ($N \sim 1/d^3$; d is the intercloud distance; D is the diameter of a typical cloud). Therefore the time between collisions is about 10^7 years. These clouds propagate randomly and collide from time to time, entailing a mechanism of momentum transfer which is characterized as a turbulent pressure, even if no eddies are present in their random motion.

Let us calculate the hydrostatic turbulence pressure as the trace of the Reynolds tensor defined in (1.170)

$$p = \frac{1}{3}MN[v'^2] \tag{7.18}$$

where M is the mass of a cloud (note $MN = mn$, m being the hydrogen mass and n being the mean value of the hydrogen number density). As $[v'^2] \approx 5 - 10$ km s^{-1} the turbulent pressure is of the order of 10^{-11}dyn cm^2. However, the 'molecular' pressure nkT, is only of the order of 10^{-13}dyn cm^2. It is therefore negligible and p calculated from (7.18) is the quantity to be considered in the MHD-equations. There is also a pressure due to cosmic rays, that is, relativistic particles, probably generated in supernova explosions, which is slightly lower than the turbulence pressure, but for simplicity it may be safely ignored.

'Molecular' viscosity is calculated to be about 3×10^{-3} g cm^{-1} s^{-1}. For the viscosity characteristic time, t_{vis}, order of magnitude arguments easily give

$$t_{vis} = \frac{\rho L^2}{\eta} \tag{7.19}$$

The value of t_{vis} obtained is very high, about 10^{14} years. The conclusion could be that viscosity is negligible, since its characteristic time is much larger than the Galaxy lifetime. Such a conclusion, however, would be wrong. It would be very difficult indeed to explain the formation of the disc with such a low viscosity coefficient. A turbulent viscosity may also play a role. To estimate it, let us apply (1.141), that is, $\eta = \rho \lambda v'$, but substituting clouds for atoms. The mean-free-path λ has been calculated above at about 100 pc. v' is a typical cloud-to-cloud velocity, of the order of 10 km s^{-1}. A value for the turbulent viscosity coefficient of 500 g cm^{-1} s^{-1} is obtained. Taking $L \approx 10$ kpc the turbulent viscosity characteristic time is 10^9 years. This is less than the galactic lifetime, hence viscosity is by no means negligible.

Eliminating the 1-subscript for the mean quantities we may write the following MHD equations for conditions prevailing in galaxies:

Continuity

$$\frac{\partial \rho}{\partial t} + \nabla \cdot (\rho \vec{v}_0) = 0 \tag{7.20}$$

Motion

$$\rho \frac{\partial \vec{v}_0}{\partial t} + \rho \vec{v}_0 \cdot \nabla \vec{v}_0 + \nabla \left(p + \frac{B^2}{8\pi} \right) = -\rho \nabla \mathcal{F} + \frac{1}{4\pi} \vec{B} \cdot \nabla \vec{B} + \frac{4}{3} \eta \nabla \nabla \cdot \vec{v}_0$$
$$- \eta \nabla \times \nabla \times \vec{v}_0 \tag{7.21}$$

Induction

$$\frac{\partial \vec{B}}{\partial t} = \nabla \times (\vec{v}_0 \times \vec{B}) + \alpha \nabla \times \vec{B} - \beta \nabla \times \nabla \times \vec{B} \tag{7.22}$$

where we have adopted the same simplifying hypothesis used in deriving (7.9), to take both the dynamo generation of magnetic fields and the turbulence diffusivity erasing them into account. The arguments given in Section 7.2.1 are still reasonable in the case of galaxies.

The equation of thermal balance has not been included, as transport terms become unimportant, production mechanisms (such as supernovae energy input) being balanced by radiation.

7.3.1   The galactic dynamo

Turbulence consists of a hierarchy of eddies or subsystems with different length scales. For the smallest ones, the magnetic Reynolds number is no longer large and the field is dissipated, with the magnetic energy in these small eddies being converted into heat. Therefore, turbulence rapidly destroys magnetic fields. However, they are observed in spiral galaxies, so that a dynamo mechanism must be present, which paradoxically is also due to turbulence. The dynamo was considered in Section 7.2.1 from the macroscopic point of view. Let us now consider how a mean magnetic field can be generated by apparently completely disordered motions, from the microscopic point of view.

Let us assume an ascending eddy which was initially a cylinder, rotating or not, in the northern hemisphere of a galaxy. (We arbitrarily call north that hemisphere from which the galactic rotation can be seen to be counterclock-wise.) As the pressure decreases with z, the ascending column expands, so that radial motions appear. These radial motions are subject to a coriolis force, so that our ascending column acquires a rotation (or an excess rotation) with respect to the surroundings (Figure 7.1). Therefore, ascending cells have a tendency to rotate clockwise in this hemisphere. The same reasoning tells us that descending cells rotate counterclockwise. Thus $\vec{v}' \cdot (\nabla \times \vec{v}')$ has a tendency to become negative in this hemisphere, irrespective of the direction of each vector. The mean value $[\vec{v}' \cdot (\nabla \times \vec{v}')]$ will therefore not be zero, that is, there is a mean non-vanishing helicity. This regularity is able to produce poloidal magnetic fields.

Assume an initial azimuthal seed field (Figure 7.2) frozen-into the disc. Assume an ascending vortex V. Because of the effect discussed above, it will rotate, generating small-scale poloidal fields. Reconnection and coalescence will produce net, closed,

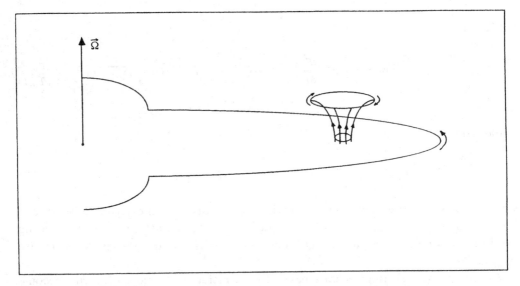

Figure 7.1   Rotation of an expanding gaseous cell moving in the vertical direction due to the coriolis forces.

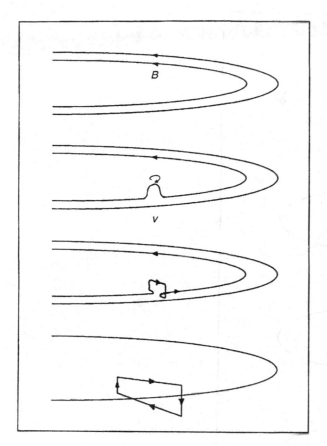

Figure 7.2 Poloidal magnetic fields generated by toroidal magnetic fields.

poloidal magnetic field lines, as all small-scale generated fields are coherent. Therefore, a seed azimuthal field produces a mean poloidal field. This is called the α-effect.

But poloidal fields in turn, under the action of differential rotation and under the frozen-in condition, may produce large azimuthal magnetic fields, as illustrated in Figure 7.3.

A point such as O has completed 0, 1, 2, and 3 turns due to galactic rotation. Another point such as O' has a lower angular velocity (as there is generally a differential rotation the angular velocity depends on r) and the magnetic field line acquires a progressive spirality. Azimuthal fields have been generated. This is called the Ω-effect. The result is that both components of the magnetic field grow. This type of dynamo is called the $\alpha\Omega$-dynamo.

The increase of the azimuthal component by differential rotation – which also plays an important role in the 11-year cycle of the Sun – is adequately described by the term $\nabla \times (\vec{v}_0 \times \vec{B})$. The effect of net helicity is a microscopic phenomenon and must be present in the α-term of the dynamo. There exists a non-vanishing mean value for $[\vec{v}' \cdot (\nabla \times \vec{v}')]$, and α in (7.22) must be proportional to it:

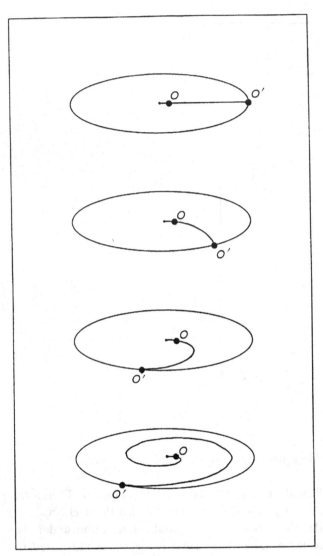

Figure 7.3   Toroidal magnetic fields generated by poloidal magnetic fields.

$$\alpha = -\frac{\tau}{3}\left[\vec{v}' \cdot (\nabla \times \vec{v}')\right] \tag{7.23}$$

The proportionality constant τ has dimensions of time, and should be identified with the characteristic time during which the vortex goes up (or down) whilst it expands. This is called the correlation time. Let us obtain an approximate value of the magnitude of α in the absence of knowledge of the order of the relation B'/B_1. The angular momentum $r^2|\nabla \times \vec{v}'|$ per unit mass of the ascending cell increases due to the coriolis torque (r is the radius of the cell). v' is now the vertical velocity of the cell. The coriolis force is proportional to $v'_r\Omega$, where v'_r is the expansion velocity of the cell and Ω is the rotational velocity of the galaxy. Hence,

$$r(v'_r \Omega) \approx \frac{r^2 |\nabla \times \vec{v}'|}{\tau} \qquad (7.24)$$

Probably $v'_r < v'$ and $r < l$ (l would be the height reached by the cell, about, say, 100 pc). If the density does not change very much in the process, $\nabla \cdot (\rho \vec{v}') = 0$, suggesting that $v'_r/r \approx v'/l$. Using $l \approx v'\tau$, we obtain $\tau \sim 1/\Omega$ and

$$\alpha \approx \frac{l\Omega}{3} \qquad (7.25)$$

As $\Omega \approx 10^{-15} \, \text{s}^{-1}$ at the galactocentric radius of the Sun, we obtain $\alpha \approx 1 \, \text{km} \ \text{s}^{-1}$. More sophisticated calculations yield $\alpha \approx 0.2 \, \text{km} \ \text{s}^{-1}$. We also noted (Section 7.2.1) that α is of the order of $v' \frac{B'}{B}$, which would imply that the fluctuating magnetic field B' is of the order of the mean field B, which is indeed observed in the Milky Way. Other models provide slightly more precise means of estimating α, although the dynamo theory as a chapter of the turbulence theory is at present severely limited in its ability to calculate coefficients. We stated in (7.10) that β is of the order of $v'B'l/B$; as $B'/B \sim 1/3$, a rough expression would be

$$\beta \approx \frac{1}{3} v'l \qquad (7.26)$$

Therefore, β is approximately $10^{26} \text{cm}^2 \text{s}^{-1}$ and more sophisticated calculations yield $\beta = 6 \times 10^{25} \text{cm}^2 \text{s}^{-1}$. Thus the turbulent diffusion time of the magnetic field, which is obtained by replacing $c^2/4\pi\sigma_0$ by β in (4.79), becomes

$$\tau_{diffusion} = \frac{l^2}{\beta} \approx 3 \times 10^7 \text{years} \qquad (7.27)$$

which is by no means negligible.

7.3.2 Large-scale hydrodynamics

Let us write down the MHD equations (7.20), (7.21) and (7.22) in cylindrical coordinates, assuming axisymmetry ($\partial/\partial\varphi = 0$) and assuming that the mean motion consists of pure rotation ($\vec{v}_0 = (0, \theta, 0)$), where θ is a function of r alone (this is suggested by observations, taking the flatness of the disc into account).

Continuity

$$\frac{\partial \rho}{\partial t} = 0 \qquad (7.28)$$

Motion

$$\frac{\partial p}{\partial r} - \rho \frac{\theta^2}{r} + \rho \frac{\partial \mathcal{F}}{\partial r} + \frac{1}{4\pi} \left(B_\varphi \frac{\partial B_\varphi}{\partial r} + B_z \frac{\partial B_z}{\partial r} + \frac{B_\varphi^2}{r} - B_z \frac{\partial B_\varphi}{\partial z} \right) = 0 \qquad (7.29)$$

$$\rho \frac{\partial \theta}{\partial t} - \frac{1}{4\pi} \left(B_r \frac{\partial B_\varphi}{\partial r} + \frac{B_\varphi B_r}{r} + B_z \frac{\partial B_\varphi}{\partial z} \right) - \eta \frac{\partial}{\partial r} \left(\frac{\partial \theta}{\partial r} + \frac{\theta}{r} \right) = 0 \qquad (7.30)$$

$$\frac{\partial p}{\partial z} + \frac{1}{4\pi} \left(B_r \frac{\partial B_r}{\partial z} + B_\varphi \frac{\partial B_\varphi}{\partial z} - B_r \frac{\partial B_z}{\partial z} \right) + \rho \frac{\partial \mathcal{F}}{\partial r} = 0 \qquad (7.31)$$

Induction

$$\frac{\partial B_r}{\partial t} = -\alpha \frac{\partial B_\varphi}{\partial z} + \beta \frac{\partial}{\partial z}\left(\frac{\partial B_r}{\partial z} - \frac{\partial B_z}{\partial r}\right) \tag{7.32}$$

$$rl\frac{\partial B_\varphi}{\partial t} = \alpha\left(\frac{\partial B_r}{\partial z} - \frac{\partial B_z}{\partial r}\right) + \beta\left(\frac{\partial^2 B_\varphi}{\partial z^2} + \frac{\partial}{\partial r}\left(\frac{\partial B_\varphi}{\partial r} + \frac{B_\varphi}{r}\right)\right) + B_r\left(\frac{\partial\theta}{\partial r} - \frac{\theta}{r}\right) \tag{7.33}$$

$$rl\frac{\partial B_z}{\partial t} = \alpha\left(\frac{\partial B_\varphi}{\partial r} + \frac{B_\varphi}{r}\right) - \beta\left(\frac{\partial}{\partial r}\left(\frac{\partial B_r}{\partial z} - \frac{\partial B_z}{\partial r}\right) + \frac{1}{r}\left(\frac{\partial B_r}{\partial z} - \frac{\partial B_z}{\partial r}\right)\right) \tag{7.34}$$

Maxwell

$$\frac{\partial B_r}{\partial r} + \frac{B_r}{r} + \frac{\partial B_z}{\partial z} = 0 \tag{7.35}$$

These equations are usually treated separately: the group (7.28)–(7.31) on the one hand, and the group (7.32)–(7.35) on the other. To integrate the first group it is usually assumed that the magnetic field is negligible.

This is a dangerous assumption, as the characteristic time of magnetic forces is of the order of $(L\rho^{1/2}B^{-1})$, as short as 5×10^7 years. The importance of magnetic fields in galactic dynamics may be large, especially considering the fact that the dominant terms in (7.29), the centrifugal term $-\rho\theta^2/r$ and the gravitational term $\rho\partial\mathcal{F}/\partial r$, may partially cancel. In fact, the two terms are usually considered to be in nearly perfect balance for the stellar system. The gas and the stellar systems usually rotate in unison, so that θ is the same for both. The gravitational potential $\mathcal{F}$ is also clearly the same for both, $\theta^2/r = \partial\mathcal{F}/\partial r$ for the stellar system, and probably for the gas, too. The two terms may therefore be suppressed in (7.29), which emphasizes the decisive role of magnetic fields. In the radial direction, it could be assumed that magnetic fields are balanced by the pressure gradient force. The maintenance of gaseous rings, such as those in the Milky Way and M31, and in many other spirals, could then be related to galactic magnetism.

Similarly, in the vertical direction, magnetic forces have an influence on the scale height of the gas. The magnetic field strength is limited by the so-called Parker instability. A highly magnetized fluid disc with horizontal magnetic fields is unstable. Gravity is balanced by thermal, cosmic, and magnetic pressures. Imagine a small perturbation in the field lines so that they become twisted with respect to the plane. The magnetic pressure decreases and matter falls down along the lines of force to the plane. The gravitational burden in the perturbed region is in part reduced and the lines of force become increasingly twisted outwards, and so on. Finally, the magnetic field, together with cosmic rays, may be ejected perpendicular to the plane out of the galaxy. The magnetic field is limited in such a way and cannot reach very high values. Parker instabilities and turbulent dissipation avoid the gradual increase of magnetic fields by dynamo effects. They limit not only the magnetic field but also the cosmic ray energies, this effect being predicted on the basis of the so-called equipartition condition, which postulates that magnetic and cosmic ray energy densities should be the same. There is,

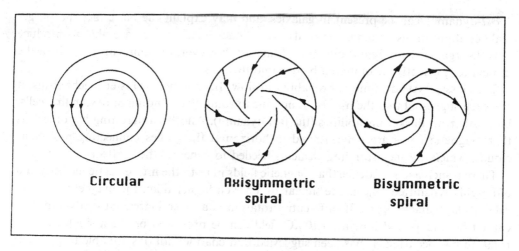

Figure 7.4 Different magnetic field patterns as predicted by the dynamo theory in spiral galaxies.

however, a somewhat controversial argument at present as to whether this condition is matched in galaxies.

The integration of the isolated second group of equations (7.32)–(7.35) constitutes the basis of dynamo theories. For this task it is usually assumed that θ is given (and hence is not influenced by $\vec{B}$). This is the so-called kinematic $\alpha\Omega$-dynamo approximation. Magnetic fields are then calculated. Observations of our own Galaxy and of other spirals, mainly by Faraday rotation and polarized synchrotron techniques, reveal two basic types of structure for the large-scale galactic fields (plotted in Figure 7.4).

All structures can reasonably be reproduced by the dynamo equations. The first corresponds to the excitation of the lowest even ($m = 0$) mode. M31 gives a clear example of circular magnetic field lines. The bisymmetric spiral structure corresponds to the excitation of the lowest uneven ($m = 1$) mode. M33 and NGC6946 both have such a structure.

The axisymmetric mode is the most favoured mode by the linear $\alpha\Omega$-dynamo theory. Magnetic field strengths should present a ring structure, having higher values at radii when an inner rotation curve $\theta \propto r$ changes to an outer rotation curve $\theta \sim$ constant.

The bisymmetric pattern closely resembles that expected after a few galactic rotations if the field were of primordial origin and then amplified by gravitational collapse and rotation, and this renewed interest in the early hypothesis of the primordial origin of galactic fields, without invoking any dynamo action. But the existence of the turbulent diffusion of the magnetic field creates serious difficulties for this hypothesis. Primordial fields could still be important as seed fields. (The dynamo amplifies only pre-existing magnetic fields.) Circular magnetic lines may also be obtained under steady-state dynamic conditions, yielding profiles for M31, that are similar to those obtained by the full dynamo approach.

$\alpha\Omega$-dynamos can be present in galaxies, and may explain the basic fact that large-scale configurations of magnetic fields exist. Other local sources are able to produce only disorganized configurations. It is of interest, however, to comment on some of the difficulties presently encountered by dynamo theories.

The classical $\alpha\Omega$-dynamo is probably too slow. In looking for faster mechanisms, it has been suggested that the α-effect is not due to rotating ascending or descending cells, but, rather, to Parker instabilities (the fast dynamo). Another interesting hypothesis is that magnetic fields are concentrated in horizontal flux tubes connecting molecular clouds, which are the ascending elements needed to generate the α-effect.

There is increasing evidence that magnetic fields in both the intergalactic medium and old highly redshifted objects are similar to, or even higher than, the magnetic fields in present-day galactic discs. If differential rotation is a basic ingredient of dynamos, it cannot be understood how a $1-10\ \mu G$ field can be produced before a single galactic rotation has occurred. It has been suggested that chaotic motions are able to generate chaotic fields if the magnetic Reynolds number is large enough. This is the fluctuation dynamo, which is very fast. Then the $\alpha\Omega$ galactic dynamo would organize rather than amplify these early chaotic fields at a later stage.

A seed field is necessary in any case. This can clearly be seen from equation (7.22). If $\vec{B}$ is initially zero, $\partial\vec{B}/\partial t = 0$ and no magnetic field can be generated. The origin of this seed field is not known at present. A very attractive hypothesis considers primeval origin at the electroweak epoch when electromagnetism was first manifested once the electroweak symmetry was broken. The differential action of certain effects on electrons and ions (notably the cosmic background radiation and gravity) could provide a weak battery. The ejection by galactic jets of magnetic fields into the intergalactic medium must also be important.

7.3.3   Cloud collapse and star formation

Molecular clouds are regions of potential gravitational collapse and subsequent star formation. Magnetic fields play an important role in this process, which is, however, rather complicated. Let us introduce the topic, taking the Virial theorem (4.109) into account. As a first example non-magnetic collapse will be considered. This provides a complementary introduction to the concept of Jeans mass (the minimum mass able to collapse for a given external pressure), which is of great importance in astrophysics and which will be examined later.

Collapse of unmagnetized clouds
Let us assume, first, a spherical unmagnetized molecular cloud in equilibrium, with an external pressure p. The cloud does not rotate. Isothermal conditions are further assumed, as contractions or expansions will normally be sufficiently slow in practice that the optically thin cloud will be able to maintain a constant temperature via radiation. The cloud is homogeneous and the internal pressure nkT may not coincide with

the external pressure p. We obtain the expression to be used in this specific problem from the Virial theorem (4.109):

$$3\Pi + \mathcal{W} - \oint_S p\vec{r} \cdot d\vec{\sigma} = 0 \tag{7.36}$$

where we calculate

$$\Pi = nkT\frac{4}{3}\pi R^3 = \frac{MkT}{m} \tag{7.37}$$

M is the cloud's mass, and R its radius, and m is the mass of a molecule, typically the mass of H_2. The potential energy is

$$\mathcal{W} = -\frac{3}{5}\frac{M^2 G}{R} \tag{7.38}$$

and

$$\oint_S p\vec{r} \cdot d\vec{\sigma} = 4\pi R^3 p \tag{7.39}$$

where $d\vec{\sigma}$ is the surface element.

Then (7.36) is written as

$$p = \frac{Y}{R^3} - \frac{Z}{R^4} \tag{7.40}$$

where

$$Y = \frac{3MkT}{4\pi m} \tag{7.41}$$

$$Z = \frac{3}{20}\frac{GM^2}{\pi} \tag{7.42}$$

Y and Z are considered to be constants, so that (7.40) should provide the radius R of the cloud in equilibrium with an external pressure, p. If p varies, for instance due to the passage of a spiral wave, R must vary according to (7.40). The $[p, R]$ curve has a maximum for

$$R_M = \frac{4}{15}\frac{GMm}{kT} \tag{7.43}$$

$$p_M = \frac{\pi}{M^2 G^3}\left(\frac{kT}{m}\right)^4 \tag{7.44}$$

It can be shown that the equilibrium becomes unstable for $R < R_M$. If the external pressure suddenly increases in a stable situation, represented by A in a $[p, R]$-diagram (Figure 7.5), the star will contract, reaching the new equilibrium at A'. But if p increases at a point such as B, the induced contraction will be unable to reach a new equilibrium. The star will gravitationally collapse as its self-gravitation will increase for decreasing values of R, and R will decrease with increasing self-gravitation.

These formulae have important consequences for the gravitational collapse of masses of gas to generate stars. In this case the value of the external pressure will be precisely the maximum pressure p_M for a critical mass M_J deduced from (7.44). This critical mass M_J is called the Jeans mass:

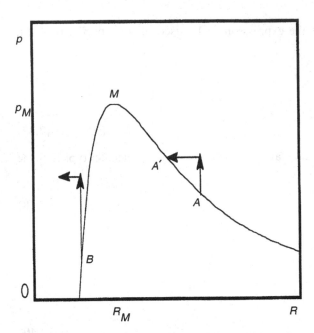

Figure 7.5 Equilibrium curve of the radius of a spherical cloud as a function of the external pressure. The equilibrium of $R < R_M$ is unstable.

$$M_J = \left(\frac{\pi}{pG^3}\right)^{1/2} \left(\frac{kT}{m}\right)^2 \tag{7.45}$$

where p is the external ambient pressure. Here T is also the ambient temperature because the temperature has been kept constant and no contraction of the critical mass has yet occurred. Therefore, we can now use $p = \frac{\rho}{m}kT$ to obtain

$$M_J = \pi^{1/2} \left(\frac{kT}{mG}\right)^{3/2} \rho^{-1/2} \tag{7.46}$$

The numerical coefficient $\pi^{1/2}$ may differ from other similar expressions (usually it is $\pi^{3/2}$) but in practice only the order of magnitude of M_J is required. The Jeans mass is a limiting mass: a mass $M < M_J$ will correspond to a critical pressure lower than the ambient pressure, as deduced from (7.44); hence it will collapse. Therefore, the interpretation of the Jeans mass is that masses lower than M_J cannot collapse to yield star formations. R_M is also called the Jeans wavelength, because the variations in external pressure may be induced by Jeans waves, that is, sound waves in which self-gravitation cannot be ignored. Wavelengths larger than R_M are not able to propagate, are unstable, and induce gravitational collapse.

We take numerical values in the interstellar medium. For $T \approx 100$ K and $n \approx 7$ cm^{-3} the order of magnitude of the Jeans mass is $3 \times 10^4 M_\odot$. This means that masses lower than this value cannot collapse. Of course, this contradicts the simple observation that our own star is less massive. The interstellar Jeans mass does coincide with the characteristic mass of an open cluster. It is well known that stars are not born in isolation but, rather, are born as members of an open cluster. The above calculation is therefore correct, but some fragmentation of the collapsing cloud must take place in the process in order to permit the existence of stars with ~ 1 $M_\odot$. If we take into account the fact that stars are formed in molecular clouds with $T \sim 10$ K and $n \sim 1000$ cm^{-3}, then the Jeans mass obtained is similar, that is, $10^5 M_\odot$.

Fragmentation

Suppose that the collapse has been initiated. As a consequence, the density increases. But we see in (7.46) that $M_J \propto \rho^{-1/2}$. Therefore, M_J will decrease during the first phase of the process. As a result, smaller and smaller masses become gravitationally unstable and collapse. Subcollapses will take place within the collapse. M_J will then decrease further and this fragmentation process will permit smaller and smaller masses to collapse. The question is: when will the fragmentation cease?

For a sufficiently large density the collapse can no longer be an isothermal process, as photons cannot escape. When the cloud becomes optically thick, the collapse tends to become adiabatic. It is easy to calculate the Jeans mass for adiabatic collapse. In (7.46) we substitute the adiabatic relation, $T \propto \rho^{\gamma-1}$, which gives $M_J \propto T^{3/2}\rho^{-1/2} \propto \rho^{1/2}$. The adiabatic Jeans mass is an increasing function of the density. Therefore, fragmentation should stop before adiabatic conditions are reached. In orders of magnitude, the isothermal–adiabatic transition will correspond to a photon mean-free-path of the order of the radius of the last fragment, that is, the so-called last fragment Jeans wavelength. As shown in (3.62), the photon mean-free-path is $(\rho\kappa_\nu)^{-1}$, where κ_ν is the opacity of the medium. As the cloud is cool, let us adopt a low value for κ_ν, of the order of 0.1 g^{-1}cm^2, as infrared wavelengths will dominate the radiative flux. From $(0.1\rho)^{-1} = R_J$, (7.43), (7.44), and (7.46), we obtained for the mass of the last fragment

$$M = \frac{4\pi}{15}\left(\frac{kT}{Gm}\right)^2 \kappa_\nu \tag{7.47}$$

The order of magnitude given by this formula should be identified with the typical mass of a star. We find that $M = 0.7$ $M_\odot$. Taking the exploratory character of the argument into account and the fact that T will increase even before the adiabatic regime is reached, the result is satisfactory. Equations (7.46) and (7.47) provide a general view of the stellar formation process that is fully borne out by experience: stars of about 1 $M_\odot$ are born in groups of about 3×10^4 stars. These groups are the open clusters. The fragmentation process must in practice produce stars with different masses. It generates the so-called initial mass function $\varphi(M)$, defined so that $\varphi(M)dM$ gives the relative number of stars born with a mass between M and $M + dM$. As massive stars die early, the observed mass function of a cluster differs from the initial one.

Equation (7.43) may be interpreted simply in a way that permits us to understand the concept of Jeans mass better. Gravitational collapse will take place when the gravitational potential energy of a particle equals its thermal energy.

Collapse of magnetized clouds

This problem is too complex to be handled analytically, requiring numerical models to follow the overall picture. Let us begin to approach the problem with a very simplified calculation. We still assume a spherical contraction of a cloud with frozen-in field. The Virial theorem must now include two new terms:

$$\mathcal{B} = \frac{1}{24} B^2 R^3 \tag{7.48}$$

obtained from (4.98) if B remains homogeneous in the contraction, and $\oint_S (\mathcal{M} \cdot \vec{r}) \cdot d\vec{\sigma}$. To estimate the latter, a hypothesis must be assumed as to the way in which the unperturbed homogeneous external field is connected, through the cloud surface, to the compressed internal field. The internal field must be kept parallel to the external. A rough hypothesis consists of assuming that the connecting surface field is radial in one of the hemispheres of the spherical cloud, and antiradial in the other hemisphere (see Figure 4.1). As for the strength, we will explore the situation using the average between the external and the internal values, which is of the order of the higher internal value. With this type of field we find a contribution of $\frac{1}{4} B^2 R^3$. From (7.48) we may conclude that the magnetic contribution is $(7/24) B^2 R^3$. Obviously, our estimate is so crude that the numerical coefficient is unimportant. We will, then, assume a magnetic contribution of the order of $B^2 R^3$. The Virial theorem becomes

$$4\pi R^3 p = \frac{3MkT}{m} - \frac{1}{R}\left(\frac{3}{5}GM^2 - B^2 R^4\right) \tag{7.49}$$

Obviously, B increases when R decreases. The exact relation should be deduced from the frozen-in condition, which implies conservation of the magnetic flux. As BR^2 is proportional to the magnetic flux, the quantity $\varphi^2 = B^2 R^4$ in (7.49) can be assumed to be constant during the contraction process. Equation (7.40) is still valid where Y is still given by (7.41) and Z is now replaced by

$$Z = \frac{3}{20\pi}GM^2 - \frac{\varphi^2}{4\pi} \tag{7.50}$$

We go on to obtain

$$R_M = \frac{4}{15}\frac{GMm}{kT}\left(1 - \frac{5}{3}\frac{\varphi^2}{GM^2}\right) \tag{7.51}$$

$$p_M = \frac{\pi}{M^2 G^3}\left(\frac{kT}{m}\right)^4 \frac{1}{\left(1 - \frac{5}{3}\frac{\varphi^2}{GM^2}\right)^3} \tag{7.52}$$

The quantity φ^2/GM^2 is proportional to the relation between the magnetic energy of the cloud ($B^2/8\pi$ times R^3) and its potential energy (GM^2/R). When the magnetic field is high, the ambient pressure required to bring about collapse is higher. Hence, we can conclude that magnetic fields are an impediment to gravitational collapse. Even when

$$1 = \frac{5}{3}\frac{\varphi^2}{GM^2} \tag{7.53}$$

collapse is impossible. Therefore, there is a limiting mass M which will collapse for any given ambient magnetic field strength. From (7.53) and $M = \frac{4}{3}\pi R^3 \rho$, with ρ being the ambient density, the limiting mass is

$$M_B = 0.26 \frac{B^3}{G^{3/2}\rho^2} \tag{7.54}$$

For $B \approx 2 \times 10^{-6}$ G and a density of ten particles per cm^3, we obtain $M_B \approx 3 \times 10^4 M_\odot$.

From (7.52) (or similarly from (7.51)), with $p_M = nkT$ and $M = \frac{4}{3}\pi R^3 \rho$, it is possible to obtain the critical mass, M_{CR}, for any value of the magnetic field. After an easy derivation we obtain

$$M_{CR} = M_J \left(1 + \frac{v_A^2}{v_S^2}\right)^{3/2} \tag{7.55}$$

where v_A is the Alfvén speed and v_S is the sound speed. It is also easy to show that there is a limiting value of M_{CR} when $v_A \gg v_S$, that is, when $M_{CR} \to M_B$.

Though the above discussion is instructive, it is partially incorrect, for several reasons. The cloud cannot remain spherical, as contraction along magnetic lines of force is favoured. The contraction is non-homologous, that is, the different equilibrium configurations are not self-similar. The condition of frozen-in magnetism is not met, particularly in the central parts. There, the density is very high, so that the central parts become opaque for cosmic radiation, which is responsible for most of the ionization of the cloud. For high densities, ambipolar diffusion breaks the frozen-in condition. Gravitational collapse is not impeded in the central parts of the clouds, where star formation takes place. The extended envelope is destroyed by winds from new stars or by supernova explosions. Because of these complex processes, numerical simulations show that the limiting mass M_{CR} is only about 15% of that deduced previously.

There is also ohmic dissipation at some stages of the contraction, also avoiding the frozen-in condition of the cloud. However, this condition is again restored when a temperature of 1600 K is reached. Then dust grains are destroyed, H_2 dissociates, and H and some metals become ionized. A high degree of ionization ensures the fulfillment of the frozen-in condition once more.

There is another action of the magnetic field which favours the collapse process. Obviously, another important impediment for collapse is angular momentum, which the cloud initially possesses because of galactic turbulence, or just as a result of the differential rotation of the galaxy. Lines of force connect the rotating inner parts with the static distribution outside, establishing a magnetic friction which brakes the cloud. Magnetic disturbances, similar to Alfvén waves, are generated and transport angular momentum outwards. Magnetic braking not only permits the collapse of rotating clouds, but is also present in the last fragmentation. Also, protostars lose angular momentum, probably in connection with the formation of planetary systems.

7.3.4   Accretion discs and jets

A large variety of astrophysical objects show evidence of accretion processes into a central compact body. Some of them are related to protostellar clouds but our attention will now be focussed on accretion in galactic nuclei. Large amounts of energy are involved, either as emitted radiation, for example in quasars and Seyfert galaxies, or in jet outflows, for example in radio galaxies.

Active galactic nuclei (AGN) show variability on small time scales. Causality restrictions allow us to estimate an upper limit to the size as $c\Delta\tau$, $\Delta\tau$ being the characteristic time of variability. X-ray emission presents very small values of $\Delta\tau$, indicating that AGN must be smaller than 10^{15} cm. The best resolution provided by VLBI is much poorer, corresponding to about 1 pc at typical distances. The dimensions of 10^{15} cm are more remarkable when the high energy output of, say, $10^{14}L_{\odot}$ in such compact objects is taken into account. The motion of the surrounding gas clouds and stars provides an estimate of the AGN mass, of the order of $10^{8}M_{\odot}$. The nuclei of all galaxies probably have some degree of activity. Even our Galaxy has a moderately active nucleus with a central mass of only $10^{6}M_{\odot}$.

A $10^{8}M_{\odot}$ object has a Schwarzschild radius of about 4×10^{6} cm, which is much less than 10^{15} cm. Nevertheless, the small size of the unresolved nuclei led to the initial conclusion that the central object was a black hole. This assumption has become widely accepted, although alternative hypotheses cannot be excluded.

Falling matter will have an increasing orbital velocity and a thin disc is expected to be formed. This is called an accretion disc and extends to about 10^{15} cm, and is thus not directly observable. Keplerian orbital velocities produce differential rotation and hence shear which is particularly high for small radii close to the Schwarzschild radius. Molecular viscosity has a very large characteristic time, as we showed for galactic discs. But other viscosity mechanisms generated by turbulence or magnetic fields play an important role as they permit angular momentum losses and therefore radial infall, and also heating in the accretion disc. This heating gives rise to thermal emission at UV and optical wavelengths, which constitutes a component of the AGN continuum, also called the ionizing continuum. Other components of the continuum spectrum are synchrotron and inverse Compton scattering.

The disc luminosity may be so high as to produce a radiation pressure preventing accretion. This critical luminosity is called the Eddington luminosity L_E. From the discussion of equations (3.172) and (5.182), we know that this can be calculated using

$$L_E = \frac{4\pi GMmc}{\sigma_T} \approx 30\,000 \left(\frac{M}{M_{\odot}}\right) L_{\odot} \tag{7.56}$$

where M is the black hole mass and σ_T is the Thomson scattering cross-section, as this is the dominant process producing the opacity. The Eddington accretion rate $\dot{M}_E$ is defined by $\dot{M}_E = L_E$, that is, if all the accreted matter were converted into radiation, $\dot{M}_E$ would be the limiting matter accretion rate. Very luminous AGN have $L \approx L_E$ and $\dot{M} \approx \dot{M}_E$, but not all AGN obey this relation. For instance, radio galaxies invert the mass energy input $\dot{M}_E$, producing jets of rapidly moving particles instead of radiation.

At about 1 pc from the central black hole is situated the so-called broad line region. The ionizing continuum emitted by the accretion disc ionizes gas clouds with high densities ($\sim 10^9 \mathrm{cm}^{-3}$) and high temperatures. This effect produces broad lines in the spectrum ($1000 - 10000$ km s^{-1} wide). The time variability of the broad line region is of the order of one week. Within a radius of 1 kpc there is a 'narrow line region' with forbidden emission lines (densities in the range $10^3 - 10^6 \mathrm{cm}^{-3}$). These lines are only 100–1000 km s^{-1} wide. Variability times are of the order of years. Seyfert galaxies of type 2 do not present broad lines in their spectra. Infrared thermal emission from dust, probably heated by the active nucleus, is also present in most AGN. Nuclear non-thermal emission also contributes to the infrared spectra, and in fact dominates in type 1 Seyferts. Reradiating dust clouds probably lie in the host galaxy, away from the nucleus. By contrast, the X-ray emission region must be very close to the black hole.

The most intractable questions about AGN are the interrelation between different types (with the angle of observation and the time of evolution being explanatory factors), the mechanism of jet production, and the physical processes in the jet itself. An optically thick, geometrically thin accretion disc may explain the high luminosity of quasars, which are the farthest AGN. During this phase the hole increases in mass and rotation velocity. Then the available supply of plasma is reduced. With low values of $\dot{M}$ other processes come into action in radio galaxies. If $\dot{M}$ is low, a low luminosity is to be expected. The problem then is the formation of jets. This process is probably a problem of magnetohydrodynamics in a highly curved space-time, which cannot be discussed here because the physics is rather complex and because present theoretical models are far from fully satisfactory. Thick discs are probably formed, supported by the ion pressure (called 'ion supported tori'), with an ion temperature much higher than the electron temperature. The energy to produce the beams along the black hole rotation axis must be extracted from the black hole spin energy. This is impossible in Schwarzschild black holes but not in Kerr–Newman black holes, where magnetic fields may provide a mechanism for the extraction of energy which can be a large fraction of the total mass of the black hole. The mass and energy of the jets greatly exceeds $\dot{M}c^2$. Here, it should be borne in mind that a large variety of hypotheses and models exist, those based on relativistic magnetohydrodynamics being the most promising.

Both jets, perpendicular to the disc or the torus, that is, along the black hole rotation axis have relativistic velocities and are extremely well collimated. They may be observed, mainly on pc-scales, by VLBI techniques and travel a long way, of the order of 100 kpc, to feed the observed lobes of radioemission. These are huge structures, much larger than the parent elliptical galaxy. Sometimes the jet meets the ambient medium and the transition shock produces a bright spot in the outer edge of the lobe. On other occasions, the jet decelerates in the turbulent central parts of the lobes, and the central parts are then brighter. Mechanisms for accelerating electrons and for producing dynamos to maintain high magnetic fields are required, and these are still largely unknown.

8 The Newtonian cosmic fluid

Some naïve thoughts provide a first insight into Newtonian cosmology. Olbers' paradox, introduced in Chapter 6 (6.7), suggests that the Universe is finite, either in space or in time, or both. If the Universe were finite in space and infinite in time, the whole Universe would have collapsed gravitationally. Therefore, the Universe is finite in time. Are there other simple arguments with which to assess the finiteness of space, that is, of matter content of the Universe? The cosmological principle may provide such an argument.

The cosmological principle is widely accepted as a reasonable, basically compatible with observations, philosophically attractive principle. From the Newtonian point of view its statement and interpretation does not present any difficulty: the point in space at which we are in the Universe is not a special point; all points in the Universe are similar, or, more precisely, all thermodynamic parameters have the same value at any point in the Universe: the Universe is homogeneous. The cosmological principle also ensures the isotropy of the Universe, that is, all directions are equivalent. Isotropy implies homogeneity, but homogeneity does not imply isotropy. For example, if the Universe were embedded in a constant magnetic field, it could be homogeneous but anisotropic. However, the Newtonian interpretation of the cosmological principle usually states that the Universe is both homogeneous and isotropic. This would only be true on a very large scale in the Universe, that is, larger than characteristic sizes of superclusters.

Once this principle is accepted, we might conclude that the Universe is infinite in space. Otherwise, a centre and an edge of the Universe would exist, in clear contradiction with the cosmological principle. An observer at the edge could observe galaxies in one direction and none in the opposite direction. This conclusion, which seems intuitive from a Newtonian perspective, cannot be drawn from a relativistic perspective: a non-Minkowskian geometry may provide a finite Universe with neither a centre nor an edge. This shortcoming of Newtonian cosmology might lead us to exclude it from modern treatments of cosmology. However, it is often used as an introduction. The main reason is that the differential equations provided by Newton's mechanics are the same as those provided by general relativity. Therefore, although having a very different language, both descriptions are locally identical. This noticeable coincidence, which must subsequently be justified, illustrates the introductory aspect of Newtonian cosmology, despite the fact that it was developed after relativistic cosmology. In fact, Newtonian cosmology is one of the simplest applications of classical fluid dynamics.

The perfect cosmological principle assumes not only homogeneity and isotropy of space, but also a similar property of time: the Universe today is as it was and will be at any time, that is, all thermodynamic parameters, such as density and temperature, are independent of time. While philosophically attractive, the perfect cosmological principle does not appear to be compatible with observations, and is not adopted in most current cosmological theories. This principle implies that the Universe is infinite in time, a possibility previously rejected. The evolution of stars and galaxies is also most easily interpreted in terms of irreversible processes, so it is difficult to believe that the Universe has not changed over time. Our arguments were certainly rather naïve, but an observational fact is difficult to reconcile with the perfect cosmological principle: the expansion of the Universe.

Spectral lines of galaxies have a positive redshift, defined as

$$z = \frac{\lambda - \lambda_0}{\lambda_0} = \frac{\nu_0 - \nu}{\nu} \tag{8.1}$$

where λ_0 and ν_0 are the rest wavelength and frequency, and λ and ν are those of the observed spectral line. Hubble's law states that

$$z = H_0 r \tag{8.2}$$

with r being the distance to the Galaxy and H_0 being Hubble's constant. The most straightforward interpretation of this redshift is a doppler effect, produced by an expansion velocity. For moderate velocities compared with unity, the change in wavelength is given by

$$\lambda = \lambda_0 (1 + v) \tag{8.3}$$

Therefore, with this interpretation,

$$\vec{v} = H_0 \vec{r} \tag{8.4}$$

where $\vec{v}$ is the macroscopic velocity. The usual subindex 0 has been suppressed because in cosmology it denotes present-day values. The value of H_0 is still subject to observational improvement. It is commonly expressed in km $s^{-1}Mpc^{-1}$, having a magnitude in the range 50–100. When not otherwise specified, an intermediate value of 75 km $s^{-1}Mpc^{-1}$ is adopted here. This is equivalent to $2.5 \times 10^{-18} s^{-1}$, or to $8.33 \times 10^{-29} cm^{-1}$. If the motion of galaxies is described by Hubble's law in the form of (8.4), all galaxies were here H_0^{-1} years ago; as $H_0^{-1} = 4 \times 10^{17} s \approx 1.3 \times 10^{10}$ years, this is a first rough estimate of the age of the Universe, called the Hubble time.

Let us recall the equation of continuity (1.51):

$$\frac{\partial \rho}{\partial t} + \nabla \cdot (\rho \vec{v}) = 0 \tag{8.5}$$

where ρ may represent the HI density, or the mass of superclusters per unit volume. Because ρ is homogeneous, $\nabla \rho = 0$. Substituting (8.4):

$$\frac{\partial \rho}{\partial t} = -\rho \nabla \cdot \vec{v} = -\rho H_0 \nabla \cdot \vec{r} = -3\rho H_0 \tag{8.6}$$

Partial or dry observer derivatives, and convective or wet observer derivatives now coincide as

$$\frac{d\rho}{dt} = \frac{\partial \rho}{\partial t} + \vec{v} \cdot \nabla \rho \tag{8.7}$$

and $\nabla \rho$ is now zero. Therefore:

$$\frac{d\rho}{dt} = -3\rho H_0 \tag{8.8}$$

at present. The perfect cosmological principle would imply that $\frac{d\rho}{dt} = \frac{\partial \rho}{\partial t} = 0$ and therefore $3\rho H_0 = 0$. This cannot be true as neither ρ nor H_0 vanish. The expansion requires a time variation of density which is incompatible with the perfect principle.

Historically, the 'steady-state' cosmology surmounted this difficulty by assuming continuous mass creation (Bondi, Gold, and Hoyle in 1948). If (8.5) is modified, including a production term P, that is, mass created per unit time and unit volume, and assuming that $\partial \rho / \partial t = 0$, $\nabla \rho = 0$,

$$3\rho H_0 = P \tag{8.9}$$

For a current density of about 5×10^{-30} g cm^{-3}, we derive a production rate of only $P = 3 \times 10^{-47}$ g cm^{-3} s^{-1}, which is far too low to be detected in the laboratory. Observational restrictions made the steady state cosmology to be rejected, and we can infer that the perfect cosmological principle is incompatible with observations.

8.1   The continuity equation

Hubble's law is derived from observations, but it can also be derived from the cosmological principle and the continuity equation. Let us start with (8.5) and assume that $\nabla \rho = 0$, not only at present but at any time in the history of the Universe. We then find that

$$\frac{1}{\rho} \frac{d\rho}{dt} = -\nabla \cdot \vec{v} \tag{8.10}$$

where ρ is a function of time only; therefore the left-hand side depends only on time. This function of time will be called $-3H(t)$, so we may write

$$\nabla \cdot \vec{v} = 3H(t) \tag{8.11}$$

As $\nabla \cdot \vec{r} = 3$, and $H(t)$ does not depend on the space coordinates, (8.11) is equivalent to

$$\nabla \cdot \vec{v} = \nabla \cdot (H(t)\vec{r}) \tag{8.12}$$

The integration provides

$$\vec{v} = H(t)\vec{r} + \nabla \times \vec{\phi}(\vec{r}, t) \tag{8.13}$$

where $\vec{\phi}(\vec{r}, t)$ is an arbitrary function. In order not to violate the isotropy of the Universe, $\vec{\phi} = a(r, t)\vec{r}$, where $a(r, t)$ is another arbitrary function, which depends on r but not on $\vec{r}$. However, the curl of such a function vanishes and therefore

$$\vec{v} = H(t)\vec{r} \tag{8.14}$$

This is valid for any time, while (8.4) was valid only for the present time, $H_0 = H(t_0)$. As H_0 is known from observations, but the function $H(t)$ remains unknown, other fluid equations will be needed in order to solve this problem. Indeed, note that H may also be negative or zero.

Instead of solving the differential equation for $H(t)$, it is conventional to use another time function $R(t)$, which is loosely termed the radius of the Universe, or, more correctly, the cosmological scale factor, defined by

$$H = \frac{1}{R}\frac{dR}{dt} \qquad (8.15)$$

Note that this definition is not complete. If R is a solution, kR, where k is any constant, is also a solution. This constant remains undetermined and will be chosen later (equation 8.24). From (8.10),

$$\frac{1}{\rho}\frac{d\rho}{dt} + 3\frac{1}{R}\frac{dR}{dt} = 0 \qquad (8.16)$$

therefore

$$\rho R^3 = \text{constant} \qquad (8.17)$$

Again using subindex 0 to denote present quantities,

$$\rho = \frac{\rho_0 R_0^3}{R^3} \qquad (8.18)$$

equation (8.17) gives the reasoning behind the name 'radius of the Universe'. Note that the possibility of an infinite Universe has not been excluded.

8.2   The motion equation

Our purpose is now to determine $H(t)$, or $R(t)$ which is equivalent. From (1.78) and neglecting viscosity, the equation of motion may be written as

$$\frac{\partial \vec{v}}{\partial t} + \vec{v}\cdot\nabla\vec{v} = -\frac{\nabla p}{\rho} - \nabla\mathcal{F} \qquad (8.19)$$

where p is the pressure and $\mathcal{F}$ is the gravitational potential. As p is a thermodynamic parameter $\nabla p = 0$. From (8.14):

$$\frac{\partial \vec{v}}{\partial t} = \frac{dH}{dt}\vec{r} \qquad (8.20)$$

$$\vec{v}\cdot\nabla\vec{v} = (H\vec{r})\cdot\nabla(H\vec{r}) = H^2\vec{r}\cdot\delta = H^2\vec{r} \qquad (8.21)$$

By applying ∇ in (8.19), taking (8.20) and (8.21) and Poisson's equation into account,

$$3\frac{dH}{dt} + 3H^2 + 4\pi\rho = 0 \qquad (8.22)$$

If H and ρ are expressed as functions of R:

$$\frac{d^2 R}{dt^2} = -\frac{4\pi}{3}\frac{\rho_0 R_0^3}{R^2} \qquad (8.23)$$

which is the equation of motion of the Universe. A first integral is obtained in a straightforward way:

$$\left(\frac{dR}{dt}\right)^2 = \frac{8\pi\rho_0 R_0^3}{3R} - k \qquad (8.24)$$

where k is an integration constant. As a scale factor is still required to define R, we may choose k to have a value of 1, -1, or zero. Depending on the value of k, we obtain three different families of model Universes.

The inclusion of viscosity in the above derivation does not modify the result, as a velocity field given by (8.14) does not produce any shear.

The simplest model is the Einstein–de Sitter model, with $k = 0$. In this case (8.24) is easily integrated:

$$R = R_0 (6\pi\rho_0)^{\frac{1}{3}} t^{\frac{2}{3}} \qquad (8.25)$$

with a proper choice of initial conditions ($R = 0$ at $t = 0$). Once R is known, it is easily calculated from (8.15) that

$$H = \frac{2}{3}\frac{1}{t} \qquad (8.26)$$

For $t = t_0$, $H = H_0$, that is, the Hubble constant. Therefore, the age of the Universe in this model is

$$t_0 = \frac{2}{3}\frac{1}{H_0} \qquad (8.27)$$

which is shorter than the Hubble time, H_0^{-1}. From (8.18) we obtain the density

$$\rho = \frac{1}{6\pi t^2} \qquad (8.28)$$

For the present-day density we then obtain $\rho_0 = (6\pi t_0^2)^{-1} = 3H_0^2/8\pi \approx 7.5 \times 10^{-37}\,\text{s}^{-2} \approx 10^{-29}\,\text{g cm}^{-3}$ (see Appendix). However, ρ_0 depends on H_0^2, which is known to be uncertain by half an order of magnitude. This value would only correspond to the present density if our Universe were an Einstein–de Sitter Universe. Whether this is the case is open to debate, and we will comment on it in the following two chapters. Observations seem to indicate lower densities, but important theoretical ideas support the hypothesis that we are in an Einstein–de Sitter Universe and that the present density has the critical value $3H_0^2/8\pi$. The density parameter Ω is defined by

$$\Omega = \frac{8\pi\rho}{3H^2} \qquad (8.29)$$

and takes the value unity for the critical density of the Einstein–de Sitter model. Its value at present, $\Omega_0 = \frac{8\pi\rho_0}{3H_0^2}$, is an important cosmological parameter, which may in principle be obtained from observations: in practice this is a notoriously difficult challenge.

A $k = -1$ Universe is said to be open. An analytical solution of (8.24) exists (for $k = -1, 0, 1$) which will be discussed in the following chapter. A $k = 1$ Universe is said to be closed, or cyclic. In this case $dR/dt = 0$ for the value of the radius of the Universe, R_M:

$$R_M = \frac{8\pi}{3}\rho_0 R_0^3 \qquad (8.30)$$

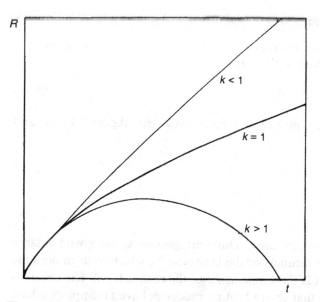

Figure 8.1 The expansion of open ($k < 1$), flat ($k = 1$), and closed ($k > 1$) universes. R is the radius or cosmological scale factor and t is the time.

After attaining this radius, the Universe would contract, instead of expanding, and would eventually undergo a 'Big-Crunch' (i.e. an inverse 'Big-Bang'), with $R = 0$ (see Figure 8.1). All such universes have $R = 0$ at some time in the past that is taken as the origin. In fact, we know that at present $dR/dt > 0$ (expansion) and that $d^2R/dt^2 < 0$ always (8.23). Therefore, all these models have a 'Big-Bang' origin. At that moment the density was infinite.

We also define a deceleration parameter q by

$$q = \frac{-R(d^2R/dt^2)}{(dR/dt)^2} \tag{8.31}$$

Some simple relations between the key parameters in cosmological models can easily be demonstrated:

$$\frac{d}{dt}\frac{1}{H} = q + 1 \tag{8.32}$$

$$\Omega = 2q \tag{8.33}$$

$$k = R^2 H^2(\Omega - 1) \tag{8.34}$$

The last equation clearly shows that if $k = 1$, $\Omega > 1$; if $k = 0$, $\Omega = 1$; if $k = -1$, $\Omega < 1$. The density of the Einstein–de Sitter model is critical.

If $\Omega > 1$, the self-gravitation of the Universe as a whole will be able to stop the expansion and produce a collapse in a finite time. If the density is low, so that $\Omega < 1$, the expansion will continue for ever.

8.3   The heat balance equation

Let us first carry out a simple derivation as in the preceding section, now taking the heat balance equation (1.79) into account in the form

$$\frac{3}{2}\frac{\rho}{m}k\left(\frac{\partial T}{\partial t}+\vec{v}\cdot\nabla T\right)+p\nabla\cdot\vec{v}=0 \tag{8.35}$$

It is implicitly assumed that the Universe is a monatomic gas. Again $\nabla T = 0$ and therefore

$$\frac{1}{T}\frac{dT}{dt}=-2H \tag{8.36}$$

or, taking (8.15) into account,

$$TR^2 = \text{constant} \tag{8.37}$$

However, this calculation may be too naïve. Different gaseous systems with a large variety of temperatures are found throughout the Universe. To which of them does the temperature in (8.37) refer? What is the temperature of the cosmic fluid? The observation of stars and galaxies reveals that severe heating processes have taken place which have not been taken into account in (8.35). Equation (8.37) would probably have been valid for the pregalactic Universe, but we know that early epochs were radiation dominated, and the above derivation would then have been valid for photons rather than for monatomic molecules. Such a calculation would also be applicable to the cosmic background radiation at ~ 2.7 K primordial radiation with very little interaction with matter at present.

In the background radiation, the number of photons per unit volume decreases as R^{-3}. Their individual energies, as a consequence of redshift, decrease as R^{-1}. Therefore, the energy density decreases as R^{-4}. But the radiative energy density is proportional to T^4. Therefore, instead of (8.37), we can write

$$TR = \text{constant} \tag{8.38}$$

or

$$T = \frac{T_0 R_0}{R} \tag{8.39}$$

where T_0 would be the present temperature of the background black body microwave radiation, that is, slightly less than 3 K. Another derivation of (8.39), maintaining, of course, the same level of simplicity as given above, is obtained as follows. Assume that matter and radiation are in equilibrium, but that the radiation pressure dominant, so that $p = \frac{1}{3}aT^4$ (equation (3.87)). If the Universe cools as it expands, and with the radiation pressure being proportional to T^4 and the matter pressure being proportional to T, this domination by radiation must have occurred at high values of T, that is, at the earliest epochs. Let us, however, solve the heat balance equation for matter. Instead of (8.35) we have

$$\frac{3}{2}\frac{\rho}{m}\frac{dT}{dt}+3H\frac{1}{3}aT^4=0 \tag{8.40}$$

Using (8.15) and (8.18)

$$R^3 T^3 = \frac{3}{2} \frac{\rho_0 R_0^3}{ma} \tag{8.41}$$

obtaining (8.38) again with suitable boundary conditions. We could go on to find a value of T_0 from this equation, but this would represent the hypothetical temperature of present-day matter if it had remained in equilibrium with the background radiation. We know that this is not the case, so that the constant in (8.41) cannot provide the current temperature.

8.4   Epochs of the Universe

The above analysis was really just an introduction. The equations may not be valid for some early epochs of the Universe, for instance when the cosmic fluid was relativistic. Two valid conclusions should be stressed: the existence of a Big-Bang, and the cooling from the Big-Bang until the present, so that $dT/dt < 0$. Looking back in the past at a cosmic fluid with increasing temperature, we should be aware that sudden changes in composition and physical properties may have modified the scenario substantially. These changes are associated with sudden changes in the equation of state, which in turn influence the expansion and the temperature evolution.

The present Universe is characterized by the existence of atoms. Most of them were produced in stars, but H, He and other minor light constituents are primordial and precede galaxy formation. If we run the clock of the Universe backwards this fluid is heated and when the temperature reaches a value of about 3000 K ionization begins to be important. This epoch is called the recombination epoch. Before this time, at 3000 K, a large quantity of photons existed, and Thomson scattering by free electrons provided an effective interaction between matter and radiation. Matter and radiation were in thermal equilibrium. After recombination, free electrons suddenly disappeared and the photons were 'decoupled' from the matter. The Universe became transparent, so that the non-interacting photons conserved their black-body distribution, their temperature becoming lower and lower as the Universe continued to expand. The cosmic background microwave radiation at ~ 2.7 K is identified with the photon system that was in equilibrium before the recombination epoch. According to (8.38) this event occurred when the Universe was about 3000/3 times smaller than it is today, since 3000 K is a slightly lower temperature than the ionization potential of hydrogen. Near recombination the Universe suffered a transition from radiation domination to matter domination, and the matter already contained slight irregularities in its density distribution which later led to the formation of galaxies.

Formation of pregalactic structures is considered in detail in Chapter 10. Here a very rough estimation of the time when this event took place is obtained as follows. The present relation of the intergalactic distances and galactic diameters is of the order of 1000. When the Universe was 1000 times smaller than it is now, galaxies were in contact with one another; thus matter was more or less homogeneously distributed.

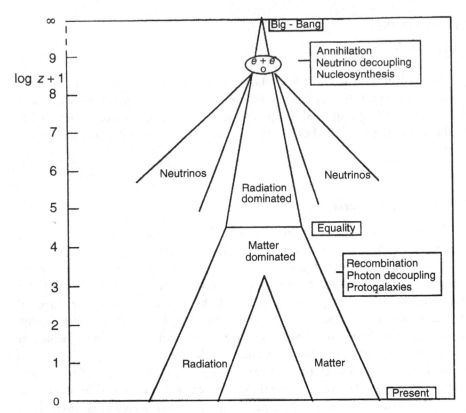

Figure 8.2 Schematic representation of the post-one-second universe. *z* is the redshift. The *x*-axis has no quantitative meaning.

Before that, an epoch existed during which the radiative energy density was higher than the matter energy density. The transition took place for

$$aT^4 = mn \qquad (8.42)$$

It will be seen in Chapter 10 that this event took place when the Universe was 10^4 times smaller than its present size. An overall picture is depicted in Figure 8.2. Before the recombination epoch, about 10^6 years after the Big-Bang, there was the 'plasma era', and afterwards the 'atomic era'. The chemical composition then was similar to the present composition: electrons, nuclei of H and He, photons, neutrinos, and dark matter particles, still unidentified (considered in Chapter 10).

We will now go back to several seconds after the Big-Bang, to find another epoch rich in decisive events. This may be called the annihilation epoch, in which three notable events took place: annihilation of electron–positron pairs, nucleosynthesis of helium, and neutrino decoupling. Particles with $m = T$ are created in particle–antiparticle pairs in important amounts when their appropriate temperature is reached. For electrons, this temperature is about 6×10^9 K, reached when the Universe was 10^9 times smaller than it is today, about ten seconds after the Big-Bang. Before annihilation the Universe contained large amounts of electrons and positrons, which annihilated in pairs to produce photons, thus enhancing the temperature. Electrons were slightly in excess with respect to positrons, and this excess prevailed, fortunately for us! It was during this epoch that the temperature was sufficiently low but the density sufficiently high to produce nucleosynthesis of helium, as well as other heavier constituents, from free protons and neutrons. The proportion of 87 protons to 14 neutrons that exists at present was determined then. The details of this nucleosynthesis provide a tool for checking cosmological models. Before about 1s, the high density enabled neutrinos to remain thermally coupled with matter particles and photons. They then decoupled, and must have a lower present-day temperature (~ 2 K) than photons because of the rise in photon temperature due to the annihilation process. Before the annihilation epoch there was the 'particle era', during which Π and K mesons were produced in large quantities.

Classical cosmology can deal with events from about 1 s after the Big-Bang. The study of earlier epochs requires the inclusion of modern elementary particle physics: the standard model of strong and electroweak interactions, grand unification, and super unification. At energies of about 1 GeV ($\sim 3 \times 10^{12}$ K) hadrons and mesons were produced out of quarks. This baryogenesis took place at about 10^{-5} s, when the Universe was about 10^{12} times smaller than it is today. Before the baryogenesis epoch the cosmic fluid was a quark–gluon plasma, and this is called the 'quark era'. At about 300 GeV a spontaneous symmetry breaking took place, corresponding to the electroweak phase transition. This marked the end of the 'electroweak unification era'. Prior to this, at $\sim 10^{15}$ GeV, the grand unification phase transition marked the end of the 'grand unification era'. The origin of matter–antimatter asymmetry during this epoch, and of possible topological defects, such as massive magnetic monopoles, must be investigated. The expansion rate then was characterized by inflation, a quantum effect that may be reproduced by adopting an equation of state of the type $p = -\epsilon$, which produced a fast exponential expansion, rendering the Universe flat. The grand-unification era begins at the Planck epoch, that is, at 10^{-43} s, with a temperature of 10^{32} K and a size for the Universe of 4×10^{-33} cm (the Planck diameter). Quantum gravity must be developed to investigate this epoch in which all four forces of nature were unified.

8.5 The Milne universe

The Milne universe is based on concepts of special relativity, and not on the field equations of general relativity. It is, then, a universe that is half-way between Newtonian and full relativistic cosmologies, sharing with the former its introductory and didactic interest. It can serve here as a bridge between the two.

Let us assume a point-like Big-Bang, observed from a Minkowskian frame. Particles flow with different radial velocities $u = r/t$, where r and t are radius and time for our Minkowskian observer. Of course, $u \leq 1$. Let us identify ourselves with this observer, placed at the centre of a sphere expanding at the speed of light. Assume that u is a constant for each particle, with no gravity perturbing its initial velocity. We see that the number density of particles decreases when our time t increases, following a law of the type $n \propto t^{-3}$. Consider now another observer placed on one of the particles that was ejected at the Big-Bang. Is it possible for this observer, in agreement with the cosmological principle, to obtain the same picture of the Universe that we obtain?

Our observer must also observe $n_0 = K\tau^{-3}$, where n_0 and τ are now the proper number density and proper time of the observer and K is a constant of proportionality. Obviously, $\tau\gamma = t$, where $\gamma = (1 - r^2/t^2)^{-1/2}$. This observer does not observe n_0 particles per unit volume, but rather n, where $n = \gamma n_0$. Thus

$$n = \gamma n_0 = \frac{\gamma K}{\tau^3} = \frac{\gamma^4 K}{t^3} = K\frac{t}{(t^2 - r^2)^2} \tag{8.43}$$

Only if we also observe this density distribution $n(r, t)$ can we affirm that all inertial observers see the same as we do. Note that for $r = 0$ we obtain our own number density and, in fact, using $r = 0$ in (8.43) we also obtain $n = Kt^{-3}$. Once the solution has been obtained, we find the density distribution rather paradoxical. As we must observe the Universe at a given time $t = constant$, and with all possible radii $r \le t$, n increases with r, and even becomes infinite for $r = t$! In this model we also find an edge to the Universe. However, an observer near what we consider to be the edge, does not see an edge, which is possible because there is an infinite number of galaxies there.

As we are in a Minkowskian system we should write for the line element

$$dS^2 = -dt^2 + dr^2 + r^2(d\theta^2 + \sin^2\theta\, d\phi^2) \tag{8.44}$$

Let us introduce some variable changes. As a radial coordinate we shall use

$$\rho = u\gamma \tag{8.45}$$

As $u = r/t$, and hence ρ is constant for any particle ejected in the Big-Bang, ρ is a coordinate which characterizes a given particle. Each ejected particle always has the same ρ. Such a coordinate is said to be a comoving one. As a time coordinate, the proper time τ of each ejected particle will be adopted, the transformation law yielding $\tau\gamma = t$. Angular coordinates θ and ϕ remain unchanged. We may obtain $t = t(\tau, \rho)$ and $r = r(\tau, \rho)$:

$$r = \rho\tau \tag{8.46}$$

$$t = \tau(1 + \rho^2)^{1/2} \tag{8.47}$$

and substitute in (8.44) to obtain

$$dS^2 = -d\tau^2 + \frac{\tau^2\, d\rho^2}{1 + \rho^2} + \rho^2\tau^2(d\theta^2 + \sin^2\theta\, d\phi^2) \tag{8.48}$$

Therefore, the spatial line element is

$$dl^2 = \tau^2\left(\frac{d\rho^2}{1 + \rho^2} + \rho^2(d\theta^2 + \sin^2\theta\, d\phi^2)\right) \tag{8.49}$$

The expression inside the large bracket is geometrically identified with the line element of a three-dimensional space with constant curvature, with $k = -1$. Actually, a constant curvature three-dimensional space has a line element of the form

$$dl^2 = \frac{d\rho^2}{1 - k\rho^2} + \rho^2(d\theta^2 + \sin^2\theta\, d\phi^2) \tag{8.50}$$

The best way to check this is to calculate the scalar curvature of a three-dimensional space where the metric tensor $g_{\alpha\beta}$ is of form

$$g = \begin{bmatrix} \frac{1}{1-k\rho^2} & 0 & 0 \\ 0 & \rho^2 & 0 \\ 0 & 0 & \rho^2\sin^2\theta \end{bmatrix} \tag{8.51}$$

But the term in parentheses in (8.49) is multiplied by τ^2, which means that the curvature is constant in the three-dimensional space, but is time-dependent. Indeed, with the coordinate change $d\rho' = \tau\, d\rho$, $\rho' = \tau\rho$, (8.49) can be rewritten as

$$dl^2 = \frac{d\rho'^2}{1 + \frac{\rho'^2}{\tau^2}} + \rho'^2(d\theta^2 + \sin^2\theta\, d\phi^2) \tag{8.52}$$

The curvature index $(-1/\tau^2)$ decreases with time. Some properties of the Milne line element are very useful from the cosmological point of view and will be commented upon further in the following chapter.

9 The relativistic cosmic fluid

Cosmology is a general relativistic topic. General relativity must replace Newtonian mechanics for systems in which the mass and spatial dimensions have similar orders of magnitude, $M \sim r$. If the density of the Universe were a constant, $M/r \approx \rho r^3/r \approx \rho r^2$, and for large enough length scales, the condition $M \approx r$ would eventually be met. Let us take the current density of 10^{-29}g cm^{-3} ($= 7.4 \times 10^{-58}$cm^{-2} in geometrized units). Then, for length scales of $r \approx 3.7 \times 10^{28}$cm ($\approx 10^4$ Mpc), calculations must be carried out using general relativity. This length is of the order of magnitude of the observable Universe, and corresponds to objects about 10^{10} years old, near the commonly accepted age of the Universe.

9.1 The cosmological principle

The Newtonian interpretation of the cosmological principle must now be reconsidered. It stated that at a given time all thermodynamic parameters are homogeneously and isotropically distributed. But what kind of time? Now every observer has his proper time and thus the time that the cosmological principle takes as a reference must be specified without invoking privileged observers or systems with peculiar characteristics. General relativity assures us that at any point there is a free-falling observer, for whom nature is explained in terms of the laws of special relativity. Observers in free-fall are called fundamental observers in cosmology and the particle of cosmic fluid that they ride is called a fundamental particle, or it could be called a fundamental galaxy, taking galaxies as the pieces from which the Universe is built. Each fundamental observer has his proper time. At a given instant of his proper time, a fundamental observer sees the Universe as homogeneous and isotropic; this is now the cosmological principle, which means that not only thermodynamic parameters, but also all metric properties, must be the same at any point, although they may be time dependent. A fundamental observer is able to obtain $T(\tau)$, $\rho(\tau)$, . . ., where T is the temperature of the cosmic fluid, ρ is its density and τ is the proper time. Another fundamental observer must obtain the same functions $T(\tau')$, $\rho(\tau')$ . . .. This gives observers the opportunity to synchronize their clocks. An event observed by the first fundamental observer is simultaneous with another event observed by the second fundamental observer when both observe the same temperature, density, and so on, of the cosmic fluid. Thermometers and barometers may be used as clocks. All fundamental observers, with an adequate choice of the origin, have a common time, which is called the cosmic time. The cosmic time is the proper time of any fundamental observer.

Non-fundamental observers have a different picture of the Universe. For them, the Universe is no longer homogeneous and isotropic. These observers have a peculiar velocity with respect to the net formed by all fundamental particles, which is called the substratum.

We should recall that a perfect fluid is that fluid for which, at any point, an observer exists who sees the local microstate as isotropic. Homogeneity is a consequence of isotropy, as discussed in Chapter 8. Therefore, the cosmic fluid is a perfect fluid. An observer exists – the fundamental observer – who sees the whole Universe as isotropic, including the local microstate.

9.2   The Robertson–Walker metric

As seen by a fundamental observer, three-dimensional space must possess constant curvature. Therefore a three-dimension line element may be of the form (8.50), except that this line element may be a function of time. If we consider ourselves as fundamental observers we may write our line element as

$$dl^2 = R^2(t) \left[\frac{dr^2}{1 - kr^2} + r^2(d\theta^2 + \sin^2\theta \, d\phi^2) \right] \tag{9.1}$$

where $R(t)$ is an unknown function of time t, called the 'radius' of the Universe, or the cosmological scale factor. k has the possible values 1, 0, or -1. For $r = 0$, we see that this line element meets the condition of isotropy, as dl does not depend on either ϕ or θ. The three-dimensional line element of Milne's universe (8.49) had a similar expression, where the unknown function $R(t)$ was precisely t^2.

The line element in four-dimensional space-time takes the form

$$ds^2 = g_{00} \, dt^2 + 2g_{0r} \, dt \, d\rho + 2g_{0\theta} \, dt \, d\theta + 2g_{0\phi} \, dt \, d\phi + dl^2 \tag{9.2}$$

For $r = 0$ events with $dr = 0, d\phi = 0, d\theta = 0$, and therefore $dl = 0, ds^2 = -d\tau^2 = g_{00} \, dt^2$. As $d\tau$ and dt must coincide in this case, we find that $g_{00} = -1$. Suppose that at a given position (r, θ, ϕ) two light signals are emitted in opposite directions in both cases, with $dr = 0$ and $d\phi = 0$. The first signal travels $d\theta$ in a time dt. In this time the second one travels $-d\theta$. In the first case we have $ds_1^2 = -dt^2 + 2g_{0\theta} \, dt \, d\theta + dl^2$. In the second case $ds_2^2 = -dt^2 - 2g_{0\theta} \, dt \, d\theta + dl^2$. Homogeneity implies that $ds_1^2 = ds_2^2$, therefore $g_{0\theta} = 0$. The same argument may be invoked to deduce $g_{0\phi} = 0$. Suppose, now, that at a given position (r, θ, ϕ) (excluding the origin (i.e. $r \neq 0$)), two light signals are emitted in opposite directions, one in the direction observer–source, and the other in the opposite direction. Again, homogeneity implies that $ds_1^2 = -dt^2 + 2g_{0r} \, dt \, dr + dl^2 = -dt^2 - 2g_{0r} \, dt \, dr + dl^2$; therefore $g_{0r} = 0$ also. Hence,

$$ds^2 = -dt^2 + dl^2 \tag{9.3}$$

From (9.1) the standard form of the Robertson–Walker metric is established as

$$ds^2 = -dt^2 + R^2(t)\left[\frac{dr^2}{1-kr^2} + r^2(d\theta^2 + \sin^2\theta\, d\phi^2)\right] \tag{9.4}$$

where $R(t)$ is to be determined.

9.2.1   The radius of a closed universe

Consider a fixed time instant in order to study some geometrical properties of three-dimensional space. Let us calculate the geodesic equation in this three-dimensional space. We can derive the following directly:

$$g_{ij} = R^2 \begin{pmatrix} \frac{1}{1-kr^2} & 0 & 0 \\ 0 & r^2 & 0 \\ 0 & 0 & r^2\sin^2\theta \end{pmatrix} \tag{9.5}$$

$$g^{ij} = R^{-2} \begin{pmatrix} 1-kr^2 & 0 & 0 \\ 0 & \frac{1}{r^2} & 0 \\ 0 & 0 & \frac{1}{r^2\sin^2\theta} \end{pmatrix} \tag{9.6}$$

and can show that only the following affine connection coefficients are non-zero:

$$\Gamma^r_{rr} = \frac{kr}{1-kr^2} \tag{9.7}$$

$$\Gamma^r_{\theta\theta} = -r(1-kr^2) \tag{9.8}$$

$$\Gamma^r_{\phi\phi} = -r(1-kr^2)\sin^2\theta \tag{9.9}$$

$$\Gamma^\theta_{r\theta} = \Gamma^\phi_{r\phi} = \frac{1}{r} \qquad r^\phi_{\phi\theta} = \frac{\cos\theta}{\sin\theta} \tag{9.10}$$

$$\Gamma^\theta_{\phi\phi} = -\sin\theta\cos\theta \tag{9.11}$$

Clearly, $\Gamma^\theta_{\theta r}$, $\Gamma^\phi_{\phi r}$, and $\Gamma^\phi_{\theta\phi}$ are also non-zero.

Therefore, the geodesic equations are

$$\frac{d^2r}{ds^2} + \frac{kr}{1-kr^2}\left(\frac{dr}{ds}\right)^2 - r(1-kr^2)\left(\frac{d\theta}{ds}\right)^2 - r(1-kr^2)\sin^2\theta\left(\frac{d\phi}{ds}\right)^2 = 0 \tag{9.12}$$

$$\frac{d^2\theta}{ds^2} + \frac{2}{r}\frac{dr}{ds}\frac{d\theta}{ds} - \sin\theta\cos\theta\left(\frac{d\phi}{ds}\right)^2 = 0 \tag{9.13}$$

$$\frac{d^2\phi}{ds^2} + \frac{2}{r}\frac{dr}{ds}\frac{d\phi}{ds} + 2\frac{\cos\theta}{\sin\theta}\frac{d\theta}{ds}\frac{d\phi}{ds} = 0 \tag{9.14}$$

Assume that $d\phi = 0$ initially. Equation (9.14) tells us that in this case $d^2\phi/ds^2 = 0$. Therefore, $d\phi/ds$ will always be zero. We are thus allowed to assume $d\phi/ds = 0$ in the above equations. Then, if $d\theta = 0$ initially, (9.13) tells us that $d^2\theta/ds^2 = 0$, so that $d\theta/ds = 0$ always. We are therefore allowed to set $d\theta/ds = 0$, too. This step is rather intuitive: pure radial geodesics exist. Then (9.12)

$$\frac{d^2r}{ds^2} + \frac{kr}{1-kr^2}\left(\frac{dr}{ds}\right)^2 = 0 \tag{9.15}$$

Taking $dt = 0$, $d\phi = d\theta = 0$ in (9.4)

$$\left(\frac{dr}{ds}\right)^2 = \frac{1-kr^2}{R^2} \tag{9.16}$$

Therefore,

$$\frac{d^2r}{ds^2} + \frac{k}{R^2}r = 0 \tag{9.17}$$

The solution is of the type

$$r = r_0 \sin\left(s\frac{k^{1/2}}{R}\right) \tag{9.18}$$

when $k = 1$, we see that for $s = 2\pi R$ the value of r is again zero, and that for an interval $\Delta s = 2\pi R$, the value of r remains unchanged. This is the reason why a $k = 1$ universe is called closed, and why the function R is called the radius of the Universe. If $k = -1$, the solution is of the type sinh (s/R), but R retains its name.

9.2.2 Comoving coordinates

Fundamental galaxies are free-falling objects and must therefore follow geodesics in four-dimensional space-time. From the above section we know that radial geodesics exist in three dimensions and that this result could easily be generalized to four dimensions. (In any case we may adopt this statement as intuitive.) We may therefore accept that fundamental galaxies follow radial geodesics. Suppose that t, r, θ, ϕ are coordinates of a fundamental galaxy, where we have accepted that θ and ϕ do not vary, that is, $d\theta = d\phi = 0$, which implies that $d\theta/d\tau = d\phi/d\tau = 0$. The time and radial components of the equation of the geodesic followed by a fundamental galaxy will then be

$$\frac{d^2t}{d\tau^2} + \Gamma^0_{00}\frac{dt}{d\tau}\frac{dt}{d\tau} + 2\Gamma^0_{r0}\frac{dt}{d\tau}\frac{dr}{d\tau} + \Gamma^0_{rr}\frac{dr}{d\tau}\frac{dr}{d\tau} = 0 \tag{9.19}$$

$$\frac{d^2r}{d\tau^2} + \Gamma^r_{00}\frac{dt}{d\tau}\frac{dt}{d\tau} + 2\Gamma^r_{r0}\frac{dt}{d\tau}\frac{dr}{d\tau} + \Gamma^r_{rr}\frac{dr}{d\tau}\frac{dr}{d\tau} = 0 \tag{9.20}$$

These are the equations of four-dimensional geodesics, in contrast to those in the previous section which were three-dimensional equations. As all fundamental observers can adjust their clocks to the cosmic time, when a given fundamental galaxy is observed by us at different times (if our own Galaxy is a fundamental one) we may set $t = \tau$. Therefore, for events taking place in the Galaxy $d^2t/d\tau^2 = 0$. We can readily derive that $\Gamma^0_{00} = \Gamma^0_{r0} = 0$ and that

$$\Gamma^0_{rr} = \frac{-RR}{1-kr^2} \tag{9.21}$$

(where $\dot{R}$ is dR/dt, as usual).

Thus (9.19) yields $dr/d\tau = 0$. Equation (9.20) reinforces this result, as, if $dr/d\tau = 0$ at only a given instant, taking $\Gamma^r_{00} = 0$ into account, we deduce that $d^2r/d\tau^2 = 0$. Hence $dr/d\tau = 0$ always. For any of the three spatial coordinates

$$\frac{d^2x^i}{d\tau^2} + \Gamma^i_{00}\left(\frac{dt}{d\tau}\right)^2 = 0 \tag{9.22}$$

and $\Gamma^i_{00} = 0$. Therefore, $d^2x^i/d\tau^2 = 0$.

Hence, the cosmological principle enables us to use coordinates which do not vary for a fundamental galaxy. These coordinates are called comoving, and will be used from now on. Even if a fundamental galaxy moves with respect to us, it has constant comoving coordinates. But motion is a relative concept and we may choose to state that a fundamental galaxy does not move. The observed expansion of the Universe may then be interpreted as a purely metric property.

9.3   The cosmological redshift

Let us adopt comoving coordinates and observe an electromagnetic wave emitted in a fundamental galaxy, with coordinates (r_1, θ_1, ϕ_1) and arriving at the observer at $(0, \theta_1, \phi_1)$. A first maximum of the wave was emitted at t_1 and the next one at $t_1 + \delta t_1$, with δt_1 being the wave period in the emitting galaxy. The first maximum arrives here after some time, at t_0, and the second one at $t_0 + \delta t_0$, δt_0 being the observed period. As we are dealing with electromagnetic signals, $d\tau = 0$. Thus, from (9.4),

$$dt^2 = R^2 \frac{dr^2}{1 - kr^2} \tag{9.23}$$

which, integrated for the first maximum ray, becomes

$$\int_{t_1}^{t_0} \frac{dt}{R} = \int_0^{r_1} \frac{dr}{\sqrt{1 - kr^2}} = f(r_1) \tag{9.24}$$

The second integral, denoted $f(r_1)$, does not depend on time. For the second maximum ray

$$\int_{t_1 + \delta t_1}^{t_0 + \delta t_0} \frac{dt}{R} = f(r_1) \tag{9.25}$$

as the distance r_1 is the same. Therefore,

$$\int_{t_1}^{t_0} \frac{dt}{R} = \int_{t_1 + \delta t_1}^{t_0 + \delta t_0} \frac{dt}{R} \tag{9.26}$$

and

$$\int_{t_1}^{t_1 + \delta t_1} \frac{dt}{R} + \int_{t_1 + \delta t_1}^{t_0} \frac{dt}{R} = \int_{t_1 + \delta t_1}^{t_0} \frac{dt}{R} + \int_{t_0}^{t_0 + \delta t_0} \frac{dt}{R} \tag{9.27}$$

Finally,

$$\int_{t_1}^{t_1+\delta t_1} \frac{dt}{R} = \int_{t_0}^{t_0+\delta t_0} \frac{dt}{R} \tag{9.28}$$

$t_1 + \delta t_1$ is so close to t_1 that R must be considered constant in the left-hand side integral, equal to $R(t_1)$. Similarly, R must be taken as a constant on the right-hand side, equal to $R(t_0)$. Hence,

$$\frac{t_1 + \delta t_1 - t_1}{R(t_1)} = \frac{t_0 + \delta t_0 - t_0}{R(t_0)} \tag{9.29}$$

and

$$\frac{\delta t_1}{R(t_1)} = \frac{\delta t_0}{R(t_0)} \tag{9.30}$$

$R(t_0) \equiv R_0$, the present value of the cosmological scale factor. As long as $R(t)$ is still an unknown function, we are allowed to conclude only that the period of the emitter will be different from that of the receiver. Observations indicate that δt_1 is shorter than δt_0; therefore $R(t)$ is an increasing function of time, at least on the present epoch. Note that no doppler effect has been explicitly invoked, so that the redshift is not necessarily a doppler shift. Hubble's law is interpreted here as a metric property, as the metric is time dependent and the expansion of the Universe is an expansion of the metric rather than an outflow of galaxies. However, the presence and contribution of a conventional doppler effect depends on our own choice of coordinates.

The redshift z was defined in (8.1) by

$$z = \frac{\lambda_0 - \lambda_1}{\lambda_1} = \frac{\delta t_0 - \delta t_1}{\delta t_1} = \frac{\nu_1 - \nu_0}{\nu_0} \tag{9.31}$$

(Note that the subindex 0 denotes observed wavelengths at present, whilst in (8.1) it denoted rest wavelengths.) Therefore,

$$z = \frac{R_0}{R} - 1 \tag{9.32}$$

9.4   The proper distance to a galaxy

Clearly, the elementary proper distance is obtained by setting $dt = 0$

$$ds = R \frac{dr}{\sqrt{1 - kr^2}} \tag{9.33}$$

Therefore, using (9.24), the proper distance to a galaxy is

$$d_1 = R \int_0^{r_1} \frac{dr}{\sqrt{1 - kr^2}} = R f(r_1) \tag{9.34}$$

The solution of the integrals gives

$$f(r_1) = \int_0^{r_1} \frac{dr}{\sqrt{1 - kr^2}} = \begin{array}{ll} \sin^{-1} r_1 & \text{for} \quad k = 1 \\ r_1 & \text{for} \quad k = 0 \\ \sinh^{-1} r_1 & \text{for} \quad k = -1 \end{array} \tag{9.35}$$

If the actual Universe has $k = 0$:

$$d_1 = Rr_1 \qquad (9.36)$$

The product Rr_1 has dimensions of length (or time) and is measured in cm (or in seconds), but note that $[R] = L$, whilst $[r] = 1$. Therefore, r_1 is a non-dimensional coordinate, indicating the position of an object but not its distance. R is a function converting r_1 in real distance. However, it is clear that (9.36) is not true for all values of k. The present proper distance will obviously be $R_0 r_1$, if $k = 0$.

Taking the logarithmic derivative in (9.34), we find

$$\dot{d}_1 = \frac{\dot{R}}{R} d_1 = H d_1 \qquad (9.37)$$

which has a formal similarity to Hubble's law. However, $\dot{d}_1 \neq z$, except for such very small distances that the following approximation may be used:

$$\dot{d}_1 = H d_1 \approx \frac{1}{R(t_1)} \frac{R_0 - R(t_1)}{t_0 - t_1} d_1 = z \qquad (9.38)$$

where (9.32) and the fact that $d_1/(t_0 - t_1) = 1$. Therefore, (9.37) is exact, but Hubble's law (8.2) is an approximate law, valid only for short distances.

9.4.1   Luminosity distance

From the observational point of view, the most basic method of determining distances to distant sources consists of measuring the flux at the Earth due to sources whose intrinsic luminosity is only known indirectly. The distance obtained in this way does not in general coincide with the proper distance.

Assume, first, that $k = 0$. A telescope of area A at the Earth forms a solid angle Ω at the object given by

$$\Omega = \frac{A}{4\pi (R_0 r_1)^2} \qquad (9.39)$$

taking as the distance to the object the present proper distance. If $k \neq 0$, this formula must be corrected. Figure 9.1 shows how the same angle at the source corresponds to different collecting areas in universes with $k = 1$ and $k = -1$. If $k = 1$ for the same Ω, the collecting area is smaller. Let us obtain the correction factor by means of a rough argument. The correction factor for $k = 1$ would be OA_0/OA_1 if we were dealing with plane angles. This is approximately equal to the relation of distances for $k = 0$ and $k = 1$, as can be seen in Figure 9.1, that is, equal to $(\sin^{-1} r_1)/r_1$. As we are dealing with solid angles the correcting factor should be $(\sin^{-1} r_1/r_1)^2$, or, in general, $(f(r_1)/r_1)^2$. On the other hand, the distance $(R_0 r_1)$ in (9.39) must be replaced by $R_0 f(r_1)$; therefore

$$\Omega = \frac{A \frac{f^2(r_1)}{r_1^2}}{4\pi R_0^2 f^2(r_1)} = \frac{A}{4\pi R_0^2 r_1^2} \qquad (9.40)$$

which is an important result. The solid angle Ω is calculated in the same way, irrespective of the value of k, as if the Universe were plane.

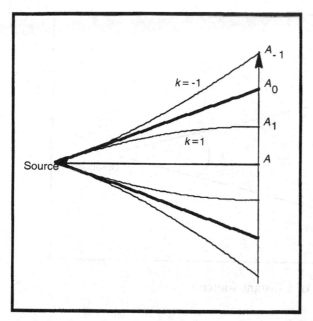

Figure 9.1 Propagation of light rays between source and detector depending on the value of k.

The flux at the collector should in principle be

$$l = \frac{L\Omega}{A} = \frac{L}{4\pi R_0^2 r_1^2} \tag{9.41}$$

However, more corrections are needed. Photons arrive redshifted and thus their energy is reduced by a factor of $R(t_1)/R_0$, but the frequency of the arrival of photons is also reduced by the same factor. At the source, a photon is emitted every Δt_1 seconds; at the Earth a photon is received every Δt_0 seconds. Clearly, $\Delta t_1/\Delta t_0 = R(t_1)/R_0$, as in (9.30). Therefore, instead of (9.41) we must use

$$l = L\frac{R^2(t_1)}{R_0^2}\frac{1}{4\pi R_0^2 r_1^2} \tag{9.42}$$

A 'luminosity distance' d_L is then defined by

$$l = \frac{L}{4\pi d_L^2} \tag{9.43}$$

Hence,

$$d_L = \frac{R_0}{R}R_0 r_1 = (1+z)R_0 r_1 \tag{9.44}$$

Other more detailed calculations provide the same result.

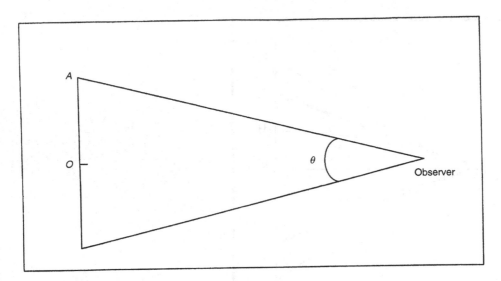

Figure 9.2   The angular extension of a distant source.

9.4.2   Diameter distance

Consider the proper distance OA in Figure 9.2, that is, a length perpendicular to the radial direction. In (9.4) we set $dt = 0$, $dr \approx 0$, and $d\phi = 0$, and obtain

$$ds^2 = R^2(t)r^2 d\theta^2 \tag{9.45}$$

Integrating, with $t = t_1$, the proper diameter D of the source is obtained:

$$D = R(t_1)r_1\theta \tag{9.46}$$

A 'diameter distance' d_D can then be defined as

$$D = d_D\theta \tag{9.47}$$

and this can be observed, provided that D is known by other means. Hence,

$$d_D = R(t_1)r_1 = (R_0 r_1)\frac{R(t_1)}{R_0} = \frac{R_0 r_1}{1 + z} \tag{9.48}$$

For small distances, the proper, luminosity and diameter distances coincide, but may differ by large amounts for large distances. The proper distance cannot be obtained directly from observations, and depends on the value of k. The distance parameter no longer has a clear interpretation and the redshift z is the preferred parameter for ordering the distances of very distant sources.

9.5   Approximate relations

Several quantities related to an object's distance have been introduced: d_1, d_L, d_D, z, r_1, t_1, $R(t_1)$. The exact relation between them requires the integration of Einstein's field equations, which will be carried out later (Section 9.7). As an example, it was shown in

(9.38) that Hubble's law is only an approximation, being strictly valid for small z only. For very large z it could be completely inappropriate, but no individual astrophysical body has been observed with very large redshifts. Known distant objects have redshifts up to about 5, and some small deviation from Hubble's law can be detected, taking observational difficulties into account. This small deviation may, however, be an important clue to determining the kind of Universe we live in. From the observational point of view, the z-range in which second-order terms in Taylor expansions become appreciable is very important. Coefficients of these Taylor expansions may subsequently be compared with the predictions of theoretical cosmological models.

As a first example, a relation $[R, t]$ must exist, which we can write in the form

$$R = R_0 \left(1 + H_0(t - t_0) - \frac{1}{2} q_0 H_0^2(t - t_0)^2 + \ldots \right) \tag{9.49}$$

H_0 and q_0 are the same quantities as were introduced in (8.15), (9.37), and (8.31), for the present epoch since

$$\dot{R} = R_0(H_0 - q_0 H_0^2(t - t_0) + \ldots) \tag{9.50}$$

$$\dot{R}_0 = R_0 H_0 \tag{9.51}$$

$$\ddot{R} = -R_0 q_0 H_0^2 + \ldots \tag{9.52}$$

$$\ddot{R}_0 = -R_0 q_0 \frac{\dot{R}_0^2}{R_0^2} = -\frac{q_0 \dot{R}_0^2}{R_0} \tag{9.53}$$

The relation between z and $(t - t_0)$ may then be obtained from (9.32). It is easily deduced that

$$\frac{R_0}{R} = 1 + H_0(t_0 - t) + \left(1 + \frac{q_0}{2}\right) H_0^2(t_0 - t)^2 + \ldots \tag{9.54}$$

Therefore,

$$z = H_0(t_0 - t) + \left(1 + \frac{q_0}{2}\right) H_0^2(t_0 - t)^2 + \ldots \tag{9.55}$$

and

$$t_0 - t = \frac{1}{H_0}\left(z - \left(1 + \frac{q_0}{2}\right) z^2 + \ldots \right) \tag{9.56}$$

In (9.24),

$$f(r) \approx r = \int_t^{t_0} \frac{dt}{R} = \frac{1}{R_0}\left((t_0 - t) + \frac{1}{2}H_0(t_0 - t_1)^2 + \ldots \right)$$

$$= \frac{1}{R_0 H_0}\left(z - \frac{1}{2}(1 + q_0)z^2 + \ldots \right) \tag{9.57}$$

Hence using (9.44), the luminosity distance is given by

$$d_L = \frac{1}{H_0}\left(z + \frac{1}{2}(1 - q_0)z^2 + \ldots \right) \tag{9.58}$$

As both z and d_L are in principle observational quantities, this relation has a special interest. H_0 and q_0 could be obtained by fitting observations and making a comparison with the theoretical $R(t)$. If $z \ll 1$, Hubble's law is found and H_0 is determined, as we know. Observations over a larger range of z may also provide q_0, a quantity that is different for different signs of k, as we will see later (equation (9.110)).

9.6   The Universe as a perfect fluid

In Section 9.1 it was shown that the cosmological principle implies that the Universe is a perfect fluid. We may therefore benefit from the perfect fluid equations for a cosmic fluid which obey the Robertson–Walker metric. By using comoving coordinates, the four-velocity is $U^0 = 1$, $U^i = 0$, which simplifies this particular application considerably.

9.6.1   The equation of continuity

Clearly, the vector $J^\alpha \equiv (n, 0, 0, 0)$. Let us consider the continuity equation in the form of (2.91). In this case:

$$g = R^6 r^4 (1 - kr^2)^{-1} \sin^2 \theta \tag{9.59}$$

As only R is time dependent, we obtain

$$\frac{d}{dt}(nR^3) = 0 \tag{9.60}$$

For the equation of continuity to be applicable, we need conservation of mass, or, more generally, conservation of the number of particles of a given mass. Therefore n in (9.60) is the number density of conserved particles. For instance, n may be the baryonic number density (after the baryogenesis epoch of course). If it is found that the number density of galaxies is conserved after their initial formation, n may also be related to the number of galaxies.

9.6.2   The energy-momentum equation

Let us now use the equation of motion in the form of (2.98). In (9.22) it was shown that $\Gamma^\mu_{00} = 0$ and therefore the third term in (2.98) always vanishes. From the cosmological principle we also know that $\partial p / \partial x^i = 0$, thus the only non-trivial equation is

$$-\frac{\partial p}{\partial t} + \frac{\sqrt{1 - kr^2}}{R^3 r^2 \sin \theta} \frac{\partial}{\partial t} \left(\left(\frac{R^3 r^2 \sin \theta}{\sqrt{1 - kr^2}} \right)(\epsilon + p) \right) = 0 \tag{9.61}$$

which is equivalent to

$$R^3 \frac{dp}{dt} = \frac{d}{dt}\left(R^3(\epsilon + p)\right) \tag{9.62}$$

This is an equation which can be used to obtain the temperature as a function of R, although it does not provide $R(t)$ itself. Let us assume as a first example the radiation-dominated era in which ϵ and p are due to photons. As $\epsilon = 3p$ (see equation (2.24)) in (9.62), we find

$$pR^4 = \text{constant} \tag{9.63}$$

As we know $p \propto T^4$ for photons, the relation $[T, R]$ in this epoch will be

$$TR = \text{constant} \tag{9.64}$$

The law $T \propto R^{-1}$ in the radiation-dominated era, and even for the present microwave radiation taken in isolation, has already been derived in (8.38) by means of an intuitive argument.

As a second example, consider the matter-dominated era, for which $\epsilon = mn + \frac{3}{2}nT$. The second term in (9.62) is

$$\frac{d}{dt}\left(R^3\left(mn + \frac{3}{2}p + p\right)\right) = \frac{5}{2}\frac{d}{dt}(R^3 p) \tag{9.65}$$

using (9.60). We then obtain

$$pR^5 = \text{constant} \tag{9.66}$$

As $p = nT$, and as $n \propto R^{-3}$, it is clear that

$$TR^2 = \text{constant} \tag{9.67}$$

which was also suggested in (8.37).

Note, however, that this equation does not inform us about the motion of the Universe itself, as the i-components of the energy-momentum equation are redundant.

9.6.3   Entropy

A general expression for the variation of the entropy, which is valid for any system, is

$$\begin{aligned}
dS &= \frac{1}{T}d(\epsilon V) + \frac{1}{T}p\,dV \\
&= \left(\frac{\epsilon + p}{T} + \frac{V}{T}\frac{\partial \epsilon}{\partial V}\right)dV + \frac{V}{T}\frac{\partial \epsilon}{\partial T}dT
\end{aligned} \tag{9.68}$$

where we assume that $p = p(V, T)$ and that $\epsilon = \epsilon(V, T)$. From $\partial^2 S/\partial V \partial T \equiv \partial^2 S/\partial T \partial V$, we obtain

$$\frac{\partial}{\partial T}\left(\frac{\epsilon + p}{T} + \frac{V}{T}\frac{\partial \epsilon}{\partial V}\right) = \frac{\partial}{\partial V}\left(\frac{V}{T}\frac{\partial \epsilon}{\partial T}\right) \tag{9.69}$$

which may be written after transformation as

$$\frac{\partial p}{\partial T} = \frac{\epsilon + p}{T} + \frac{V}{T}\frac{\partial \epsilon}{\partial V} \tag{9.70}$$

Substituting (9.70) in (9.68) gives

$$dS = \frac{\partial p}{\partial T}dV + \frac{V}{T}\frac{\partial \epsilon}{\partial T}dT \tag{9.71}$$

Equations (9.70) and (9.71) are very general. It is usual in astrophysics to replace the thermodynamic variable V by n. As we know that $nV = \text{constant}$, $n\,dV + V\,dn = 0$ and (9.70) may be rewritten as

$$\frac{\partial p}{\partial T} = \frac{\epsilon + p}{T} - \frac{n}{T}\frac{\partial \epsilon}{\partial n} \tag{9.72}$$

Instead of using (9.71) it is usual to divide by (nV) to obtain the specific entropy per baryon σ:

$$d\sigma = -\frac{\partial p}{\partial T}\frac{1}{n^2}\,dn + \frac{1}{Tn}\frac{\partial \epsilon}{\partial T}\,dT \tag{9.73}$$

As $d\sigma = 0$ in the Universe, we have

$$\frac{\partial p}{\partial T}\frac{dn}{n} = \frac{\partial \epsilon}{\partial T}\frac{dT}{T} \tag{9.74}$$

It is easy to check the validity of (9.72) for the present system, with $p = nT$ and $\epsilon = mn + \frac{3}{2}nT$. Consider a relativistic system, where we know that $\epsilon = 3p$ and we may assume that $\epsilon = \epsilon(T)$ is a function of T, but not of n (baryons may in fact be absent). Then if p is also a function of T only, but not of n, equation (9.72) gives

$$\frac{dp}{dT} = \frac{4p}{T} \tag{9.75}$$

and therefore

$$p \propto T^4 \tag{9.76}$$

and

$$\epsilon \propto T^4 \tag{9.77}$$

We know that this is true for systems of photons, but it is also valid for other relativistic systems such as for neutrinos and for relativistic positrons and electrons, present in early epochs of the universe. The proportionality constant in (9.77) is simply the constant a for a system of photons, given by (3.85), but it may differ for other systems. Indeed, it is $(7/16)a$ for neutrinos and $(7/8)a$ for electrons and for positrons.

For a relativistic system (9.74), using (9.76) and (9.77), and $\epsilon = 3p$, gives

$$\frac{1}{3}\frac{dn}{n} - \frac{dT}{T} = 0 \tag{9.78}$$

$$\frac{n^{1/3}}{T} = \text{constant} \tag{9.79}$$

As $n \propto R^{-3}$

$$T \propto R^{-1} \tag{9.80}$$

This temperature decrease when the Universe expands has already been found for photons in (9.64). Now we see that it is also valid for earlier epochs characterized by neutrinos and positron–electron pairs.

If (9.74) is now applied to the present baryonic component of the Universe, we find that $T \propto n^{2/3}$ and hence $T \propto R^{-2}$, assuming a model in equilibrium. We know, however, that the present baryonic system is far from thermodynamic equilibrium and T has a very large range of values in galaxies and stars.

Let us return to the relativistic systems. For such systems, a very simple expession may be found for their entropy. Substituting (9.70) in (9.71), and setting $\partial \epsilon/\partial V = 0$ and $\partial p/\partial V = 0$, it is easy to deduce that

$$S = \frac{V(\epsilon + p)}{T} \tag{9.81}$$

Therefore, the specific entropy per baryon is

$$\sigma = \frac{\epsilon + p}{nT} = \frac{4}{3} \frac{\epsilon}{nT} \tag{9.82}$$

Equation (9.82) may be applied to the following eras of the evolving Universe: pre-neutrino-decoupling (neutrinos + electrons + positrons + photons), pre-annihilation (electrons + positrons + photons), photon-dominated era (photons) and present microwave background radiation (photons).

In the pre-neutrino-decoupling era, there were three types of neutrino and the corresponding antineutrinos (six types), electrons and positrons (two), and photons (one). Therefore,

$$\epsilon = \left(6 \times \frac{7}{16} + 2 \times \frac{7}{8} + 1\right) aT^4 = \frac{43}{8} aT^4 \tag{9.83}$$

In the pre-annihilation era,

$$\epsilon = \left(2 \times \frac{7}{8} + 1\right) aT^4 = \frac{11}{4} aT^4 \tag{9.84}$$

and subsequently (matter-dominated era excluded), ϵ is simply given by

$$\epsilon = aT^4 \tag{9.85}$$

For the specific entropy per baryon for the three different eras considered:

$$\sigma = \frac{43}{6} \frac{aT^3}{n}, \quad \text{or} \quad \frac{11}{3} \frac{aT^3}{n}, \quad \text{or} \quad \frac{4}{3} \frac{aT^3}{n} \tag{9.86}$$

As the Universe is a perfect fluid, σ is a constant. Therefore, during neutrino-decoupling, neutrinos carried away an entropy $(43/6 - 11/3)aT^3/n = (21/6)aT^3/n$, and the temperature of neutrinos subsequently decreased steadily as a result of expansion. However, the temperature of photons increased abruptly during annihilation. As σ must be a constant, if T_b is the temperature before annihilation and T_a the temperature shortly afterwards:

$$\frac{11}{3} \frac{aT_b^3}{n} = \frac{4}{3} \frac{aT_a^3}{n} \tag{9.87}$$

and thus

$$\frac{T_a}{T_b} = \left(\frac{11}{4}\right)^{1/3} \approx 1.4 \tag{9.88}$$

As the neutrinos had the same temperature as the photons immediately prior to annihilation, the relation (9.88) must give the approximate relation between the present photon temperature (~ 3 K) and the present primordial neutrino background temperature. Primordial neutrinos must have a temperature of about 2 K.

We have seen that entropy analysis for highly relativistic systems can easily be performed via the integrated expression (9.82). Let us now consider the critical epoch in which the transition from a photon-dominated universe to a matter-dominated universe took place.

Considering now that $\epsilon = mn + \frac{3}{2}nT + aT^4$ (taking the contributions of baryons and photons into account) and that $p = nT + \frac{1}{3}aT^4$, and with the help of equation (3.91), where now σ, the specific photon entropy per baryon, is called σ_γ, we readily obtain

$$(1 + \sigma_\gamma)\frac{dn}{n} = 3(2 + \sigma_\gamma)\frac{dT}{T} \tag{9.89}$$

σ_γ is also a constant throughout this era. For $T_\gamma \approx 2.7$ K, $n_\gamma \approx 4 \times 10^3 \text{cm}^{-3}$. However, for $mn \approx 10^{-29}\text{g} \ \text{cm}^{-3}$, $n \approx 6 \times 10^{-6}\text{cm}^{-3}$. Therefore, the present photon entropy per baryon is of the order of $10^9 - 10^{10}$, and this value has been constant since the neutrino-decoupling epoch, so that $\sigma_\gamma \gg 1$. Hence $n \propto T^3$, and we obtain $T \propto R^{-1}$ as in previous epochs. Therefore, the law governing the temperature decrease $T \propto R^{-1}$ is also valid for that period before recombination which was matter-dominated.

9.7   Einstein's field equations

Our next objective is to obtain the function $R(t)$ by means of Einstein's field equations. The alternative expression (2.86) will be adopted for this task. The energy-momentum tensor will be of the form

$$\tau^{\mu\nu} = pg^{\mu\nu} + (p + \epsilon)U^\mu U^\nu \tag{9.90}$$

with $U^0 = 1$, $U^i = 0$. Then

$$\tau_\lambda{}^\lambda = 3p - \epsilon \tag{9.91}$$

From (2.94),

$$R_{\mu\nu} = -8\pi\left(pg_{\mu\nu} + (p + \epsilon)U_\mu U_\nu - \frac{1}{2}g_{\mu\nu}(3p - \epsilon)\right)$$
$$= -8\pi\left(\frac{(\epsilon - p)}{2}g_{\mu\nu} + (p + \epsilon)U_\mu U_\nu\right) \tag{9.92}$$

Hence,

$$R_{00} = -4\pi(\epsilon + 3p) \tag{9.93}$$

$$R_{rr} = -4\pi(\epsilon - p)\frac{R^2}{1 - kr^2} \tag{9.94}$$

$$R_{\theta\theta} = -4\pi(\epsilon - p)R^2 r^2 \tag{9.95}$$

$$R_{\phi\phi} = -4\pi(\epsilon - p)R^2 r^2 \sin^2\theta \tag{9.96}$$

The Ricci tensor must now be calculated. This has already been partly carried out in (9.7)–(9.11), (9.21), etc. After completing the derivation, we obtain

$$3\frac{\ddot{R}}{R} = -4\pi(\epsilon + 3p) \tag{9.97}$$

and

$$R\ddot{R} + 2\dot{R}^2 + 2k = 4\pi R^2(\epsilon - p) \tag{9.98}$$

These equations enable us to find both $R(t)$ and $\epsilon(t)$, provided the equation of state $\epsilon = \epsilon(p)$ is known. However, other combinations are preferred. From (9.97) and (9.98) and eliminating ϵ:

$$2\frac{\ddot{R}}{R} + \frac{\dot{R}^2}{R^2} + \frac{k}{R^2} = -8\pi p \tag{9.99}$$

The pair of equations which will replace (9.97) and (9.98) are obtained as follows. Let us first eliminate $\ddot{R}$:

$$\frac{8\pi\epsilon}{3}R^3 = \dot{R}^2 R + kR \tag{9.100}$$

which is the Einstein–Friedmann equation. Our second equation is obtained from the time derivative of (9.100), using (9.99):

$$\frac{d}{dt}(\epsilon R^3) = -3R^2\dot{R}p \tag{9.101}$$

which is the equation of energy conservation. We will take (9.100) and (9.101) as our fundamental equations. Note that (9.101) is not a new equation for us. It is the same as (9.62), as may be quickly confirmed. As well as (9.100) and (9.101) we need the equation of state, which has changed from epoch to epoch. There are, however, some basic conclusions that are independent of the equation of state.

Equation (9.97) tells us that $(\ddot{R}) < 0$, always. Present observations indicate that $\dot{R}_0 > 0$. Therefore, $R = 0$ at some time in the past. This time is adopted as the origin (the Big-Bang).

One possibility, if ϵ and p were both 0, would be that $\ddot{R} = 0$. In this case $\dot{R} = $ constant, $R = (R_0/t_0)t$, $\dot{R} = \dot{R}_0 = R_0/t_0$, $t_0 = R_0/\dot{R}_0 = 1/H_0$, the lifetime of the Universe would coincide with the Hubble time. But if $\ddot{R} < 0$, then $t_0 < 1/H_0$: so the Hubble time is in fact an upper limit to the time interval between the Big-Bang and the present.

If in equation (9.101), $p > 0$, and $\dot{R}_0 > 0$, then $d(\epsilon R^3)/dt < 0$, that is, ϵ decreases faster than R^3 increases. Eventually the left-hand side in (9.100) will become negligible. Then $\dot{R}^2 = -k$. For $k = -1$, $\dot{R}^2 = 1$; $\dot{R} = 1$; $R = t$ which means expansion for ever. Therefore, an open universe will expand to infinity. Suppose, on the other hand that $k = 1$ (closed universe). Then $\dot{R}^2 = -k$ provides imaginary values for R. Of course, this situation cannot in fact exist. Before reaching 0, ϵ will take the value $\frac{3}{8\pi}R^{-2}$, so that $\dot{R} = 0$, which corresponds to a maximum. As $\ddot{R} < 0$ a decrease in R will follow this maximum, and an inverse Big-Bang, or Big-Crunch will eventually take place. Lastly, assume that $k = 0$. For $\epsilon \approx 0$, $\dot{R}^2 R = 0$, $\dot{R} = 0$, stopping the expansion. However, for $\dot{R} = 0$, $\epsilon R^3 = $ constant (as deduced from (9.101)), so that $\epsilon \approx 0$ corresponds to $R \approx \infty$. Expansion ceases only asymptotically for $t = \infty$. We started this discussion by noting that $\dot{R}_0 > 0$, but we don't know if $\dot{R} > 0$ always. If $\dot{R} < 0$, then $d/dt(\epsilon R^3) > 0$. But $\dot{R} < 0$ would hold for only a time after R reached its maximum value, with $\dot{R} = 0$. Equation (9.100) shows that this maximum requires $k = 1$. Only a closed universe possesses this maximum.

Let us divide (9.100) by R^3:

$$\frac{8\pi\epsilon}{3} = H^2 + \frac{k}{R^2} \tag{9.102}$$

For $k = 0$, a critical density ϵ_c is obtained

$$\epsilon_c = \frac{3}{8\pi} H^2 \tag{9.103}$$

If $k = 1$, $\epsilon > \epsilon_c$. If $k = -1$, $\epsilon < \epsilon_c$ as deduced from (9.102). By defining the density parameter

$$\Omega = \frac{\epsilon}{\epsilon_c} \tag{9.104}$$

if $\Omega > 1$, the universe is closed and will eventually recollapse. If $\Omega < 1$, the universe is open, and will expand forever. The condition (9.103) is the closure condition. It is also clear that

$$\Omega = 1 + \frac{k}{R^2 H^2} \tag{9.105}$$

9.8   The matter-dominated era

As mentioned in Section 8.4 this era began about 10^3 years after the Big-Bang. The problem here of course is our need to know the equation of state, which clearly requires a knowledge of the dominant type of particles. Classical cosmology was developed assuming that the Universe mainly consists of nuclei, having as an equation of state $\epsilon = mn + \frac{3}{2}p$ (see, for instance, (2.22)), where $p = nT$. Under conditions prevailing when matter is dominant $p \ll mn$; therefore we are allowed to adopt the simplifying hypothesis $p \approx 0$. A direct consequence of (9.101) is that in a matter-dominated universe ϵR^3 is a constant, equal to $\epsilon_0 R_0^3$. However, the reader must be warned about the possible dominance of dark matter with a very different equation of state. For instance, if light neutrino dark matter were predominant, we should use $\epsilon = 3p$ as the equation of state. The evolution of the universe would then be similar to that in the radiation-dominated era. Once this warning has been heeded, let us present the classical cosmology for an atomic universe. The identification of the nature of dark matter will be reconsidered in the next chapter. The following formulae are correct for most of the possible dark matter particles.

Solutions may be given in terms of the boundary-values R_0 and ϵ_0, but it is more useful to express them in terms of the observational values H_0 and q_0. For this transformation we use (9.102) for the present time:

$$\frac{8\pi\epsilon_0}{3} = H_0^2 + \frac{k}{R_0^2} \tag{9.106}$$

and a relation which may be obtained from (9.101) with the definition of q_0 (8.31):

$$q_0 = \frac{-R_0 \ddot{R}_0}{\dot{R}_0^2} \tag{9.107}$$

From (9.99), with $p = 0$,

$$-2q_0 H_0^2 + H_0^2 + \frac{k}{R_0^2} = 0 \tag{9.108}$$

Combining (9.103), (9.104), (9.106), and (9.108),

$$\Omega_0 = 2q_0 \tag{9.109}$$

relating ϵ_0 to q_0. From (9.108),

$$R_0^2 = \frac{k}{H_0^2(2q_0 - 1)} \tag{9.110}$$

Note that for $k = 0$, $q_0 = 1/2$, R_0 is indeterminate, but we will see that in practice that does not introduce any problems.

Another interesting equation can be obtained:

$$\left(\frac{\dot{R}}{R_0}\right)^2 = H_0^2 \left(1 - 2q_0 + 2q_0 \frac{R_0}{R}\right) \tag{9.111}$$

which is in practice the equation to be integrated. With the substitution

$$x = \frac{R}{R_0} \tag{9.112}$$

equation (9.111) can be rewritten as

$$\dot{x} = H_0 \left(1 - 2q_0 + \frac{2q_0}{x}\right)^{\frac{1}{2}} \tag{9.113}$$

From this, we find t as a function of R/R_0:

$$t = \frac{1}{H_0} \int_0^{R/R_0} \left(1 - 2q_0 + \frac{2q_0}{x}\right)^{-\frac{1}{2}} dx \tag{9.114}$$

and the age t_0 of the Universe is

$$t_0 = \frac{1}{H_0} \int_0^1 \left(1 - 2q_0 + \frac{2q_0}{x}\right)^{-\frac{1}{2}} dx \tag{9.115}$$

9.8.1 Friedmann models

Friedmann models provide $R(t)$ by integrating (9.111) for $k = 0$ (Einstein–de Sitter or flat model), $k = 1$ (closed or cyclic model), and $k = -1$ (open model). Let us begin with the simplest one, which is also the possibility favoured by theoretical cosmologists, that is, let us first set $k = 0$.

Einstein–de Sitter universe

From (9.108) we have

$$q_0 = \frac{1}{2} \tag{9.116}$$

Therefore (9.113) is simply

$$\dot{x} = H_0 x^{-1/2} \tag{9.117}$$

which gives

$$x = \left(\frac{3}{2} H_0\right)^{2/3} t^{2/3} \tag{9.118}$$

a familiar result (8.25). Therefore, in this case,

$$t_0 = \frac{2}{3} \frac{1}{H_0} \tag{9.119}$$

which has already been obtained in (8.27). Other familiar formulae are

$$H = H_0 x^{-3/2} = \frac{2}{3} t^{-1} \tag{9.120}$$

$$\epsilon = mn = \frac{3 H_0^2}{8\pi} x^{-3} = \frac{1}{6\pi} t^{-2} \tag{9.121}$$

Lastly, from (9.67) the temperature of the matter is

$$T = T_0 x^{-2} = T_0 \left(\frac{3}{2} H_0\right)^{-4/3} t^{-4/3} \tag{9.122}$$

In this case, the meaning of the matter temperature is not clear, as it does not seem to be in thermodynamic equilibrium. The meaning of T_0 is also very unclear. Equation (9.122) should be applicable at the recombination epoch: $T_R = T_0 x_R^{-2}$. As it was estimated that $T_R \approx 3 \times 10^3$ K and $x_R \approx 10^{-3}$, we have $T_0 \approx T_R 10^{-6} \approx 3 \times 10^{-3}$ K, which does not correspond to a measurable physical temperature. Also note that T in (9.122) has nothing to do with the 2.7 K black-body temperature, which should follow a $TR =$ constant law, as stated in (9.64).

The closed universe

To integrate (9.114) for $k = 1$, that is, $q_0 \rangle 1/2$, it is best to use the variable transformation

$$1 - \cos\theta = \frac{2q_0 - 1}{q_0} x \tag{9.123}$$

Then

$$tH_0 = q_0 (2q_0 - 1)^{-3/2} \int_0^\theta \frac{\sin\theta}{\sqrt{-1 + \frac{2}{1-\cos\theta}}} d\theta \tag{9.124}$$

$$= q_0 (2q_0 - 1)^{-3/2} \int_0^\theta (1 - \cos\theta) d\theta = q_0 (2q_0 - 1)^{-3/2} (\theta - \sin\theta)$$

This equation is a cycloid (the curve described by a point on the rim of a rolling wheel of radius A) plotted in Figure 9.3. The cycloid has parametric equations $X = A\theta$; $Y = A - A\cos\theta$, where A is the wheel's radius and θ is the parameter of the curve. In this case $X \equiv t$; $A \equiv q_0 (2q_0 - 1)^{-3/2} H_0^{-1}$ and $Y \equiv R$. This last identity must be justified:

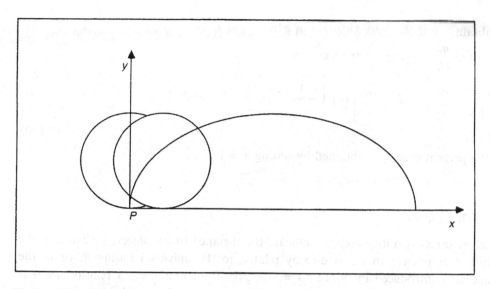

Figure 9.3 Definition of a cycloid.

$$Y = A - A\cos\theta = \frac{q_0}{H_0}(2q_0 - 1)^{-3/2}(1 - \cos\theta)$$

$$= \frac{q_0}{H_0}(2q_0 - 1)^{-3/2}\frac{2q_0 - 1}{q_0}\frac{R}{R_0} = R \tag{9.125}$$

where we have used (9.110). If there is a second turn to the universal wheel it is beyond the scope of this mathematical derivation. The maximum value of R is given by

$$R_{max} = 2A = \frac{2q_0(2q_0 - 1)^{-3/2}}{H_0} \tag{9.126}$$

and the time of the 'Big-Crunch' is $2\pi A$. Lastly, the solution may be written as

$$t = \frac{q_0}{H_0}(2q_0 - 1)^{-3/2}\left[\cos^{-1}\left(1 - \frac{2q_0 - 1}{q_0}x\right) - \sqrt{1 - \left(1 - \frac{2q_0 - 1}{q_0}x\right)^2}\right] \tag{9.127}$$

and the present time, t_0, is obtained by setting $x = 1$.

The open universe
The integration is similar, with

$$\theta = i\psi \tag{9.128}$$

As $\cos(i\psi) = \cosh\psi$, let us introduce the variable transformation

$$1 - \cosh\psi = \frac{2q_0 - 1}{q_0}x \tag{9.129}$$

We obtain

$$t = \frac{q_0}{H_0}(1 - 2q_0)^{-3/2}(\sinh\psi - \psi)$$

$$= \frac{q_0}{H_0}(1 - 2q_0)^{-3/2}\left[\sqrt{\left(1 + \frac{1 - 2q_0}{q_0}x\right)^2 - 1} - \cosh^{-1}\left(1 + \frac{1 - 2q_0}{q_0}x\right)\right]$$

$$(9.130)$$

and the present time t_0 is obtained by setting $x = 1$.

9.8.2   Distances

The most direct quantity used to indicate the distance to an object is obviously the redshift z. It is clear that z is directly related to the universal radius R, or to the parameter x introduced in (9.112): $x = (1 + z)^{-1}$ from (9.32). As a relation (x, t) is already available – (9.118), (9.127) and (9.130) – we know how old the Universe was when the light of the object was emitted. Let us find the relation between the radial coordinate r_1 and the redshift, first using $k = 1$. In (9.24) the second integral is $\sin^{-1} r_1$. The first integral is

$$\int_{t_1}^{t_0}\frac{dt}{R} = \int_{R_1}^{R_0}\frac{dR}{R\dot{R}} = \int_{(1+z)^{-1}}^{1}\frac{dx}{\dot{x}}\frac{1}{x}\frac{1}{R_0}$$

$$= \frac{1}{H_0 R_0}\int_{(1+z)^{-1}}^{1}\frac{dx}{\sqrt{1 - 2q_0 + \frac{2q_0}{x}x}}\frac{1}{x} \qquad (9.131)$$

$$= \frac{1}{H_0 R_0}\int_{(1+z)^{-1}}^{1}\frac{dx}{\sqrt{(1 - 2q_0)x^2 + 2q_0 x}}$$

where we have used (9.113). Integration then gives

$$\sin^{-1}r_1 = -\frac{1}{H_0 R_0}\frac{1}{\sqrt{2q_0 - 1}}\left[\sin^{-1}\frac{2(1 - 2q_0)x + 2q_0}{\sqrt{4q_0^2}}\right]_{(1+z)^{-1}}^{1} \qquad (9.132)$$

$$= \sin^{-1}\left(1 + \frac{1 - 2q_0}{q_0}\frac{1}{1 + z}\right) - \sin^{-1}\left(1 + \frac{1 - 2q_0}{q_0}\right)$$

using (9.110). This expression may be rearranged:

$$r_1 = \frac{zq_0 + (q_0 - 1)(-1 + \sqrt{2q_0 z + 1})}{H_0 R_0 q_0^2(1 + z)} \qquad (9.133)$$

which is the formula required.

Equation (9.133) also applies when $k = -1$. For $k = 0$:

$$r_1 = \frac{1}{R_0 H_0}\int_{(1+z)^{-1}}^{1}x^{-1/2}dx = \frac{2}{R_0 H_0}\left(1 - \frac{1}{\sqrt{1 + z}}\right) \qquad (9.134)$$

but this expression may also be deduced from (9.133) for $q_0 = 1/2$. Therefore, (9.133) is valid for any of the three possible universes. In (9.134), the problem is that R_0, and thus also r_1, remain undetermined. However, the distance $R_0 r_1$ does not.

The luminosity distance, as well as the diameter distance, can easily be found from the above formula.

9.8.3 The horizon

The horizon is defined as the surface beyond which light has not had time to reach our Galaxy. As time increases, the horizon moves away; at time t its distance is r_H. From (9.24) r_H is a function of t given by

$$\int_0^{r_H} \frac{dr}{\sqrt{1 - kr^2}} = \int_0^t \frac{dt'}{R(t')} \tag{9.135}$$

For the simplest case $k = 0$, the first integral gives r_H. By solving the second integral:

$$r_H = \frac{3}{R_0 \left(\frac{3}{2} H_0\right)^{2/3}} t^{1/3} \tag{9.136}$$

and for the proper distance of the horizon, using (9.34):

$$d_H = \frac{2}{H_0} \left(\frac{R}{R_0}\right)^{3/2} = 3t \tag{9.137}$$

A horizon also exists for a closed universe, with $k = 1$; using (9.34) and (9.132):

$$d_H = R \sin^{-1} r_H = -\frac{R}{R_0 H_0 \sqrt{2q_0 - 1}} \left[\sin^{-1}\left(\frac{1 - q_0}{q_0} x + 1\right)\right]_0^x$$

$$= \frac{x}{H_0 \sqrt{2q_0 - 1}} \left[\sin^{-1}(1) - \sin^{-1}\left(1 - \frac{2q_0 - 1}{q_0} x\right)\right] \tag{9.138}$$

$$= \frac{x}{H_0 \sqrt{2q_0 - 1}} \cos^{-1}\left(1 - \frac{2q_0 - 1}{q_0} x\right)$$

and for $k = -1$:

$$d_H = \frac{x}{H_0 \sqrt{1 - 2q_0}} \cosh^{-1}\left(1 + \frac{1 - 2q_0}{q_0} x\right) \tag{9.139}$$

For $k = -1$, as x increases, the argument of $\cosh^{-1}$ increases and hence d_H increases faster than x, and we will eventually see all fundamental galaxies. This is also true for $k = 0$, as the horizon recedes with the first power of t, whilst R grows with $t^{2/3}$. For $k = 1$, the life of the Universe is finite and we therefore do not have to wait indefinitely. Even here, though, the whole Universe will eventually be observed. In this case, there is a circumference $L = 2\pi R$, and it is interesting to calculate

$$\frac{d_H}{L} = \frac{1}{2\pi} \cos^{-1}\left[1 - \frac{(2q_0 - 1)R}{q_0 R_0}\right] \tag{9.140}$$

As R increases, the ratio d_H/L increases. When R reaches its maximum value given by (9.126), or

$$\frac{R_{max}}{R_0} = \frac{2q_0}{2q_0 - 1} \tag{9.141}$$

then $d_H/L = 1/2$, we see half the maximum length scale of the Universe. (By completing the observation in the opposite direction, all fundamental galaxies could be observed.) To obtain $d_H/L = 1$, the function R must vanish. Therefore the whole Universe can be observed only at the Big-Bang and at the final collapse.

9.8.4   Numerical examples

Let us adopt $H_0 = 75$ km s^{-1} Mpc^{-1} ($\approx 2.5 \times 10^{-18}$s^{-1}), and three different values of q_0: 0.1 (open), 0.5 (Einstein–de Sitter) for which the Hubble time is 13 Gyr, and 1 (closed).

The present value of the radius of the Universe is calculated from (9.110):

q_0	R_0(Mpc)
0.1	4500
0.5	*indeterminate*
1.0	4000

The Universe's circumference for $q_0 = 1$ would be $2\pi R_0 = 25\,000$ Mpc.

Let us consider five objects at redshifts $z = 0.01$, $z = 0.1$, $z = 1$, $z = 4$, and $z = 1000$ (the 2.7 K radiation). The coordinate r_1 is calculated from (9.133):

Values of r_1

q_0/z	0.01	0.1	1	4	1000
0.1	9×10^{-3}	8.5×10^{-2}	0.63	1.66	7.87
0.5		*indeterminate*			
1.0	9.9×10^{-3}	9.1×10^{-2}	0.50	0.80	0.99

The luminosity distance is calculated from (9.44):
Values of luminosity distance, d_L (Mpc)

q_0/z	0.01	0.1	1	4	1000
0.1	40	419	5600	37000	35×10^6
0.5	40	409	4700	22000	8×10^6
1.0	40	396	4000	16000	4×10^6

The lifetime of the Universe is calculated from (9.119), (9.127), and (9.130) by setting $x = 1$, and the present proper distance of the horizon, d_H, from (9.137), (9.138), and (9.139) by setting $x = R/R_0 = 1$:

q_0	t_0(Gyr)	d_H(Mpc)
0.1	10.7	13000
0.5	8.5	8000
1.0	7.2	6000

9.8.5 The curved Universe

There is an increasing tendency among theorists to assume that our Universe is plane, in agreement with inflationary models. But even if $k \neq 0$, the Universe behaves as a plane at early epochs. Let us consider the Einstein–Friedmann equation (9.100) in the form

$$\dot{R}^2 = \frac{8\pi\epsilon_0 R_0^3}{3R} - k \tag{9.142}$$

At early epochs, with small values of R, the first term on the right-hand side must be larger than the second, which is constant throughout the whole history of the Universe. Let us estimate over what time interval the hypothesis of $k = 0$ is reasonable. From (9.100) and (9.116),

$$\frac{8\pi\epsilon_0}{3} = 2q_0 H_0^2 \tag{9.143}$$

So the gravitational term in (9.142) may be written as

$$\frac{8\pi\epsilon_0}{3} \frac{R_0^3}{R} = 2q_0 H_0^2 \frac{k}{H_0^2 (2q_0 - 1)} \frac{R_0}{R} = \frac{2q_0}{|2q_0 - 1|} \frac{R_0}{R} \tag{9.144}$$

(except for $k = 0$).

If this term is greater than $|k|$ (i.e. greater than unity), curvature effects may be assumed to be negligible. For a plane universe, $q_0 = 1/2$, it is ∞. Obviously no curvature effects at all are expected in an Einstein–de Sitter universe. For $q_0 = 0.1$, curvature effects become of the same order of magnitude as gravitational ones, for $R_0/R = 4$ (a small value compared with R_0/R at the recombination epoch of about 1000). For $q_0 > 0.5$, the transition takes place for $R_0/R < 1$, that is, in the future. At recombination, curvature effects would be slightly perceptible for $q_0 < 5 \times 10^{-4}$. Therefore, curvature becomes important, if at all, only just before the present time. For calculations involving early epochs, curvature may be set equal to zero.

9.9   The radiation-dominated era

If radiation dominates, $\epsilon = 3p$. It has already been seen that $p \propto R^{-4}$ (from (9.63)), therefore, and that $\epsilon \propto R^{-4}$, that is, $T \propto R^{-1}$, as has already been pointed out in (9.64). This important relation was again obtained in (9.80), and could also be obtained from (9.101). Let us now obtain $R(t)$ from (9.100). The equation is easily integrated, yielding

$$R^2 = 2\sqrt{\frac{8\pi}{3} [\epsilon R^4]} \; t - kt^2 \tag{9.145}$$

where $[\epsilon R^4]$ is obviously a constant.

For $k = 0$, we find $R \propto t^{1/2}$, which means a slower expansion than for the matter-dominated era (with $R \propto t^{2/3}$, equation (9.118)). For $k = 1$, R vanishes at the Big-Bang and also at a time given by

$$t(R = 0) = 2\sqrt{\frac{8\pi}{3}[\epsilon R^4]} \tag{9.146}$$

The maximum value of R is reached in half this time. (Before reaching the maximum, however, the Universe will in practice no longer be dominated by radiation.) Instead of (9.110), we now have

$$R^2 = \frac{k}{H^2(q-1)} \tag{9.147}$$

Therefore, for a plane universe, with $k = 0$, we obtain $q = 1$ instead of $q = 1/2$ as at present. It is also easy to obtain

$$\Omega = q \tag{9.148}$$

instead of (9.109).

We need not consider the problem of distances, because no objects are to be detected during this era. We may calculate the position of the horizon at a given t. Let us assume that $k \approx 0$ and apply (9.135):

$$d_H(t) = R(t)r_H(t) = R(t)\int_0^t \frac{dt'}{R(t')} = 2t \tag{9.149}$$

Thus the horizon recedes at a lower speed for smaller values of t ($d_H = 3t$; see equation (9.137)).

9.10   The microwave background radiation

As often stated throughout this chapter and the preceding one, the $\sim$3 K background radiation corresponds to the black-body photon system which becomes decoupled from matter at $z \approx 1100$. This black body cools as a result of expansion. As the system is isolated it maintains its thermodynamic equilibrium, so that a black-body spectrum at a lower temperature is observed at present. The most precise determination of its spectrum and temperature has been obtained by COBE. It is a perfect black-body within the limits of measurement error, at 2.735±0.006 K. Anisotropies in the angular distribution are connected with the problem of protogalactic inhomogeneities, and will be discussed in the following chapter. Another type of anisotropy may arise from the proximity of the horizon, at the decoupling epoch. A region which is more than one horizon distance away from another one is causally disconnected from it. Therefore, anisotropy on an angular scale, corresponding to the horizon when decoupling took place, is to be expected. In any case there is an interesting transition between super- and sub-horizon inhomogeneities. We must calculate the angle formed by the horizon scale at the decoupling epoch.

In the Robertson–Walker metric (9.4) let us set $dt = 0$, $dr = 0$, $d\varphi = 0$

$$ds = Rr\,d\theta \tag{9.150}$$

Here $R = R_0/(1 + z)$ and z is about 1100. To calculate r we use (9.133) or, more simply, (9.134) if we assume that the Universe was plane at this early epoch. From (9.134),

$$r \approx \frac{2}{R_0 H_0} \tag{9.151}$$

Therefore,

$$ds \approx \frac{R_0}{1+z} \frac{2}{R_0 H_0} d\theta \approx \frac{2}{H_0} \frac{1}{z} d\theta \tag{9.152}$$

is integrated, giving

$$s = \frac{2}{H_0 z} \theta \tag{9.153}$$

where s is the horizon scale length, which was given by (9.137) (this epoch is matter-dominated and with vanishing curvature), that is,

$$d_H \approx \frac{2}{H_0} z^{-3/2} \tag{9.154}$$

Then θ is the angle characterizing the present angular scale of the horizon at decoupling:

$$\theta = z^{-1/2} \approx \frac{1}{40} \text{rad} \approx 2° \tag{9.155}$$

The 2°-angular scale thus divides anisotropies into super- and sub-horizon anisotropies and density inhomogeneities. The interpretation of the different angular scales is considered in the following chapter.

10 The fluid of galaxies

10.1 Main properties

The fluid of stars in a galaxy closely resembles the fluid of molecules. The fluid of galaxies does not. Let us examine three main differences.

(a) The distribution of galaxies is not chaotic. There are groups, clusters, superclusters, and there is a large-scale structure, which we are now beginning to realize. The distinction between clusters and superclusters is not sharp, and as superclusters are elements of larger structures their limits and sizes are difficult to establish. The largest observed structures are as large as the limits of the deepest surveys. A continuous spectrum of inhomogeneities describes structures larger than a galaxy better than it describes discrete objects such as stars and galaxies.

The relative increase in the density with respect to the mean density δ is about $10^2 - 10^3$ for clusters. Typical intercluster distances of about 5 Mpc and inter-supercluster distances of about 25 Mpc are revealed by cross-correlation studies. The value of δ for a galaxy is about 10^5.

The large-scale structure of the Universe is complex. Many clusters are aligned in huge filaments. Others seem to form sheets. There are also voids which are apparently deplete of galaxies. A simplified picture of the large-scale structure might consist of an ensemble of large polyhedral voids. In the limiting sheets separating two adjacent voids there are clusters. Along the limiting line intersections there are more clusters. At the limiting vertices there are still more clusters. The linear dimensions of the voids are typically $20 - 50$ Mpc. The picture is probably oversimplified and other authors see a forest with trees with interconnected branches and roots. In this case, the higher density of galaxies corresponds to the different parts of the tree. A 'Great Wall' of dimensions $60 \times 170 \times 10$ Mpc^3 has possibly been identified, and the density in the Hydra–Centaurus direction seems to be much higher (in the ~ 1000-Mpc-neighbourhood). This is the direction of the so-called Great-Attractor, with a possible mass of about $10^{16} M_\odot$. A super-attractor 200 Mpc away in the same direction, with a mass ten times larger, may exist.

On qualitative grounds the question of how large-scale structure can be generated is not difficult to answer. Any relatively small part of the Universe behaves as the whole Universe, that is, it obeys the Einstein–Friedmann equation. A bubble with the same density as the mean density expands at the same rate as the whole Universe expands. An initially less dense bubble would expand faster, as fast as an open low density universe. Thus voids grow relatively with respect to the surroundings. This expansion of a void is limited only by the expansion of adjacent voids. In this way the polyhedral structure is achieved. However, the quantitative description is much more difficult, because the spectrum of primordial density fluctuations is unknown since the dominant elementary particle of the Universe is not known for certain (i.e. we do not know the material out of which the Universe is made), and because we do not know if light traces mass, that is, the distribution of galaxies does not necessarily coincide with the distribution of matter.

(b) We now encounter the dark matter problem. Inflation theories of the very early Universe, as well as some philosophical arguments, suggest that $\Omega_0 = 1$. However, the baryonic visible matter contributes $\Omega_{BV} = 0.01$ at most, even when the whole electromagnetic radiation spectrum is explored. Theoretical models of primordial nucleosynthesis in the early Universe are able to limit the baryonic contribution (either seen or unseen) to $\Omega_B \approx 0.1$. Therefore, there is a large amount of dark baryonic matter, may be as much as ten times the visible baryonic matter, and there is a very large amount of non-baryonic dark matter, possibly more than ten times the baryonic dark matter. The nature of this non-baryonic dark matter is unknown. From the point of view of our fluid of galaxies, we could say that our molecules flow through a sea of unknown nature, which possesses more mass than they do.

Our Galaxy, as a representative spiral galaxy, probably also has a large amount of dark matter. This conclusion is based on the flatness of the rotation curve at large radii, that is, the rotation velocity θ is nearly independent of the galactocentric radius r in the outer disc. A Keplerian rotation curve $\theta \propto r^{-1/2}$ would be expected instead if the $10^{11} M_\odot$ of visible stars were the only source of gravitational potential, and if the gravitational and centrifugal forces were in balance. Either the mass of the Galaxy is much higher, at least $10^{12} M_\odot$, or other forces are present in the balance. The first interpretation is the more accepted and will therefore be adopted here. We assume that the mass of our Galaxy is more than $10^{12} M_\odot$. From the point of view of the fluid of galaxies this is a bad starting point: we do not know the mass of our molecules.

It is thought that the galactic dark matter is possibly mainly baryonic and lies in a spherical halo. Candidates for galactic baryonic dark matter are planet-like objects, brown dwarfs, neutron stars, black holes, or even gas at a temperature low enough to emit negligible thermal X-radiation. Recent possible microlensing detections of LMC stars have been interpreted as due to halo compact objects and seem to favour the brown dwarf candidates. But other interpretations of these events cannot be ruled out.

It is not only the flat rotation curves of spiral galaxies that seem to indicate dark matter problem. The Virial theorem when applied to elliptical galaxies and to clusters also indicates that the necessary amount of matter to maintain these systems bound under gravity is much higher than that observed. This conclusion is not free from observational mis-interpretations, but it is commonly accepted. Observations on a large scale and a cosmic Virial theorem suggest a matter density corresponding to $\Omega_0 = 1$, but here assumptions tend to be self-fulfilling.

The larger the scale of the system under study, the closer to unity seems to be the value of Ω_0 obtained. One possible way to explain this is by the hypothesis of biased galaxy formation. The formation of bright galaxies could be a threshold phenomenon, so that they would be formed only in very high density fluctuations. Faint or failed galaxies associated with non-baryonic dark matter would fill space more homogeneously. The bright galaxies would reflect the structure we see, but voids would not be as empty as they appear. This possibility is far from being demonstrable, however.

A large experimental, observational, and theoretical effort is being made at present to identify dark matter particles of non-baryonic nature. There are two types of possibility: hot dark matter consisting of a hot ($\epsilon = 3p$) fluid at recombination, and cold dark matter consisting of a cold ($p = 0$) fluid at recombination. Models that assume hot dark matter usually face problems in reproducing small-scale clustering, as it is difficult to concentrate such fast particles. Light but non-massless neutrinos (10–30 eV) are the best candidates. There is a long list of candidates for WIMPs (weak interacting massive particles), that is, the particles that could constitute the cold dark matter. In this list are heavy neutrinos, magnetic monopoles, primordial black holes (of planetary masses, formed at early epochs out of non-baryonic matter), neutralinos, etc. They are usually very massive (> 1 GeV). However, one of the strongest candidates is the axion, which is very light ($\sim 10^{-5}$ eV). If the Universe is flat, there could be about 10^9 axions/cm^3, 10^{16} times the number density of protons!

(c) The mean velocity of a comoving fundamental volume element is zero. The observed mean flow is the Hubble flow. However, there may be peculiar velocities of field galaxies and clusters with respect to the Hubble flow. We will show that peculiar velocities are associated with density inhomogeneities. A well-observed value of a typical peculiar velocity is provided by that of our own Galaxy. The Milky Way moves with respect to the cosmic microwave background radiation at 622 ± 20 km s^{-1}, towards $l = 277° \pm 2°$, $b = 30° \pm 2°$, as deduced from measurements of the dipole anisotropy by the satellite COBE. $\delta T/T$ of this dipole anisotropy is 3.2 ± 0.2 mK (see Section 3.3.3). We suspect that this is the actual peculiar velocity because the speed and direction coincide with those of the Milky Way with respect to the X-ray background. The X-ray background is emitted by distant galaxies, so that it can be identified with the unperturbed Hubble frame. It also coincides with the velocity of the local group with respect to a sample of elliptical galaxies. For this latter determination distances have been obtained using the observed correlation between diameter and internal velocity dispersion (Faber–Jackson effect). The peculiar velocities of clusters may be obtained if their distances can be determined by methods other than the redshift–distance relation. This is of considerable potential interest as peculiar velocities are due to the distribution of matter, either dark or visible, and may be used to determine the mass distribution, which could well differ from the light distribution.

With typical velocity dispersions of the order of 600 km s^{-1}, a galaxy has a path of only ~ 10 Mpc in one Hubble time. That means that forces such as the pressure gradient force have little influence on the establishment of the large-scale structure. Initial conditions are important in any non-equilibrium system, but they are particularly important in this case. What we see is the result of initial density fluctuations, modified only by the action of gravity rather than that of the balance of intervening forces.

These fluctuations must have originated during inflation. We need to know the primordial spectrum in order to determine future evolution. One of the most attractive possibilities consists of the so-called adiabatic (or curvature) fluctuations. Harrison and Zeldovich deduced from inflation a δ-spectrum independent of the rest mass involved in a fluctuation. COBE has obtained a good agreement with the Harrison–Zeldovich spectrum, although the more refined experiment on Tenerife appears to give results in disagreement with this. Cosmic strings, that is, line-like effects originated during phase transitions, might also constitute the seeds of structure formation.

10.2   The birth of galaxies

10.2.1 The galactic Jeans mass

The concept of Jeans mass, developed in Chapter 7 to deal with the birth of stars, now requires some generalization in order to be applied to the birth of galaxies. Although there are some important differences between the interstellar and the intergalactic media, we are also interested in the formation of clusters and large-scale structures.

Adiabatic fluctuations

Protostellar fluctuations were considered to be isothermal, as photons were allowed to escape from the transparent protostellar cloud, thus maintaining constant temperature. In the pre-galactic phase of the Universe photons were one of the basic constituents of our cloud, even being dominant in early epochs. Therefore, the fluctuations of any model will, in general, be adiabatic. Some protogalactic isothermal fluctuations are also possible, but they will not be considered here. From annihilation to recombination we may write

$$\epsilon = mn + aT^4 \tag{10.1}$$

$$p = \frac{1}{3}aT^4 \tag{10.2}$$

$$\sigma = \frac{4aT^3}{3n} \tag{10.3}$$

as shown in Chapters 2 and 9. m and n are the mass and number density of the dominant non-relativistic particle (baryon, axion, . . .), σ is the entropy per dominant non-relativistic particle, ϵ is the energy density, p is the hydrostatic pressure, and a is a constant with a value slightly dependent on the dominant relativistic particles, which are considered here to be photons. The contribution of the kinetic energy of non-relativistic particles to ϵ and p has been neglected. For adiabatic processes σ is a constant and therefore

$$3\frac{\delta T}{T} = \frac{\delta n}{n} \tag{10.4}$$

Hence,

$$\delta\epsilon = m\delta n + 4aT^3\delta T = mn\left(3\frac{\delta T}{T}\right) + 4aT^4\frac{\delta T}{T} = (3mn + 4aT^4)\frac{\delta T}{T} \tag{10.5}$$

and

$$\delta p = \frac{4}{3}aT^3\delta T = \frac{4}{3}aT^4\frac{\delta T}{T} \tag{10.6}$$

The sound speed V_s, is given by

$$V_s^2 = \left(\frac{\partial p}{\partial \rho}\right)_Q \rightarrow \left(\frac{\partial p}{\partial \epsilon}\right)_Q \tag{10.7}$$

where subindex Q denotes adiabatic fluctuations, and the arrow represents generalization to a hot system. Then

$$V_s^2 = \frac{\frac{4}{3}aT^4\frac{\delta T}{T}}{(3mn + 4aT^4)\frac{\delta T}{T}} = \frac{T\sigma}{3(m + T\sigma)} \tag{10.8}$$

For instance, if the system is very hot $m \ll T\sigma$ and we obtain the known result $V_s^2 = 1/3$ (see Section 2.2.4). Hence (10.8) is a more general formula, which is valid for relativistic systems that are not in extreme states.

However, after recombination, with the decoupling of photons, (10.1), (10.2), and (10.3) must be substituted by

$$\epsilon = mn + \frac{3}{2}nT \tag{10.9}$$

$$p = nT \tag{10.10}$$

$$n = constant \ T^{1/(\gamma-1)} \tag{10.11}$$

with $\gamma = c_p/c_v \approx 5/3$. Following the same procedure, we now obtain

$$V_s^2 \approx \frac{\gamma T}{m} \tag{10.12}$$

if $T \ll m$. This is of course a very familiar result, and we now have expressions for the sound speed from annihilation up to recombination (10.8) and after recombination (10.12). We have made this calculation because Jeans waves are propagated at the speed of sound, so V_s appears explicitly in the formula for the Jeans mass.

Jeans mass in a relativistic fluid

Classically, the Jeans mass is a critical mass above which Jeans waves cannot propagate acoustically, grow exponentially, or collapse gravitationally. To find the classical expression for the stellar Jeans mass quickly, let us recall an interesting interpretation given in Section 7.3.3: a cloud has a typical size of a Jeans wavelength λ_J when its gravitational energy equals its thermal energy

$$\frac{(\rho\lambda_J^3)^2}{\lambda_J} \approx T\frac{\rho}{m}\lambda_J^3 \tag{10.13}$$

Hence,

$$\lambda_J \approx V_s \rho^{-1/2} \tag{10.14}$$

and thus the classical stellar Jeans mass is

$$M_J \approx V_s^3 \rho^{-1/2} \tag{10.15}$$

in agreement with (7.46). For the relativistic case, we should replace ρ by ϵ and V_s by its value given by (10.8):

$$\lambda_J = V_s \epsilon^{-1/2} \tag{10.16}$$

The total energy density is in fact involved in the Jeans oscillation. To calculate the Jeans mass, however, we must multiply λ_J^3 by the rest mass density only, as this is the mass that will collapse if gravitational instability develops and leads to galaxy formation. We then obtain

$$M_J = mn\lambda_J^3 = mnV_s^3 \epsilon^{-3/2} \tag{10.17}$$

We see in the first equality that if $n \propto R^{-3}$ and $\lambda_J \propto R$ as any other length in the expansion, M_J is invariant in the expansion, as is the rest mass. The correct formula, derived by Lifshitz using a more standard relativistic treatment, is very similar $(M_J = mnV_s^3 (\epsilon + p)^{-3/2})$. As $p < \epsilon$, the order of magnitude of (10.17) is correct, and only the order of magnitude of the Jeans mass is interesting for our purposes. Note that m is the rest mass of a non-relativistic particle predominant in the fluid which will be present in the structures that subsequently develop. For instance, it could be an axion or a neutralino mass. If we are interested in a small-scale pre-galactic baryonic structure it should be identified with a proton mass.

Jeans mass in an expanding medium

When the gravitational evolution of the cloud is considered, the expansion of the Universe is by no means negligible. This is clearly confirmed when the characteristic time of expansion is compared with the period of a Jeans wave. The former would be of the order of

$$\frac{1}{H} = \frac{R}{\dot{R}} \approx \frac{R}{\left(\frac{8\pi\epsilon R^3}{3R}\right)^{1/2}} = \left(\frac{3}{8\pi\epsilon}\right)^{1/2} \approx \epsilon^{-1/2} \tag{10.18}$$

where (9.142) with $k = 0$, and (9.62) with $p = 0$, have been used to obtain the magnitude. The period of a Jeans wave would be of the order of

$$\frac{\lambda_J}{V_s} = \frac{V_s \epsilon^{-1/2}}{V_s} = \epsilon^{-1/2} \tag{10.19}$$

the same as in (10.18). For wavelengths $\lambda < \lambda_J$, that is, for $\lambda < V_s \epsilon^{-1/2}$, there is no collapsing evolution, so that the wave will consist of pure acoustic oscillations. If we divide this inequality by V_s and use (10.18) we can also see that acoustic oscillations are propagated for wave periods less than the Hubble time. For $\lambda \gg V_s \epsilon^{-1/2}$ the formation of structures begins. How this gravitational collapse evolves will be studied below, but we can anticipate the result that it does not have the violent exponential characteristic of the proto-stellar collapse. So we should speak of a moderate density increase, rather than a collapse.

A perturbation of size λ will grow in the expansion, even if $\delta\epsilon/\epsilon$ does not, for instance when it lies in the acoustic regime. But $1/H$ also grows, that is, the horizon grows, as $1/H$ is of the order of the horizon distance, both in radiation- and matter-dominated epochs (from (9.120), (9.137), (9.145), and (9.149)). If $\lambda \propto R$ and the horizon increases either as R^2 (radiation era) or as $R^{3/2}$ (matter era), an epoch is reached when a perturbation larger than the horizon becomes sub-horizon-sized.

The evolution of the Jeans mass

M_J has varied with time thoughout the history of the Universe. To determine the function $[M_J, R]$ let us begin shortly after annihilation.

(a) Radiation-dominated Universe

As $m \ll T\sigma$ in (10.8), $V_s^2 = 1/3$. With (10.3), $\epsilon = aT^4$ and thus

$$M_J \approx mn\epsilon^{-3/2} = m\frac{4aT^3}{3\sigma}(aT^4)^{-3/2} \approx \frac{m}{a^{1/2}\sigma}T^{-3} = \frac{m}{a^{1/2}\sigma}T_0^{-3}\left(\frac{R}{R_0}\right)^3 \qquad (10.20)$$

Hence $M_J \propto R^3$ during this era.

(b) Pre-recombination matter-dominated Universe

The transition between radiation and matter domination took place for $\epsilon_{matter} \approx \epsilon_{radiation}$, that is,

$$mn_0\left(\frac{R_0}{R}\right)^3 \approx aT_0^4\left(\frac{R_0}{R}\right)^4 \qquad (10.21)$$

and thus

$$\frac{R_0}{R} = \frac{mn_0}{aT_0^4} \qquad (10.22)$$

We do not know the mass of the non-relativistic particles prevailing in the present Universe. But, irrespective of the nature of dark matter, we may assume that mn_0 is the present matter density required to close the Universe. If we adopt (as a reasonable compromise) $H_0 \approx 75$ km s^{-1}Mpc$^{-1} = 2.5 \times 10^{-18}$s, the critical density is 10^{-29}g cm^{-3}. We then obtain for $(R_0/R) \approx z$ at the equality epoch the value of 2.5×10^4, of the order of 10^4, as stated in (8.42).

Between equality and recombination (more precisely photon decoupling), $m \gg T\sigma$ and hence $V_s^2 = T\sigma/3m$. As $\epsilon \approx mn$,

$$M_J = mn\left(\frac{T\sigma}{3m}\right)^{3/2}(mn)^{-3/2} = m^{-1/2}\left(\frac{4aT^3}{3\sigma}\right)^{-1/2}\left(\frac{T\sigma}{3m}\right)^{3/2}$$

$$= \frac{1}{6}\frac{\sigma^2}{m^2}a^{-1/2} \qquad (10.23)$$

Note here that M_J does not depend on T, that is, it does not depend on R. For this constant baryonic Jeans mass we obtain $M_J = 0.4\sigma^2 M_\odot$, about $4 \times 10^{19} M_\odot$.

(c) Post-recombination Universe

Here $V_s^2 \approx (T/m)$, where T is now the matter temperature:

$$M_J = mn\left(\frac{T}{m}\right)^{3/2}(mn)^{-3/2} = m^{-2}n^{-1/2}T^{3/2} = m^{-2}\left(\frac{4aT_\gamma^3}{3\sigma}\right)^{-1/2}T^{3/2} \qquad (10.24)$$

where T_γ is the radiation temperature. As $T \propto R^{-2}$ and $T_\gamma \propto R^{-1}$, we have $T \propto T_\gamma^2$. The proportionality constant is obtained knowing that at recombination both temperatures were the same: T_R. Then

$$T \approx \frac{T_\gamma^2}{T_R} \qquad (10.25)$$

Substituting this result in (10.24),

$$M_J \approx m^{-2}\left(\frac{\sigma T_\gamma^3}{aT_R^3}\right)^{1/2} = m^{-2}\left(\frac{\sigma T_{\gamma 0}^3}{aT_R^3}\right)^{1/2}\left(\frac{R_0}{R}\right)^{3/2} \qquad (10.26)$$

Therefore, during this most recent epoch $M_J \propto R^{-3/2}$.

Summarizing the previous discussion:

Annihilation–equality

$$M_J = 1.4 \times 10^{28}(R/R_0)^3 \, M_\odot \qquad (10.27)$$

Equality–recombination

$$M_J = 4 \times 10^{19} \, M_\odot \qquad (10.28)$$

Recombination–present

$$M_J = 0.5(R/R_0)^{-3/2} \, M_\odot \qquad (10.29)$$

There is an apparent discontinuity at equality. Substituting in (10.27) for $R/R_0 = (2.5 \times 10^4)^{-1}$ at equality yields $9 \times 10^{14} \, M_\odot$, far from the value $4 \times 10^{19} \, M_\odot$ given by (10.28). This is due to our approximations. Equation (10.28) would in fact be correct for $\epsilon_{matter} \approx 10\,\epsilon_{radiation}$, that is, for an R_0/R value equal to 0.1 times the value given by (10.22), about 2.5×10^3. This is about the R_0/R value of the recombination epoch. Thus the asymptotic value (10.28) is never reached.

There was a real discontinuity at recombination, as this was a sudden event. Baryons coupled to photons were unable to participate in a collapse. This constraint suddenly disappeared at decoupling and the formation of protogalactic clouds could begin without any impediment.

The curve $[M_J, R]$ is plotted in Figure 10.1. We observe that the asymptotic value (10.28) is never reached. Note the dramatic decrease at recombination when baryons are decoupled from the photons. The Jeans mass at present is of the order of $1 M_\odot$. Let us assume a primordial structure with a baryonic mass of $10^{12} M_\odot$, that is, the mass of a galaxy such as the Milky Way (dark matter included). After annihilation it became unstable and the density increased with respect to that of the surrounding medium. The rate at which the density relatively increased must still be calculated. At about $z \approx 3 \times 10^5$ the Jeans mass became larger, and the perturbation became an acoustic crest, preventing the increase of $\delta\epsilon/\epsilon$. Recombination took place in these non-violent conditions. Our perturbation again became unstable, like any other perturbation of

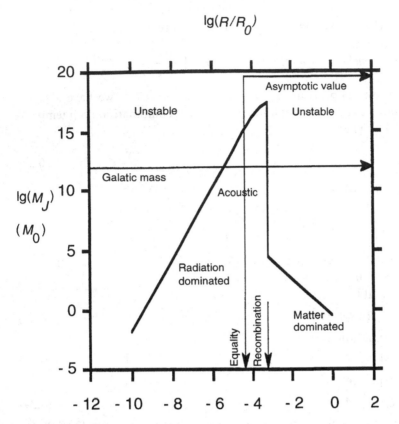

Figure 10.1 Evolution of the Jeans mass along the history of the Universe. The size of the Universe with respect to its present size is used as a time variable.

more than $3 \times 10^4 M_\odot$ (note, this is the mass of a typical globular cluster). The rate at which $\delta = \delta\rho/\rho$ increased will be calculated below. Protogalactic clouds formed. Galaxies themselves with quasar and starlight emissions appeared much later, at $z \approx 4$.

However, this plot does not answer a basic question: why galaxies (and clusters) have the mass they have.

10.2.2 The mass of a galaxy

The Silk mass

On 'small' scales, such as that of a perturbation of $10^{12} M_\odot$, the cosmic fluid is not perfect. As a result, small-scale perturbations were damped during the acoustic phase and could not survive until recombination. The damping mechanism was photon diffusion. When matter becomes denser because of the passage of waves, fast photons tend to diffuse to less dense zones, breaking down the adiabatic condition. Let us assume, first, that matter is made up of colliding particles, such as baryons, interacting with photons via Thomson scattering. Baryons are pulled to less dense regions by photons. Remember the scattering cross-section $\sigma_T = (8\pi e^4)/(3m_e^2) = 0.67 \times 10^{-24} \text{cm}^{-2} \equiv 7.4 \times 10^{-46}$ s and that the photon mean-free-path from (3.62) is $\lambda_\gamma = (\sigma_T n)^{-1}$. Imperfect fluids are characterized by viscosity and thermal conduction, in contrast with ideal fluids. Therefore, as the fluid becomes imperfect, photon diffusion may be interpreted both as a viscosity and as a thermal conduction. When the photon mean-free-path becomes of the order of $\sim 0.1\lambda$, where λ is the perturbation wavelength, the dissipative effects of photon diffusion become important, that is, for

$$\lambda \approx 10\frac{(R/R_0)^3}{\sigma_T n_0} = 10\frac{m(R/R_0)^3}{\sigma_T(mn_0)} \tag{10.30}$$

which corresponds to a mass of

$$M = mn\lambda^3 = 10^3\frac{m^3(R/R_0)^6}{(mn_0)^2\sigma_T^3} \tag{10.31}$$

For a present baryon density of 10^{-30}g cm^{-3} ($\Omega_B = 0.1\Omega_0$), at recombination we find that masses less than $2 \times 10^{12}\text{M}_\odot$ cannot collapse. Rather satisfactorily, this is the order of a typical galactic mass. Larger structures would be unaffected by photon diffusion.

Note that $\lambda_\gamma \propto R^3$, whilst an acoustic perturbation grows as $\lambda \propto R$, and thus damping is most effective near recombination. This is intuitively understandable because the photon mean-free-path becomes larger for a less dense universe, so that photons can then escape more easily from the cloud.

To improve our grasp of the process, let us calculate how many oscillations took place during the acoustic phase. Just before recombination $V_s = (T\sigma/3m)^{1/2} \approx 0.02$. In any case it must be lower than 0.58. The period of the wave was λ/V_s. For a $10^{12}\text{M}_\odot$ perturbation, λ is given by (10.30), so the period is about 10^5 years. This figure is close to the time at recombination. Although smaller masses would have shorter periods, it could be said that these acoustic waves oscillated just once!

The process does not much resemble a wave. Rather, we would say that primordial local concentrations of mass slightly lower than that of a galaxy could grow until $z = 10^6$. Then, an acoustic trend to convert an overdense region into an underdense one was impeded by flowing photons. Subsequently photons dragged along baryons and the relic high density zone disappeared. Only masses greater than about one galactic mass were not smoothed in this way by photons. For these large perturbations the rather sudden process of recombination interrupted the oscillations.

A more precise calculation, taking the photon diffusion viscosity and thermal conduction in a relativistic imperfect fluid into account, was first carried out by Silk.

Free streaming

There are particles, such as neutrinos or axions, which were decoupled in the epoch of interest, and which could therefore stream freely out of an overdense cloud without dragging baryons via Thomson scattering or any other mechanism. By means of this free motion they were able to smooth out inhomogeneities. Only inhomogeneities larger than the transit path length of these particles were free from this kind of damping, which is called free streaming damping. An order of magnitude of this length is obtained as follows. Let us consider neutrinos, which are effective dampers because of their high speeds. After neutrino-decoupling they constituted a hot fluid with speed close to unity. At some epoch the temperature had dropped to $T_\nu \approx m_\nu$ and the fluid was no longer relativistic. Their speed was reduced and their ability to smooth out acoustic oscillations decreased. The length we are looking for may be identified with the maximum distance that a neutrino with a speed close to unity can travel from the epoch of neutrino-decoupling until the epoch when neutrinos lost their relativistic regime, t_{NR}. Therefore the length must be approximately t_{NR}. The neutrino-decoupling time and the additional length traversed by slow neutrinos can be assumed to be negligible. The distance t_{NR} later increased by a factor R_0/R_{NR}. It is easy to estimate t_{NR} and R_{NR} as they are the time and cosmic scale factors when the Universe had a temperature of m_ν. For $m_\nu = 30$ eV we can easily calculate $t_{NR} \approx 10^{10}$s and $R_{NR} \approx 1.7 \times 10^{-5} R_0$. Therefore the free streaming length scale is

$$\lambda_{FS} = t_{NR} \frac{R_0}{R_{NR}} \approx 6 \times 10^{14} s \equiv 6 \text{ Mpc} \tag{10.32}$$

For a present density of baryons of $\sim 10^{-30}$g cm^{-3}, we obtain a minimum mass of $3 \times 10^{12} M_\odot$. This is also a reasonable value for the mass of a galaxy. It is, however, an underestimate, and more detailed calculations yield typical masses of galaxy clusters.

10.2.3 The growth of perturbations

From (10.14) and (10.15),

$$\frac{M_J}{\lambda_J} \approx V_s^2 \tag{10.33}$$

After recombination $V_s^2 \approx T/m \ll 1$. From Chapters 5 and 9 we know that a general relativistic treatment is necessary when the length scale of a system is less than, or of the order of, its mass. Now $M_J \ll \lambda_J$, which means that a Newtonian description of the gravitational growth of a perturbation is adequate. However, before equality, $V_s^2 = 1/3$, M_J, and λ_J have similar orders of magnitude, and a relativistic treatment is required.

Post-recombination growth

After photon-decoupling, the contraction of baryonic systems is no longer impeded, and the formation of protogalactic structures is unavoidable. The quantity $\delta = \delta\rho/\rho$ characterizes the magnitude of the perturbation. This is a relative quantity. If baryons contribute one-tenth of the critical density, then $10^{-30}(1100)^3 \approx 10^{-21}$g cm$^{-3} \approx 10 M_\odot$pc^{-3} was the baryon density at recombination. This value is higher than the present density in our galaxy. These figures are certainly difficult to compare. The large volume of the halo on the one hand, and the mass variations during the Galaxy lifetime on the other, have to be taken into account. Also note that the density of a protogalactic cloud is higher than the average density at a given epoch. In any case, we can perhaps best see the process of galaxy formation as an isolation of the protogalactic cloud from the expanding surrounding medium, rather than as a real collapse. Self-gravitation impedes further expansion of the cloud, rather than producing a contraction.

Newtonian cosmology is allowed after recombination. We will assume small values of δ, $\delta \ll 1$, equivalent to considering linear perturbations. It is reasonably clear that $\delta \ll 1$ at the beginning, but the critical value $\delta = 1$, which renders the equations non-linear, was certainly reached well in the past, as we know that $\delta \approx 10^5$ today. Let us, however, consider the linear regime as characterizing the initial evolution. We will further assume zero pressure and a flat universe.

Any quantity, such as the density, may be decomposed into a mean value and a perturbed one, $\rho \rightarrow \rho + \rho_1$, where $\rho_1 \equiv \delta\rho \ll \rho$. The mean value follows the time evolution of the Einstein–de Sitter universe:

$$\rho = \rho_0 (R/R_0)^{-3} \tag{10.34}$$

$$\vec{v} = (\dot{R}/R)\vec{r} \tag{10.35}$$

$$\vec{g} = -\frac{4\pi\rho}{3}\vec{r} \tag{10.36}$$

where

$$\dot{R}^2 = \frac{8\pi\rho}{3} R^2 \tag{10.37}$$

By means of a standard analysis of perturbations of the cosmological fluid equations in Chapter 8, we easily find the following

(a) Continuity

$$\frac{\partial\rho}{\partial t} + \nabla \cdot (\rho\vec{v}) + \frac{\partial\rho_1}{\partial t} + \rho\nabla \cdot \vec{v}_1 + \vec{v} \cdot \nabla\rho_1 + 3\rho_1 \frac{\dot{R}}{R} = 0 \tag{10.38}$$

The first two terms cancel, as implied by the continuity equation for mean quantities. Therefore,

$$\dot{\rho}_1 + \rho\nabla \cdot \vec{v}_1 + \vec{v} \cdot \nabla\rho_1 + 3\rho_1 \frac{\dot{R}}{R} = 0 \tag{10.39}$$

(b) Motion

$$\dot{\vec{v}}_1 + \frac{\dot{R}}{R}\vec{v}_1 + \frac{\dot{R}}{R}\vec{r} \cdot \nabla\vec{v}_1 - \vec{g}_1 = 0 \tag{10.40}$$

where we have ignored the effect of the pressure gradient force, assuming zero pressure. We also have, for the mean quantities,

$$\nabla \times \vec{g} = 0 \tag{10.41}$$

$$\nabla \cdot \vec{g} = -4\pi\rho \tag{10.42}$$

equivalent to $\vec{g} = -\nabla\phi$ and Poisson's equation. The perturbed equations are

$$\nabla \times \vec{g}_1 = 0 \tag{10.43}$$

$$\nabla \cdot \vec{g}_1 = -4\pi\rho_1 \tag{10.44}$$

Let us assume, and then check, a solution of the type

$$\rho_1 = B\mathcal{E} \tag{10.45}$$

$$\vec{v}_1 = \vec{V}\mathcal{E} \tag{10.46}$$

$$\vec{g}_1 = \vec{G}\mathcal{E} \tag{10.47}$$

where

$$\mathcal{E} = e^{i\vec{q}\cdot\frac{\vec{r}}{R}} \tag{10.48}$$

B, $\vec{V}$, and $\vec{G}$ are the amplitudes of ρ_1, $\vec{v}_1$, and $\vec{g}_1$. They are functions of time only. The idea is that $\mathcal{E}$ is invariant in comoving coordinates. But now we use a classical Newtonian nomenclature. $\vec{r}$ is the usual position vector. If it denotes a comoving particle it should grow in proportion to R. To avoid this growth, we divide by R. Hence, $\vec{q}$ represents a 'comoving' invariant wave number. Then $R(2\pi/q)$ represents the physical size of the perturbation. $\vec{q}$ must be kept constant. In (10.39), (10.40), (10.43), and (10.44)

$$\dot{B} + i\rho\frac{1}{R}(\vec{V}\cdot\vec{q}) + 3\frac{\dot{R}}{R}B = 0 \tag{10.49}$$

$$\dot{\vec{V}} + \frac{\dot{R}}{R}\vec{V} - \vec{G} = 0 \tag{10.50}$$

$$\vec{q} \times \vec{G} = 0 \tag{10.51}$$

$$\frac{i}{R}(\vec{G}\cdot\vec{q}) = -4\pi B \tag{10.52}$$

From (10.51) we see that $\vec{G} \parallel \vec{q}$. From (10.52),

$$\vec{G} = \frac{i4\pi RB}{q^2}\vec{q} \tag{10.53}$$

Let us take

$$\vec{V} = \vec{V}_\perp + iw\vec{q} \tag{10.54}$$

where $\vec{V}_\perp$ is perpendicular to $\vec{q}$ and $iw\vec{q}$ is the $\vec{V}$-component along $\vec{q}$. Here w is a function of time only. Then, in (10.50),

$$\dot{\vec{V}}_\perp + iw\vec{q} + \frac{\dot{R}}{R}\vec{V}_\perp + \frac{\dot{R}}{R}iw\vec{q} - i\frac{4\pi RB}{q^2}\vec{q} = 0 \tag{10.55}$$

equivalent to two equations

$$\dot{\vec{V}}_\perp + \frac{\dot{R}}{R}\vec{V}_\perp = 0 \tag{10.56}$$

and

$$\dot{w} + w\frac{\dot{R}}{R} - \frac{4\pi RB}{q^2} = 0 \tag{10.57}$$

The variable

$$\delta = \frac{B}{\rho} \tag{10.58}$$

is equivalent to $\delta\rho/\rho$, and its time variation is what we are looking for. From (10.49),

$$\dot{\delta} = \frac{wq^2}{R} \tag{10.59}$$

From (10.56), $\vec{V}_\perp \propto 1/R$, hence $\vec{V}_\perp$ decreases with time and thus has no interest for us. We may set $\vec{V}_\perp = 0$. From (10.57), (10.58), and (10.59), eliminating w, we obtain

$$\ddot{\delta} + 2\dot{\delta}\frac{\dot{R}}{R} - 4\pi\rho\delta = 0 \tag{10.60}$$

which is the differential equation to be integrated. Given that $\rho = \rho_0(R/R_0)^{-3}$ and $R \propto t^{2/3}$ during the matter-dominated era, these equations have two solutions of the form $\delta \propto t^{2/3}$ and $\delta \propto t^{-1}$. The second solution would produce an expansion of the cloud and is therefore uninteresting. Our required solution is therefore

$$\delta \propto t^{2/3} \propto R \propto (1 - z)^{-1} \tag{10.61}$$

In fact, a non-linear theory of growth yields the result

$$\delta \propto R^n \qquad\qquad n \geq 3 \tag{10.62}$$

but this is complicated to obtain, and is beyond our scope here.

The peculiar velocities

Let us now consider equation (10.57). We saw in (10.61) that $\delta \propto R/R_0$ for linear perturbations. In general we may write $\delta = A(R/R_0) + E$. To determine the two constants, A and E, we are not allowed to use δ_0 as a particular point because the linear regime certainly does not apply at the present time. However, let us have a look into the problem and consider a linear analysis even at the present time, and assume that $E = 0$, that is, δ_R (at recombination) is very small. Then $\delta = \delta_0(R/R_0)$. We recall (9.118) in the form $R = Dt^{2/3}$, where $D = R_0(6\pi\rho_0)^{1/3}$ during the matter-dominated epoch. From (10.59), we then find that

$$w = \frac{\delta_0 \dot{R} R}{R_0 q^2} \tag{10.63}$$

Substituting for the function $R(t)$:

$$w = \frac{2}{3}\frac{\delta_0 R_0}{q^2}(6\pi\rho_0)^{2/3}t^{1/3} \tag{10.64}$$

At present the velocity amplitude, introduced in (10.54), is

$$V_0 = w_0 q = \frac{2}{3}\frac{\delta_0 R_0}{q}(6\pi\rho_0)^{2/3}t_0^{1/3} \tag{10.65}$$

From (10.65) or directly from (10.63), noting that R_0/q is of the order of the length of the perturbation λ_0, we obtain the interesting relation for the present epoch:

$$V_0 \approx H_0\lambda_0\delta_0 \tag{10.66}$$

where V_0 is the amplitude of the peculiar velocity, and gives an estimate of the peculiar velocities in general. This quantity can in principle be observed by subtracting the velocity of the Hubble flow from the observed velocity of a given object. For this task an independent way of determining distances without invoking the Hubble law is necessary. λ_0 is, in principle, directly observable.

One of the most important potential results of this formula is that it may provide δ_0, the density contrast of the distribution of matter, either dark or visible! Therefore, the dynamic effects of the density as shown gravitationally would permit us to draw a dark matter distribution map. This would be very important if light did not trace mass, so that δ_0 was not observationally determinable. Another interesting application lies in the fact that it could offer a way to restrict the value of the density parameter Ω_0. As we have set $\Omega_0 = 1$, we cannot see in fact that there should be a numerical coefficient different from unity in (10.66), but there is a coefficient which depends critically on Ω_0 (in fact on $\Omega_0^{0.6}$).

Other applications which are based on the preceding ideas, that is, on the use of peculiar velocities to determine mass, are potentially promising as links between observation and prediction.

Relativistic growth

As stated above and as shown in Figure 10.1, prior to recombination and the epoch of acoustic propagation, another epoch exists in which the perturbations are unstable and grow. The timing of this epoch depends on the rest mass of the perturbation; it is about $z \approx 3 \times 10^5$ for a typical galactic mass. Then V_s is close to unity, so that from (10.33) we can see that M_J and λ_J are of the same order of magnitude, hence a relativistic treatment is necessary.

As in the derivation carried out above for Newtonian perturbations, we introduce perturbations in the Einstein field equations in the form of (2.86). Let us consider $\delta g_{00} = 0$, $\delta g_{0i} = 0$, that is, pure spatial perturbations in the metric, the unperturbed metric tensor being of the Robertson–Walker type ($g_{00} = -1$, $g_{0i} = 0$, $g_{ij} = R^2 \delta_{ij}$). Let us assume that $k = 0$, which is always reasonable for early epochs. Now R is a function of time given by (9.145):

$$\frac{R}{R_0} = 2\left(\frac{2\pi}{3}\epsilon_0\right)^{1/4} t^{1/2} \tag{10.67}$$

as we assume the Universe to be radiation dominated. Also from (9.63) we find

$$\epsilon = (\epsilon_0 R_0^4) R^{-4} = \frac{3}{2^5 \pi} t^{-2} \tag{10.68}$$

and $p = \epsilon/3$. To calculate $T_{\mu\nu}$ we may set $U^i = 0$ for the unperturbed four-velocity. From equation (2.95) we know that

$$0 = \delta(g_{\mu\nu} U^\mu U^\nu) = (\delta g_{\mu\nu}) U^\mu U^\nu + g_{\mu\nu}(\delta U^\mu) U^\nu + g_{\mu\nu} U^\mu (\delta U^\nu) \tag{10.69}$$

Only δg_{ij} is non-vanishing, but $U^i = 0$, hence the first term is zero. In the second term, we have $g_{00}(\delta U^0) U^0$ but not $g_{ii}(\delta U^i) U^i$ as $U^i = 0$. The same is true for the third term. Therefore, we obtain

$$2g_{00}(\delta U^0) U^0 = 2\delta U^0 = 0 \tag{10.70}$$

that is, $\delta U^0 = 0$.

The detailed derivation is somewhat lengthy but easy, and we will simply indicate the procedure. An equation of motion

$$\frac{\partial}{\partial t}\left(R^5 U_1^i(\epsilon + p)\right) = -R^3 \frac{\partial p_1}{\partial x^i} \tag{10.71}$$

and an energy conservation equation

$$\dot{\epsilon}_1 + \frac{3\dot{R}}{R}(\epsilon_1 + p_1) = -(\epsilon + p)\left(\frac{\partial}{\partial t}\left(\frac{\delta g_{kk}}{2R^2} + \frac{\partial}{\partial x^i} U_1^i\right)\right) \tag{10.72}$$

are again found. Using the same subindex criterion as before, U_1^i is the fluctuating or perturbed velocity, and ϵ_1 and p_1 are the perturbed energy density and hydrostatic pressure, respectively.

We find directly from the continuity equation (2.91) that

$$\frac{\partial(n_1/n)}{\partial t} = -\frac{\partial U_1^i}{\partial x^i} - \frac{\partial}{\partial t}\left(\frac{\delta g_{kk}}{2R^2}\right) \tag{10.73}$$

We then assume, as for the Newtonian case, that the quantities δg_{ij}, ϵ_1, U_1^i, and n_1 are proportional to $e^{i\vec{q}\cdot\vec{r}}$. We now define

$$\delta = \frac{\epsilon_1}{\epsilon + p} = \frac{3}{4}\frac{\epsilon_1}{\epsilon} \tag{10.74}$$

and find

$$\delta g_{kk} = -2R^2\delta \tag{10.75}$$

and

$$\ddot{\delta} + \frac{\dot{\delta}}{t} - \frac{\delta}{t^2} = 0 \tag{10.76}$$

Equation (10.76) was our basic objective and has two solutions: $\delta \propto t$ and $\delta \propto t^{-1}$. The second one does not produce growth and thus

$$\delta = \text{constant} \times t \tag{10.77}$$

quite different from the dramatic growth that characterizes proto-stellar collapse. Note that we have considered perturbations in the metric, and therefore in the curvature (which are called adiabatic perturbations), but not in the equation of state (which are called isothermal perturbations). We have found that perturbations grow faster in the radiation-dominated epoch. There is an intuitive reason behind this: they grow relatively with respect to a medium that expands more slowly.

Photon anisotropies and density inhomogeneities

It is possible to make an approximate plot of $[\lg \delta, \lg R/R_0]$, sketched in Figure 10.2, in the following way. Assume a present perturbation with $\delta \approx 10^4$. We then draw the curve backwards with a slope of about 3, until the value $\delta = 1$ is reached, and smoothly change the slope to unity at the recombination epoch. In this case the acoustic propagation epoch corresponds to a vanishing slope. When the unstable zone is again reached, the relativistic growth is characterized by a value for the slope of 2.

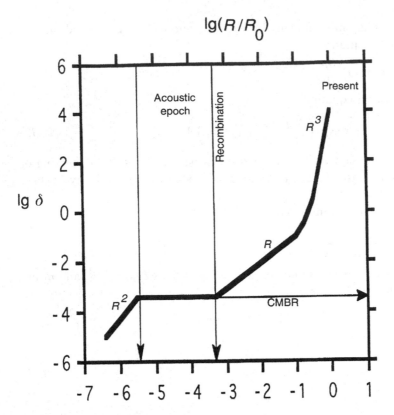

Figure 10.2   The evolution of a perturbation along the different epochs of the Universe

There is an interesting possible way to test this predicted evolution. By means of the cosmic background radiation, we may observe the density perturbation map as it was at the recombination epoch. The anisotropies can be interpreted as $\delta T/T$ at recombination, and the problem then is to interpret $\delta T/T$ in terms of $\delta \equiv \delta\rho/\rho$. From (10.4) we can set $\delta\rho/\rho$ at three times the value of $\delta T/T$. From the satellite COBE, complemented with the Tenerife experiment, we find that $\delta T/T \approx 1.1 \times 10^{-5}$, hence $\delta\rho/\rho$ would be of the order of 3.3×10^{-5} at recombination, clearly less than our expected values. It must be noted that the angular scale in these experiments was a few degrees, which corresponds to super-horizon structures, which currently are a size larger than 1000 Mpc. On angular scales of about 1 arcmin, upper limits for $\delta T/T$ of about 1.5×10^{-5} are found.

However, some dark matter models predict coefficients relating $\delta T/T$ and $\delta\rho/\rho$ which are much higher than 3. This is because baryonic structures in fact begin to grow only after recombination once they are decoupled from radiation. However, WIMPs, if they exist, would be decoupled from photons much earlier, at the equality epoch, and would therefore begin to grow much earlier. Let us assume a linear Newtonian growth of WIMP perturbations from equality to recombination. As $R(equality)/R(recombination) \approx (1.1 \times 10^3)/(2.5 \times 10^4) \approx 1/23$, then we may expect

$(\delta\rho/\rho)_{WIMP} \approx 23(\delta\rho/\rho)_{BARYON}$. Therefore, at recombination $\delta\rho/\rho \approx 3 \times 23(\delta T/T)$ $\approx 7.6 \times 10^{-4}$. This value is in agreement with our rough predictions made in Figure 10.2. It is true that the structures we observe today are baryonic, but baryons would probably have fallen into the WIMPs' potential wells at later epochs, so that today they would trace the early WIMP perturbations. Broad agreement between COBE measurements and the theoretical predictions of some dark matter models has been claimed, but we still have a long way to go, both theoretically and observationally, before we can claim a satisfactory cosmological model which provides a good explanation of the presence and properties of galaxies.

Appendix

Some conversion factors and constants in geometrized units

$$1 \text{ cm} = 3.33 \times 10^{11} \text{s}$$
$$1 \text{ g} = 2.47 \times 10^{-39} \text{s}$$
$$1 \text{ erg} = 2.75 \times 10^{-60} \text{s}$$
$$1 \text{ K} = 3.79 \times 10^{-76} \text{s}$$
$$1 \text{ gauss} = 8.61 \times 10^{-15} \text{s}^{-1}$$
$$h = 4.23 \times 10^{-87} \text{s}^2$$
$$e = 9.57 \times 10^{-36} \text{s}$$
$$m_p = 4.13 \times 10^{-63} \text{s}$$
$$m_e = 2.52 \times 10^{-66} \text{s}$$
$$1 \text{ pc} = 1 \times 10^8 \text{s}$$
$$H_0(75 \text{ km s}^{-1}\text{Mpc}^{-1} = 2.49 \times 10^{-18} \text{s}^{-1}$$
$$\text{M}_\odot = 4.92 \times 10^{-6} \text{s}$$
$$a = 2.69 \times 10^{259} \text{s}^{-6}.$$

Bibliography

Allen, C.W. (1991) *Astrophysical Quantities*. The Athalon Press. London.

Alfvén, H. (1981) *Cosmic Plasma*. Reidel Pub. Co. Dordrecht.

Audouze, J. and Tran Thanh Van, J. (eds.) (1983) *Formation and Evolution of Galaxies and Large Structures in the Universe*. NATO ASI Series, vol. 117. Reidel Pub. Co. Dordrecht.

Audouze, J. and Israel, G. (1988) *The Cambridge Atlas of Astronomy*. Cambridge University Press.

Arp, H. (1987) *Quasars, Redshifts and Controversies*. Interstellar Media. Berkeley. Ca.

Asseo, E. and Gresillon, D. (eds.) (1987) *Magnetic Fields and Extragalactic Jets*. Proc. Institut d'etudes scientifiques de CARGESE. Editions de Physique. Les Ulis. France.

Athanassoula, E. (ed.) (1983) *Internal Kinematics and Dynamics of Galaxies*. IAU Symposium N°. 100. Reidel Pub. Co. Dordrecht.

Balescu, R. (1975) *Equilibrium and Non-equilibrium Statistical Mechanics*. John Wiley and Sons, Inc. New York.

Ballester, J.L. and Priest, E.R. (eds.) (1991) *Dynamics and Structure of Solar Prominences*. Servei de publicaciones i intercanvi cientific de la Universitat de les illes Balears. Palma.

Battaner, E. (1986) *Fluidos Cosmicos*. Labor. Barcelona.

Battaner, E. (1988) *Fisica de las noches estrelladas*. Tusquets Editores. Barcelona.

Battaner, E. (1991) *Planetas*. Alianza Editorial. Madrid.

Beck, R. and Gräve, R. (eds.) (1987) *Interstellar Magnetic Fields*. Proc. Workshop Schloss Ringberg, Tegernsee. Springer-Verlag. Berlin.

Beckman, J. and Pagel, B.E.J. (eds.) (1989) *Evolutionary Phenomena in Galaxies*. Cambridge University Press.

Belvedere, G.(ed.) (1989) *Accretion Disks and Magnetic Fields in Astrophysics*. Proc. European Physical Society Study Conference in Noto (Sicily). Kluwer Academic Publishers. London.

Binney, J. and Tremaine, S. (1987) *Galactic Dynamics*. Princeton University Press. Princeton.

Bittencourt, J.A. (1986) *Fundamentals of Plasma Physics*. Pergamon Press. Oxford.

Blandford, R.D., Netzer, H., and Woltjer, L. (1990) *Active Galactic Nuclei*. Saas-Fee Advanced course. Springer-Verlag. Berlin.

Bloemen, H. (ed.) (1991) *The Interstellar Disk–halo Connection in Galaxies.* IAU Symposium N°. 144. Kluwer Academic Publishers. London.

Bohm-Vitense, E. (1989) *Introduction to Stellar Astrophysics* I and II. Cambridge University Press.

Bok, B. and Bok, P.F. (1981) *The Milky Way.* Harvard University Press. Cambridge Mass.

Börner, G. (1988) *The Early Universe.* Springer-Verlag. Berlin.

Bowers, R. and Deeming, T. (1984) *Astrophysics* I and II. Jones and Bartlett Pub. Inc. Boston.

Briggs, G.A. and Taylor, F.W. (1988) *Photographic Atlas of the Planets.* Cambridge University Press.

Burgers, J.M. (1969) *Flow Equations for Composite Gases.* Academic Press. New York.

Cabannes, H. (1970) *Theoretical Magnetofluiddynamics.* Academic Press. New York.

Canuto, V.M. and Elmegreen, B.G. (eds.) (1988) *Handbook of Astronomy, Astrophysics and Geophysics (vol. II).* Gordon and Breach Sci. Pub. New York.

Cap, F. (1976) *Handbook of Plasma Instabilities.* Academic Press. New York.

Casertano, S., Sackett, P., and Briggs, F. (eds.) (1991) *Warped Disks and Inclined Rings around Galaxies.* Cambridge University Press.

Celnikier, L.M. (1989) *Basics of Cosmic Structures.* Editions Frontieres. Gif-sur-Yvette Cedex. France.

Chamberlain, J.W., and Hunten, D.M. (1987) *Theory of Planetary Atmospheres.* Academic Press. New York.

Chandrasekhar, S. (1942) *Principles of Stellar Dynamics.* Dover Pub., Inc. New York.

Chandrasekhar, S. (1950) *Radiative Transfer.* Oxford University Press. Oxford.

Chandrasekhar, S. (1960) *Plasma Physics.* The University of Chicago Press. Chicago.

Chandrasekhar, S. (1967) *An Introduction to the Study of Stellar Structure.* Dover Pub., Inc. New York.

Chandrasekhar, S. (1970) *Hydrodynamic and Hydromagnetic Stability.* Clarendon Press.

Chandrasekhar, S. (1983) *The Mathematical Theory of Black Holes.* Oxford University Press. Oxford.

Chandrasekhar, S. (1987) *Ellipsoidal Figures of Equilibrium.* Dover Pub., Inc. New York.

Chapman, S. and Cowling, T.G. (1970) *The Mathematical Theory of Non-uniform Gases.* Cambridge University Press.

Chen, F.F. (1974) *Introduction to Plasma Physics.* Plenum Press. New York.

Chin, Hong-Yee (1967) *Stellar Physics.* Blaisdel Pub. Co. London.

Chorin, A.J. and Marsden, J.E. (1990) *A Mathematical Introduction to Fluid Mechanics.* Springer-Verlag. Berlin.

Clayton, D.D. (1968) *Principles of Stellar Evolution and Nucleosynthesis.* The University of Chicago Press. Chicago.

Collins, G.W. (1989) *The Fundamentals of Stellar Astrophysics.* Freeman and Co. New York.

Combes, F. and Casoli, F. (eds.) (1991) *Dynamics of Galaxies and their Molecular Cloud Distributions*. IAU Symp. N°. 146. Kluwer Academic Publishers. London.

Cook, A.H. (1980) *Interiors of the Planets*. Cambridge University Press.

Cornell, J. (ed.) (1989) *Bubbles, Voids and Bumps in Time: The new cosmology*. Cambridge University Press.

Cowling, T.G. (1976) *Magnetohydrodynamics*. Adam Hilger.

Cox, J.P. and Guili, R.T. (1968) *Principles of Stellar Structure*. Gordon and Breach Sci. Pub. New York.

Dalgarno, A. and Layzer, D. (eds.) (1987) *Spectroscopy of Astrophysical Plasmas*. Cambridge University Press.

Demianski, M. (1985) *Relativistic Astrophysics*. Pergamon Press. Oxford.

Dolgov, A.D., Sazhin, M.V., and Zeldovich, Ya.B. (1990) *Basics of Modern Cosmology*. Editions Frontieres. Gif-sur-Yvette. France.

Edmunds, M.G. and Terlevich, R.J. (eds.) (1992) *Elements and the Cosmos*. Cambridge University Press.

Encrenaz, T., Bibring, J.P., and Blanc, M. (1990) *The Solar System*. Springer-Verlag. Berlin.

Eriksson, K.E., Lindgren, K., and Mansson, B.A. (1989) *Structure, Context, Complexity, Organization*. World Scientific. Singapore.

Faber, S.M. (ed.) (1987) *Nearly Normal Galaxies: From the Planck time to the present*. Santa Cruz Summer Workshops in Astronomy and Astrophysics. Springer-Verlag. Berlin.

Fall, S.M. and Lynden-Bell, D. (eds.) (1981) *The Structure and Evolution of Normal Galaxies*. Cambridge University Press.

Foukal, P.V. (1990) *Solar Astrophysics*. John Wiley and Sons, Inc. New York.

Gerbal, D. and Mazure, A. (eds.) (1993) *Clustering in the Universe*. Editions Frontieres. Gif-sur-Yvette. France.

Gilmore, G.F., King, I.R., van der Kruit, P.C., and Buser, R. (1990) *The Milky Way as a Galaxy*. University Science Books. Mill Valley. California.

Glyn Jones, K. (1991) *Messier's Nebulae and Star Cluster*. Cambridge University Press.

Habing, H.J. (ed.) (1970) *Interstellar Gas Dynamics*. IAU Symp. 39. Reidel Pub. Co. Dordrecht.

Harwit, M. (1988) *Astrophysical Concepts*. Springer-Verlag. Berlin.

Haynes, R. and Milne, D. (eds.) (1991) *The Magellanic Clouds*. IAU Symp. 148. Kluwer Academic Publishers. London.

Hewitt, A., Burbidge, G., and Fang, L.Z. (eds.) (1987) *Observational Cosmology*. IAU Symp. 124. Reidel Pub. Co. Dordrecht.

Hirschfelder, J.O., Curtiss, C.F., and Bird, R.B. (1954) *Molecular Theory of Gases and Liquids*. John Wiley and Sons, Inc. New York.

Houghton, J.T. (1986) *The Physics of Atmospheres*. Cambridge University Press.

Hughes, P.A. (ed.) (1991) *Beams and Jets in Astrophysics*. Cambridge University Press.

Jones, B.J.T. and Jones, J.E. (eds.) (1981) *The Origin and Evolution of Galaxies*. NATO ASI Series 97. Reidel Pub. Co. Dordrecht.

Kaufmann, W.J. (1991) *Universe*. Freeman and Co. New York.

Kikuchi, H. (ed.) (1981) *Relation between Laboratory and Space Plasmas*. Reidel Pub. Co. Dordrecht.

Kippenhahn, R. and Weigert, A. (1990) *Stellar Structure and Evolution*. Springer-Verlag. Berlin.

Kitchin, C.R. (1988) *Astrophysical Techniques*. Adam Hilger. Bristol.

Klimishin, I.A. (1991) *Modern Asronomy*. Specktrum Akademisher Verlag. Heidelberg.

Kolb, E.W. and Turner, M.S. (1990) *The Early Universe*. Addison-Wesley Pub. Co.

Kourganoff, V. (1963) *Basic Methods in Transfer Problems*. Dover Pub., Inc. New York.

Kourganoff, V. (1970) *Introduction à la Physique des Interieurs Stellaires*. Dunod. Paris.

Krall, N.A. and Trivelpiece, A.W. (1973) *Principles of Plasma Physics*. McGraw Hill Inc. New York.

Lamb, H. (1975) *Hydrodynamics*. Cambridge University Press.

Landau, L.D. and Lifshitz, E.M. (1951) *Classical Theory of Fields*. Addison-Wesley. New York.

Landau, L.D. and Lifshitz, E.M. (1959) *Fluid Mechanics*. Pergamon. Oxford.

Lang, K. (1986) *Astrophysical Formulae*. Springer-Verlag. Berlin.

Linde, A. (1990) *Particle Physics and Inflationary Cosmology*. Harwood Academic Press.

Mandelbrot, B.B. (1983) *The Fractal Geometry of Nature*. Freeman and Co. New York.

Maran, S.P. (ed.) (1992) *The Astronomy and Astrophysics Encyclopedia*. Cambridge University Press.

Mardirossian, F., Giuricin, G., and Mezzetti, M. (eds.) (1984) *Clusters and Groups of Galaxies*. Astrophysics and Space Science Library, vol. 111. Reidel Pub. Co. Dordrecht.

Meyer, F., Duschl, W.J., Frank, J., and Meyer-Hofmeister, E. (eds.) (1989) *Theory of Accretion Disks*. Kluwer Academic Publishers. London.

Michaud, G. and Tutukov, A. (eds.) (1991) *Evolution of Stars: The photospheric abundance connection*. IAU Symp. 145. Kluwer Academic Publishers. London.

Michelson, I. (1970) *The Science of Fluids*. Van Nostrand Reinhold Co. New York.

Mihalas, D. and Binney, J. (1968) *Galactic Astronomy: Structure and kinematics*. Freeman and Co. New York.

Mihalas, D. (1978) *Stellar Atmospheres*. Freeman and Co. New York.

Mihalas, D. and Mihalas, B.W. (1984) *Foundations of Radiation Hydrodynamics*. Oxford University Press. Oxford.

Miller, J.S. (1985) *Astrophysics of Active Galaxies and Quasi-Stellar Objects*. Oxford University Press. Oxford.

Misner, C.W., Thorne, K.S., and Wheeler, J.A. (1973) *Gravitation*. Freeman and Co. New York.

Monin, A.S. and Yaglom, A.M. (1971) *Statistical Fluid Mechanics*. The MIT Press. Cambridge, Mass.

Naber, G.L. (1988) *Spacetime and Singularities. An introduction*. Cambridge University Press.

Narlikar, J.V. (1983) *Introduction to Cosmology.* Jones and Bartlett Pub. Inc. Boston.

Novikov, I. (1990) *Black Holes and the Universe.* Cambridge University Press.

Novotny, E. (1973) *Introduction to Stellar Atmospheres and Interiors.* Oxford University Press. Oxford.

Oegerle, W.R., Fitchett, M.J., and Danly, L. (eds.) (1990) *Clusters of Galaxies.* Space Telescope Science Institute. Cambridge University Press.

Osterbrock, D.E. (1989) *Astrophysics of Gaseous Nebulae and Active Galactic Nuclei.* University Science Books. Mill Walley, California.

Pacholczyk, A.G. (1977) *Radio Galaxies.* Pergamon Press. Oxford.

Parker, E.N. (1979) *Cosmical Magnetic Fields.* Clarendon Press. Oxford.

Pecker, J.C. and Schatzman, E. (1959) *Astrophysique generale.* Masson et cie. editeurs. Paris.

Peebles, P.J.E. (1971) *Physical Cosmology.* Princeton University Press. Princeton.

Peebles, P.J.E. (1980) *The Large-scale Structure of the Universe.* Princeton University Press. Princeton.

Peratt, A.L. (1992) *Physics of the Plasma Universe.* Springer-Verlag. Berlin.

Priest, E.R. (1987) *Solar Magneto-Hydrodynamics.* Reidel Pub. Co. Dordrecht.

Prigogine, I. (1980) *From Being to Becoming.* Freeman and Co. New York.

Raine, D.J. (1981) *The Isotropic Universe.* Adam Hilger. Bristol.

Ramana Murthy, P.V. and Wolfendale, A.W. (1986) *Gamma-ray Astronomy.* Cambridge University Press.

Ratcliffe, J.A. (1972) *An Introduction to the Ionosphere and Magnetosphere.* Cambridge University Press.

Reddish, V.C. (1974) *The Physics of Stellar Interiors. An introduction.* Edinburgh University Press.

Reeves, H. (1991) *The Hour of our Delight.* Freeman and Co. New York.

Ridley, B.K. (1984) *Time, Space and Things.* Cambridge University Press.

Rindler, W. (1977) *Essential Relativity.* Springer-Verlag. Berlin.

Riordan, M. and Schramm, D.N. (1991) *The Shadows of Creation: Dark matter and the structure of the Universe.* Freeman and Co. New York.

Rose, W.K. (1973) *Astrophysics.* Holt, Rinehart and Winston. New York.

Rowan-Robinson, M. (1981) *Cosmology.* Clarendon Press. Oxford.

Rowan-Robinson, M. (1985) *The Cosmological Distance Ladder. Distance and time in the Universe.* Freeman and Co. New York.

Roy, A.E. and Clarke, D. (1988) *Astronomy. Principles and practice.* Adam Hilger. Bristol.

Ruzmaikin, A.A., Shukurov, A.M., and Sokoloff, D.D. (1988) *Magnetic Fields of Galaxies.* Kluwer Academic Publishers. London.

Sánchez, F., Collados, M., and Rebolo, R. (eds.) (1992) *Observational and Physical Cosmology.* Cambridge University Press.

Sandage, A. (1961) *The Hubble Atlas of Galaxies.* Carnegie Institute of Washington. Washington, DC.

Saslaw, W.C. (1985) *Gravitational Physics of Stellar and Galactic Systems*. Cambridge University Press.

Schutz, B.F. (1985) *A First Course in General Relativity*. Cambridge University Press.

Schwarzschild, M. (1958) *Structure and Evolution of the Stars*. Princeton University Press. Princeton.

Sciama, D.W. (1973) *Modern Cosmology*. Cambridge University Press.

Sellwood, J.A. (ed.) (1989) *Dynamics of Astrophysical Discs*. Cambridge University Press.

Sersic, J.W. (1982) *Extragalactic Astronomy*. Reidel Pub. Co. Dordrecht.

Setti, G., Spada, G., and Wolfendale, A.W. (1981) *Origin of Cosmic Rays*. IAU Symp. 94. Reidel Pub. Co. Dordrecht.

Seymour, P. (1986) *Cosmic Magnetism*. Adam Hilger. Bristol.

Shapiro, S.L. and Tenkolsky, S.A. (1983) *Black Holes, White Dwarfs and Neutron Stars*. John Wiley and Sons, Inc. New York.

Shu, F.H. (1982) *The Physical Universe: An introduction to astronomy*. University Science Books. Mill Valley. California.

Shu, F.H. (1991) *The Physics of Astrophysics I and II*. University Science Books. Mill Valley. California.

Shuter, W.L.H. (ed.) (1983) *Kinematics, Dynamics and Structure of the Milky Way*. Proc. Workshop in Vancouver. Reidel Pub. Co. Dordrecht.

Silk, J. (1989) *The Big Bang*. Freeman and Co. New York.

Spitzer, L. (1978) *Physical Processes in the Interstellar Medium*. John Wiley and Sons, Inc. New York.

Stephani, H. (1990) *General Relativity*. Cambridge University Press.

Stix, M. (1989) *The Sun*. Springer-Verlag. Berlin.

Sundelius, B. (ed.) (1991) *Dynamics of Disc Galaxies*. Proc. Conference at Varberg Castle in Sweden. Goteborgs University, and Chalmers Univ. of Technology. Goteborg.

Tenorio-Tagle, G., Moles, M., and Melnick, J. (eds.) (1989) *Structure and Dynamics of the Interstellar Medium*. Springer-Verlag. Berlin.

Tully, B. (1988) *Nearby Galaxy Catalogue*. Cambridge University Press.

Unsöld, A. (1977) *The New Cosmos*. Springer-Verlag. Berlin.

Wald, R.M. (1984) *General Relativity*. The University of Chicago Press. Chicago.

Walker, G. (1989) *Astronomical Observations. An optical perspective*. Cambridge University Press.

Weedman, D.W. (1988) *Quasar Astronomy*. Cambridge University Press.

Weinberg, S. (1972) *Gravitation and Cosmology*. John Wiley and Sons, Inc. New York.

Wielen, R. (ed.) (1990) *Dynamics and Interaction of Galaxies*. Proc. International Conference in Heidelberg. Springer-Verlag. Berlin.

Wynn-Williams, C.G. and Cruikshank, D.P. (eds.) (1981) *Infrared Astronomy*. IAU Symp. 96. Reidel Pub. Co. Dordrecht.

Zeldovich, Ya B. and Novikov, I.D. (1971) *Relativistic Astrophysics*. The University of Chicago Press. Chicago.

Zeldovich, Ya B., Ruzmaikin, A.A., and Sokoloff, D.D. (1983) *Magnetic Fields in Astrophysics*. Gordon and Breach Sci. Pub. New York.

Zeytounian, R.K. (1991) *Mecanique des fluides fondamentale*. Springer-Verlag. Berlin.

Zirin, H. (1988) *Astrophysics of the Sun*. Cambridge University Press.

Zombeck, M. (1990) *Handbook of Space Astronomy and Astrophysics*. Cambridge University Press.

Index

Printed in the United States
By Bookmasters